New Techniques and Applications in Lipid Analysis

New Techniques and Applications in Lipid Analysis

Editors

Richard E. McDonald
National Center for Food Safety and Technology
Food and Drug Administration
Summit-Argo, Illinois

Magdi M. Mossoba
Food and Drug Administration
Washington, D.C.

Champaign, Illinois

AOCS Mission Statement
To be a forum for the exchange of ideas, information, and experience among those with a professional interest in the science and technology of fats, oils, and related substances in ways that promote personal excellence and provide high standards of quality.

AOCS Books and Special Publications Committee
E. Perkins, chairperson, University of Illinois, Urbana, Illinois
N.A.M. Eskin, University of Manitoba, Winnipeg, Manitoba
M. Pulliam, C&T Quincy Foods, Quincy, Illinois
T. Foglia, USDA—ERRC, Wyndmoor, Pennsylvania
L. Johnson, Iowa State University, Ames, Iowa
Y.-S. Huang, Ross Laboratories, Columbus, Ohio
J. Lynn, Edgewater, New Jersey
M. Mossoba, Food and Drug Administration, Washington, D.C.
G. Nelson, Western Regional Research Center, San Francisco, California
F. Orthoefer, Stuttgart, Arkansas
J. Rattray, University of Guelph, Guelph, Ontario
A. Sinclair, Royal Melbourne Institute of Technology, Melbourne, Australia
G. Szajer, Akzo Chemicals, Dobbs Ferry, New York
B. Szuhaj, Central Soya Co., Inc., Fort Wayne, Indiana
L. Witting, State College, Pennsylvania

Copyright © 1997 by AOCS Press. All rights reserved. No part of this book may be reproduced or transmitted in any form or by any means without written permission of the publisher.

The paper used in this book is acid-free and falls within the guidelines established to ensure permanence and durability.

Library of Congress Cataloging-in-Publication Data

New techniques and applications in lipid analysis/editors, Richard
   E. McDonald, Magdi M. Mossoba.
      p. cm.
   Includes bibliographical references and index.
   ISBN 0-935315-80-2 (alk. paper)
   1. Lipids—Analysis.   I. McDonald, Richard E., 1946–
II. Mossoba, Magdi M.
QP751.N486   1997   97-3204
547'.77—dc21   CIP

Printed in the United States of America with vegetable oil-based inks.
00 99 98 97   5 4 3 2 1

Preface

The role of lipids in human nutrition has generated a great deal of interest as well as controversy during the past several years. Accurate identification and quantitation of the myriad lipid components and isomers are essential to any study of the nutritional effects of these compounds. This monograph, with contributions written by internationally recognized experts, attempts to highlight recent advances in lipid analysis. Most of the chapters of this book were first presented at the American Oil Chemists' Society Annual Meeting in Indianapolis, Indiana, on April 29–May 1, 1996, and focus on novel analytical techniques as well as new developments and applications of established lipid methods. Although this book emphasizes state-of the-art analytical methods, new applications of some of the classical methods are also reviewed. This book will therefore serve both students and experienced researchers as an informative and comprehensive reference book covering the latest and most important analytical topics in lipid chemistry.

Richard E. McDonald
Magdi M. Mossoba

About the Editors

Richard E. McDonald attended the University of Massachusetts, earning a B.S. in 1968 and a Ph.D. in 1983. In the intervening years, he served as a navy lieutenant in Vietnam, obtained an M.S. from Michigan State University, and worked as a chemist for Miles Laboratories. After obtaining his doctorate, he worked as a group leader in analytical chemistry at Nestlé. In 1985 he accepted a position as a lipid chemist with the Food and Drug Administration (FDA), focusing on the analysis of lipids in processed edible oils. He is the author of over 35 scientific articles and 4 patents, and he has been invited to serve in leadership positions of the American Oil Chemists' Society (AOCS) and the Institute of Food Technologists (IFT). At the AOCS he has served as vice chairperson of the Uniform Methods Committee and is currently on the Executive Committee of the Analytical Division. At IFT he is chairperson of the Food Chemistry Division and chairperson-elect of the Stephen S. Chang Award Jury for Lipid or Flavor Science. Dr. McDonald is currently chief of FDA's food processing branch and is also serving as the director of scientific programs at the National Center for Food Safety and Technology/FDA in Illinois.

Magdi M. Mossoba graduated from the American University in Cairo with a B.Sc. in 1972. After receiving his Ph.D. from the Chemistry Department at Georgetown University, Washington, D.C., in 1980, Dr. Mossoba was awarded a Fogarty Fellowship at the National Institutes of Health, Bethesda, Maryland, from 1980 to 1983, where his research focused on the structural characterization of transient species in biologically important matrices using magnetic resonance, spectroscopic, and computer simulation techniques. He joined the University of Maryland Cancer Center, Baltimore, Maryland, in 1983–1984, where he investigated the chemical role of antitumor agents in photodynamic therapy. Since 1984, Dr. Mossoba has been with the U.S. Food and Drug Administration in Washington, D.C. Using hyphenated gas chromatography–Fourier transform infrared spectroscopy techniques, his research in food safety and nutrition has included the identification and quantitation of *trans* fatty acid isomers in margarine and hydrogenated vegetable and marine oils, the structural elucidation of cyclic fatty acid monomers in heated oils, the characterization of conjugated linoleic acid isomers, and the rapid determination of the total *trans* content of foods using internal reflection spectroscopy. Dr. Mossoba's work is documented in over 58 scientific publications and 4 book chapters. He is currently the vice chairperson of the Uniform Methods Committee of the AOCS, a member of the AOCS Books and Special Publications Committee, and the AOAC associate referee for hydrogenated fats.

Contents

	Preface	v
Chapter 1	Newer Methods for Fat Analysis in Foods D. Firestone and M.M. Mossoba	1
Chapter 2	Use of Stable Isotopes to Study Incorporation of Dietary Fat into Blood Lipids J.P. DeLany and T.-S. Lu	34
Chapter 3	Qualitative and Quantitative Analysis of Triacylglycerols Using Atmospheric-Pressure Chemical Ionization Mass Spectrometry W.C. Byrdwell and W.E. Neff	45
Chapter 4	Liquid Chromatography with On-Line Electrospray Mass Spectrometry of Oxidized Diphosphatidylglycerol M. Bergqvist and A. Kuksis	81
Chapter 5	Stereospecific Analysis of Docosahexaenoic Acid-Rich Triacylglycerols by Chiral-Phase HPLC With Online Electrospray Mass Spectrometry J.J. Myher, A. Kuksis, and P.W. Park	100
Chapter 6	NMR Characterization of Fatty Compounds Obtained via Selenium Dioxide–Based Oxidations G. Knothe	121
Chapter 7	Supercritical Fluid Chromatography: A Shortcut in Lipid Analysis J.W. King and J.M. Snyder	139
Chapter 8	Analysis of Unusual Triglycerides and Lipids Using Supercritical Fluid Chromatography D.G. Hayes	163
Chapter 9	Oxidation Products of Conjugated Linoleic Acid and Furan Fatty Acids M.P. Yurawecz, N. Sehat, M.M. Mossoba, J.A.G. Roach, and Y. Ku	183
Chapter 10	Analysis of Lipid Oxidation Products by Combination of Chromatographic Techniques G. Márquez-Ruiz and M.C. Dobarganes	216
Chapter 11	Analysis of *trans* Fatty Acids J. Fritsche and H. Steinhart	234

Contents

Chapter 12	Separation of Fatty Acid Methyl Esters and Triacylglycerols by Ag-HPLC: Silver-Ion and Normal-Phase Contributions to Retention ....256 *R.O. Adlof*	
Chapter 13	Near Infrared Analysis of Oilseeds: Current Status and Future Directions ....266 *J.K. Daun and P. Williams*	
Chapter 14	Application of Fourier Transform Infrared Spectroscopy in Edible Oil Analysis ....283 *J. Sedman, F.R. van de Voort, and A.A. Ismail*	
Chapter 15	Recent Applications of Iatroscan TLC-FID Methodology ....325 *R.G. Ackman, and H. Heras*	
Chapter 16	Natural Antioxidants in Lipids ....341 *T. Rathjen and H. Steinhart*	
Chapter 17	Coffee Lipids: Analysis of the Diterpene 16-*O*-methylcafestol as an Indicator of Admixing of Coffees ....356 *N. Sehat and G. Niedwetzki*	
Chapter 18	Improvements in Recovery of Petroleum Hydrocarbons from Marine Fish, Crabs, and Mussels ....380 *R.G. Ackman, H. Heras, and S. Zhou*	
	Index 395	

Chapter 1

Newer Methods for Fat Analysis in Foods

D. Firestone and M.M. Mossoba

Food and Drug Administration, Center for Food Safety and Applied Nutrition, Washington, DC 20204

Introduction

Both classical and instrumental techniques are important for analysis of fats and oils, even though the latter provide more information about lipid composition, quantity, and molecular structure. This review describes useful techniques for lipid analysis, including those involving sophisticated instrumentation, and some of their applications.

Lipid Extraction

To analyze the fat content of foods by various chemical, chromatographic, and spectroscopic methods, a preliminary qualitative or quantitative extraction step is usually required. A large number of procedures and standardized methods are available, and an analyst must choose the most appropriate extraction procedure for the lipid that needs to be identified and determined. The different fatty components obtained in an analytical extract depend on the extraction procedure—specifically, the solvent used, its polarity, boiling point, miscibility with water and other solvents, solubility properties, and volatility. Nonpolar organic solvents, such as hexane or supercritical carbon dioxide (SC-CO_2), are suitable for the neutral or simple lipids, which include esters of fatty acids, acylglycerols, and unsaponifiable matter. Complex or polar lipids (such as phospholipids, glycolipids, lipoproteins, oxidized acylglycerols, and free fatty acids) are preferentially extracted in polar solvents such as methanol. Such extractions require breaking ionic and hydrogen bonds with the proteins or carbohydrates of the food matrix. Solid-phase extraction is particularly useful for complex polar lipids.

Quantitative determination of total fat initially involves acid or base digestion of the test sample in order to free bound and complex lipids prior to extraction. The extraction procedure involves a sequence of steps. First, the food is mixed, ground, or boiled in an organic solvent to maximize the contact area between the food and solvent; next, material that is not extracted is separated by decantation, centrifugation, or filtration from the organic phase; the latter is then purified by washing, re-extraction, drying, filtration, or adsorption; finally, the solvent is eliminated by distillation or evaporation under a stream of an inert gas.

Dedicated to the memory of Dr. Thomas H. Smouse.

Three basic procedures are used to extract lipids from foods: reflux (Soxhlet, etc.), acid or alkaline digestion prior to solvent extraction, and nonheating methods (both solvent extraction and dry-column purification). A quantitative method that does not alter the structure or chemical composition of the lipids of the food product must be used. Other considerations include solvent toxicity, flammability, and cost; simplicity and speed of the procedure; and reproducibility. Less toxic solvent mixtures have also been used. Although a more polar solvent system may provide greater efficiency, it often decreases selectivity. A rapid dry-column method for total lipid extraction from animal tissues was developed for elution of 1-g test portions on calcium phosphate dihydrate columns with 90:10 methylene chloride:methanol (1). Lee et al. (2) described a simple, rapid solvent extraction method for determining total lipids in fish tissue that uses chloroform–methanol and an Eberbach blending jar.

The U.S. Food and Drug Administration, in response to the Nutrition Labeling and Education Act of 1990 (NLEA), defined fat for nutrition labeling purposes as the sum of fatty acids expressed as triglyceride equivalents and saturated fat as those fatty acids that do not contain a double bond. Carpenter et al. (3) reviewed and summarized analytical methods for fat and other lipid components that are required to be or may be placed voluntarily on the new food labels. They noted that while there are many official analytical methods for determination of fat in various food matrices, it is not always clear which methods should be used. Park and Goins (4) reported a method for *in situ* preparation of fatty acid methyl esters (FAMEs) for gas chromatographic (GC) analysis of the fatty acid composition of foods without prior lipid extraction, but did not apply the method to the determination of total fat in the foods. House et al. (5) described a method for quantitative measurement of total fat in foods that meets the requirements of the NLEA. The method involves acid or alkaline hydrolysis and the extraction of fat from various foods with ether, followed by conversion of the fatty acids to their FAMEs and quantitative measurement of the FAMEs by GC. The method, applicable to foods with 1–50% fat, has undergone collaborative study and was recommended for adoption by AOAC International as a first action method. Rader et al. (6) determined total fat and saturated fat in foods by packed-column GC after acid hydrolysis. The authors concluded that the acid hydrolysis–packed-column GC method satisfies labeling requirements for determination of total fat and saturated fat for a large number of foods. They also reported that use of a capillary column gave results comparable to those obtained with the packed column. An acid hydrolysis–capillary column GC method for determination of total, saturated, and monounsaturated fat in cereal products was adopted in 1966 by AOAC International (7). Further discussion of methodology for fat analysis as it pertains to food labeling is beyond the scope of this review. Additional background information on this topic is provided in the references cited above.

Nontoxic, inexpensive, and nonexplosive SC-CO_2 has also been used to extract nonpolar triacylglycerols, even though other polar components such as phospholipids, proteins, and carbohydrates are also somewhat soluble in it. Advantages of using supercritical fluid extraction (SFE) include ease of solvent removal from the extract, selectivity, and the possibility of on-line coupling to a chromatographic technique. Lembke and Engelhardt (8) described an SFE method as an alternative to solvent extraction for rapid determination of total fat content in food. Quantitative lipid extractions were carried out in 35 min, compared to 5–6 h with classical solvent extraction methods. Commercial instrumentation is currently available for automated SFE of foodstuffs. Physical methods have also been applied to determination of total fat in foods without prior fat extraction. These methods, including rapid and nondestructive near infrared (NI) and wide-line nuclear magnetic resonance (NMR) spectroscopy, are important because no single extraction procedure is applicable to all foods. A recently published monograph describes the principles and general applications of SFE (9).

Classical Chemical Techniques

A number of specific tests are used to measure the chemical characteristics and quality of food fats. These tests include iodine value (IV), acidity (free fatty acids), 1-monoglyceride content, color, anisidine value, thiobarbituric acid test, ultraviolet (UV) absorbance, and peroxide value (PV). The content of unsaturated fatty acid hydroperoxides, routinely determined by iodometric measurement of PV, is of toxicological interest as well as an indicator of deterioration due to processing and storage of food and food fats. An alternative iodometric method for PV determination using acetic acid–isooctane solvent (10) was developed recently to replace the standard method (11), which uses chloroform, a known carcinogen. However, with the alternative method it is somewhat difficult to determine the titration endpoint, and there is a 15–30 s delay in neutralizing the starch indicator with high-PV test samples. Other methods for PV determination reported recently include the use of microperoxidase for colorimetric measurement of hydroperoxides at 554 nm (12), coulometric determination (13), spectrophotometric measurement of the I^-_3 chromophore at 360 nm (14), chemiluminescence induced by the reaction between luminol and peroxides (15), and Fourier transform infrared (FTIR) transmission spectroscopy for measurement of the hydroperoxide band of vegetable oils centered at 3444 cm^{-1} (16).

A new method was proposed for determining acidity in vegetable oils based on the nontitration pH-metric technique using a reagent consisting of 0.20 M triethanolamine in a 1:1 solution of water and isopropanol (17). Supercritical fluid chromatography (SFC) on an ODS silica gel column (18) and capillary zone electrophoresis (19) have also been proposed as alternative methods for determination of free fatty acids.

Determination of the position of double bonds is an important aspect of elucidating the structure of fatty acids. This has been accomplished by a combination of complex chemical and separation techniques (20). Ozonolysis methods consist of reacting ozone with a fatty acid or its derivative. This reaction splits the fatty acid chain at the position of unsaturation, resulting in two stable fragments. A monoethylenic FAME yields an aldehyde and an aldehyde ester fragment by reductive ozonolysis.

However, oxidative ozonolysis gives ester and diester products that can be readily separated by GC. The complexity of the method increases with the presence of additional double bonds on the fatty acid chain. A diethylenic FAME is first reduced to an alcohol prior to oxidative ozonolysis to yield a mixture of a monoester, a diester, and an alcohol ester, which is fractionated by thin-layer chromatography (TLC). The alcohol is converted to acetate, and the three fragments are subsequently analyzed by GC. Oxidative ozonolysis plus silver ion–TLC, partial hydrazine reduction, GC–mass spectrometry (MS), and GC coupled with FTIR spectroscopy were used to establish the structure of γ-linolenic acid geometrical isomers in borage oil subjected to treatment (steam-vacuum deodorization) at elevated temperatures (21).

Triethylenic FAMEs first undergo hydrazine reduction prior to ozonolysis. This qualitative reaction is optimized such that it partially converts the triene FAME structure into a mixture of mostly the three possible monoenes and as little as possible of the dienes and the saturated compound. The position of the double bond in the different monoene products corresponds to one of the three original double bond positions in the parent triene FAME, because hydrazine reduction does not alter the double bond position or configuration. The monoene products are fractionated by TLC as mercuric adducts (whose migration is solely dependent on the degree of unsaturation) and reconverted to FAMEs. The position of double bonds in monoene FAMEs is then determined by ozonolysis and GC. The identity of linolenic acid isomers in low-calorie spreads was confirmed by GC–MS of the dimethyl esters resulting from ozonolysis of the monoenes isolated by silver ion–TLC after hydrazine reduction of linolenate isomers separated by HPLC (22). The geometrical and positional isomers of linoleic acid in partially hydrogenated vegetable oils were isolated and identified by silver ion–TLC of the FAMEs, capillary GC, partial hydrazine reduction and oxidative ozonolysis, and GC–MS of the picolinyl ester derivatives (23). Partial hydrazine reduction of linolenic acid was used to prepare a mixture of 18:2 acids used for a study of *cis-trans* isomerization of ethylenic bonds in heated linseed oil (24).

Solid-Phase Extraction (SPE)

SPE is a sample preparation technique for the isolation, concentration, purification, and fractionation of analytes from complex mixtures. A variety of solid phases are commercially available for normal-phase (NP), reverse-phase (RP), and ion exchange (IE) extraction. In addition, many newer cartridges contain mixed-mode blended or copolymerized phases for specific isolations and purifications. SPE generally allows reduced solvent consumption, shorter test sample preparation time, and high analyte recoveries, so it is an attractive alternative to liquid/liquid extraction. Sebedio et al. (25) found that SPE (Sep-Pak^{TM} cartridges) and the standard IUPAC-AOAC silica column chromatographic procedure gave similar results for the fractionation of nonpolar and polar lipids from frying oils. Kaluzny et al. (26)

developed a method for rapid separation of lipid classes in high yield and purity using aminopropyl bonded-phase columns. Wilson et al. (27) described a rapid method for preparing methyl or ethyl esters of n-3 polyunsaturated fatty acids from fish oil by SPE using aminopropyl bonded-phase columns.

Prieto et al. (28) developed a method for separating neutral lipids, glycolipids, and phospholipids from wheat flour using a combination of silica and aminopropyl bonded-phase columns. Chromatography on a silica column separated the steryl esters, triglycerides, free fatty acids, diglycerides, monoglycerides, monogalactosylglycerides, digalactosylglycerides, phosphatidylcholine, and lysophosphatidylcholine from wheat flour lipid extracts. Chromatography on an aminopropyl bonded-phase column separated N-acyl-phosphatidylethanolamine and N-acyl-lysophosphatidylethanolamine, which were coeluted on the silica column. Vaghela and Kilara (29) reported a modified Bligh and Dyer procedure for extraction of total lipids from whey protein concentrates. The extracted lipids were further separated into lipid classes (free fatty acids, phospholipids, cholesterol esters, triacylglycerols, cholesterol, diacylglycerols, and monoacylglycerols) using aminopropyl SPE columns. Schmarr et al. (30) developed a procedure for analysis of polar cholesterol oxidation products in foods involving a Soxhlet extraction, transesterification of the lipid extract, separation of the sterol oxide fraction on an aminopropyl SPE column, and GC of the trimethylsilyl (TMS) derivatives. GC peak assignment was supported by GC–MS. Perez-Camino et al. (31) devised a rapid, simple method for determination of diacylglycerol isomers in vegetable oils by SPE, using a diol bonded-phase column to isolate the diacylglycerols, followed by GC of the TMS derivatives.

Christie (32) showed that commercial SPE columns with bonded benzenesulfonic acid IE medium could be readily converted to the silver ion form for separation of FAMEs with zero to six double bonds as well as *cis* and *trans* FAME isomers. Silver ion (argentation) chromatography using SPE columns was found to be simpler to use than argentation TLC and gave clean fractions uncontaminated by silver ions. Three recent reviews discuss the mechanisms and applications of SPE for lipid separation (33–35).

Thin-Layer Chromatography (TLC)

TLC is a simple chromatographic technique that allows the separation in a single run of a mixture of lipids with widely different polarities. TLC plates are coated with an adsorbent that is suitable for a particular application. For example, silica gel G containing calcium sulfate is used for separating cholesterol esters, triacylglycerols, free fatty acids, cholesterol, diacylglycerols, monoacylglycerols, and phospholipids. Silica gel with silver nitrate is employed for separating fatty acids according to the degree of unsaturation and double bond configuration (*cis* or *trans*). Plates are developed in a lid-covered glass chamber containing an appropriate solvent system. When the solvent front reaches the top of the plate, the plate is removed from the chamber and any residual solvent is evaporated. TLC plates can also be developed in a second direction at a 90° angle with a different solvent system for additional separation or resolution.

Spots can be visualized (under ultraviolet light) after spraying the TLC plate with a fluorescent dye such as Rhodamine 6G or 2',7'-dichlorofluorescein (nondestructive) or by spraying with 50% sulfuric acid and charring at 180°C (destructive). Other reagents are specific for visualizing individual lipid classes such as phospholipids. Migration of spots is expressed by the R_f value: the ratio of the distance migrated by the lipid spot to the distance migrated by the solvent.

High-performance (HP) TLC is a rapid technique that requires little time (several min instead of 0.5 to 3 h), uses small test solution volumes (0.1 µL rather than 1 to 5 µL) to be applied, and provides improved accuracy and an order of magnitude better sensitivity (0.5 to 5 ng). Precoated aluminum sheets are commercially available for HPTLC. Two-dimensional TLC and HPTLC are particularly useful for separation of polar lipid classes and are used for analysis of phospholipids and glycolipids.

Lipid fractions on silica gel can be quantified *in situ* by densitometric or fluorimetric detection. Quantitation can also be achieved by coupling TLC with a flame ionization detector. With the nondestructive spray reagents, lipids may be recovered from the silica gel by elution with an organic solvent for further analytical evaluation.

Microcolumn TLC is performed on a silica gel–glass frit coating on a quartz rod. Quantitation is carried out by scanning with a flame ionization detector (FID) and using the commercially available Iatroscan TLC–FID system. Tanaka et al. (36) described a method for quantitative determination of mono-, di-, and triglycerides and free fatty acids by using a boric acid–impregnated Chromarod with the Iatroscan system. Sebedio et al. (37) studied the quantitative analysis of FAME geometrical isomers and triglycerides differing in unsaturation using silver nitrate–impregnated rods with the Iatroscan system. The authors observed that the system was useful for quantitation of *trans* fatty acids in margarines and partially hydrogenated oils. Przybylski and Eskin (38) used two-dimensional TLC combined with FID-Iatroscan to determine the phospholipid content of canola oil at early stages of processing. Recoveries of greater than 90% were obtained for all the phospholipids measured. De Schrijver and Vermeulen (39) reported a procedure to separate and quantitate phospholipids in animal tissues using the Iatroscan TLC–FID system and oxalic acid–impregnated Chromarods. O'Keefe et al. (40) compared the oxidative stability of high-oleic and normal peanut oils by following increases in polar compounds measured by TLC–FID with the Iatroscan Mark 4 analyzer.

Molkentin and Precht (41) reported a rapid method for determination of *trans*-octadecenoic (*trans*-18:1) acids in edible fats involving silver ion–TLC on commercially available silica plates followed by GC analysis of the isolated *trans*-18:1 acids on a highly polar 50-m or 100-m capillary column. Wolff et al. (42) determined the fatty acid composition of French butters by silver ion–TLC and capillary GC with emphasis on quantitation of the *trans*-18:1 fatty acids determined as the fatty acid isopropyl esters (FAIPE). Total FAIPE were analyzed by capillary GC. *Trans*-18:1 acids were isolated by silver ion–TLC followed by extraction from the silica gel of the combined saturated and *trans*-monounsaturated acids for GC analysis (the saturated acids were employed as internal standards). Bayard and Wolff used silver ion–TLC and capillary GC to determine the *trans*-18:1 isomer content of margarines and shortenings (43) and beef tallow (44).

Peyrou et al. (45) used the Iatroscan Mark 5 TLC–FID analyzer to separate and quantitate mono-, di-, and triacylglycerols, whereas Nakamura et al. (46) designed a rapid method to separate polyunsaturated fatty acids (PUFA) in fish and other marine lipids by TLC of the FAMEs on commercially available silica gel plates using *n*-hexane:ethyl ether:acetic acid (95:5:1 v/v) as developing solvent. Two TLC spots were observed; the lower spot contained PUFA with more than two double bonds, whereas the upper spot contained saturated, monounsaturated, and most of the diunsaturated fatty acids as determined by GC. The total PUFA content of marine lipids was estimated by densitometric measurement. Two reviews (47,48) discuss the applications of TLC and TLC–FID for lipid analysis.

Size Exclusion Chromatography (SEC)

SEC and high-performance (HP) SEC are simple separation techniques based on the relative size of lipid molecules. The stationary phase, which consists of crosslinked macromolecules, discriminates among molecules by excluding the larger ones (which will elute first) and allowing the smaller ones to diffuse mechanically, partially or completely, into pores of appropriate size (for example, 100 or 500 Å). The stationary phase usually used for fat analyses consists of copolymers of styrene–divinyl benzene. HPSEC is a powerful technique for certain applications such as the separation of hydrolytic, oxidative, and thermal products of frying fat. These alteration products include triacylglycerol dimers, oxidized triacylglycerol momomers, diacylglycerols, monoacylglycerols, and fatty acids, including cyclic fatty acids. The refractive index or evaporative light-scattering detector can be used for quantitation.

Aitzetmuller (49) showed that SEC can be used to indicate the extent of heating and polymerization of heated fats. Perrin et al. (50) investigated use of several polystyrene–divinylbenzene columns for HPSEC separation of triacylglycerol dimers and polymers.

Christopoulou and Perkins (51) developed an HPSEC method for separation of monomer, dimer, and trimer fatty acids in frying fats by using a system consisting of two styrene/divinylbenzene copolymer columns with a toluene mobile phase and refractometry for detection. Romero et al. used HPSEC to determine the amounts and distribution of polar compounds in extra virgin olive oil (52) and high-oleic sunflower oil (53) used for frying. Marquez-Ruiz et al. (54,55) used a combination of silica column adsorption chromatography and SEC to determine the distribution and quantity of fatty acid monomers, dimers, and polymers in frying fats to obtain useful information about the influence of oxygen and temperature on samples of unknown origin. Evaluation of commercial frying fats with levels of polar compounds around the limit for fat rejection (21 to 28% polar compounds, as determined by silica column chromatography) gave values of total altered fatty acids ranging from 8 to 11% (55). An HPSEC procedure has been validated by interlaboratory study as a standard method for determination of as little as 3% polymerized triglycerides in vegetable fats and oils (56).

High-Performance Liquid Chromatography (HPLC)

HPLC is used for separating nonvolatile, high-molecular-weight lipids in several modes employing either adsorption or partition chromatography. Adsorption chromatography is widely used for separating classes of lipids according to the nature and number of polar functional groups, such as ester bonds and hydroxyl and phosphate functions. For instance, good separation of triacylglycerols, diacylglycerols, sterols, free fatty acids, and monoacylglycerols can be obtained with a gradient of ethanol into hexane–chloroform (9:1). In this normal phase (NP) HPLC the adsorbent is silica gel, which is a porous material with hydroxyl groups on its surface. An on-line NPHPLC GC method was reported for determination of Δ^7- and $\Delta^{8(14)}$-sterols as a means of identifying adulterated olive oils (57). Silver ion chromatography, in which silica gel is impregnated with silver nitrate, is a form of adsorption chromatography in which the separation occurs, as mentioned earlier for argentation TLC, according to the number and configuration of double bonds in a molecule.

Silver ion HPLC was used to analyze vegetable oil triacylglycerols (58) as well as to separate geometrical and positional FAME isomers of mono-, di-, and polyunsaturated fatty acids (59). FAMEs were less well resolved than phenacyl derivatives, with which baseline resolution of the *cis*-6-, *cis*-9-, and *cis*-11 isomers was possible. Silver ion HPLC was also used to determine the monoenoic fatty acid distribution in hydrogenated vegetable oils (60). Sempore and Bezard (61) described a method for determination of the individual molecular species of vegetable oil triacylglycerols based on application of preparative silver ion TLC, RPHPLC, and chiral-phase HPLC. Myher et al. (62) also used chiral-phase HPLC for stereoscopic analysis of triglycerides rich in long-chain polyunsaturated fatty acids in a scheme involving partial deacylation with ethylmagnesium bromide or pancreatic lipase, separation of the diacylglycerols by borate TLC, resolution of the *sn*-1,2- and *sn*-2,3-enantiomers by chiral-phase HPLC; identification of the enantiomer species by chiral-phase LC-MS with electrospray; and determination of the fatty acid composition by capillary GC. Christie (63) recently reviewed the application of silver ion HPLC and chiral chromatography to the structural analysis of triacylglycerols.

Reverse-phase (RP) HPLC, which is based on partition chromatography, is used to separate individual components that belong to one lipid class. In this case, the stationary phase usually consists of nonpolar octadecylsilane (C_{18}) bonded phase, while the mobile phase is a more polar solvent system such as methanol-water. This chromatographic mode was applied to the separation of geometrical and positional FAME isomers of octadecenoic acid (64). FAMEs from partially hydrogenated fish oils and vegetable oils were separated by RPHPLC. Free fatty acids prepared from margarine were also resolved by this approach. RPHPLC was used to determine the positional distribution of fatty acids in the monounsaturated triacylglycerols of vegetable oils (65), whereas RPHPLC with amperometric detection was used for determination of tocopherols and tocotrienols in vegetable oils (66). Because HPLC separations are conducted at ambient temperatures, they also allow the RP- and NPHPLC separations of thermally labile lipid products, such as isomeric hydroperoxides generated by the oxidation of fatty acids. RPHPLC has also been used as an enrichment step in a method for determination of cyclic fatty acid monomers (CFAM) in heated fats (67). Methyl esters of heated linseed and sunflower oil were hydrogenated and fractionated on a semipreparative HPLC column. The fraction enriched with CFAM was analyzed by capillary GC, and the major hydrogenated CFAMs were identified by GC–MS.

Detectors in the ultraviolet (UV)-visible range are the most commonly used for HPLC. Because most lipids absorb in the 200 to 210-nm range, only solvents that are transparent in this absorption window, such as hexane, acetonitrile, methanol, and water, can be used. UV detection at 206 nm was used to separate lipids (hydrocarbons, sterols, mono-, di-, and triacylglycerols, etc.) from soybean oil into classes and individual species by RPHPLC (68). Derivatives of fatty acids with aromatic chromophores (i.e., phenacyl esters), which exhibit stronger UV absorptions, can be used to improve detection sensitivity, as can derivatives (i.e., anthrylmethyl) for fluorescence detection (69). Miwa and Yamamoto (70) reported determination of fatty acid composition of fats and oils with separations similar to those achieved by capillary GC by employing direct derivatization of saponified test samples with 2-nitrophenylhydrazine, followed by RPHPLC with detection at 400 nm.

Other detectors include refractive index, flame ionization, and light scattering (also known as mass detectors). The evaporative light-scattering detector (ELSD) (71) is highly reproducible and insensitive to solvent changes and polarity. Its response is related to analyte mass, thus making it useful for the quantitation of eluting species. Vaghela and Kilara (72) developed a procedure for separation and quantitation of phospholipids in whey protein concentrates by using a narrow-bore HPLC column and ELSD. Detection sensitivity was improved more than ten-fold compared to a standard analytical column, and solvent consumption was reduced by 80%. Picchioni et al. (73) applied NPHPLC with ELSD to the analysis of plant (apple and carrot) phospholipids and glycolipids. Quantitative results agreed with those obtained by TLC and lipid-P analyses. Abidi et al. (74) recommended use of *tert*-butyl methyl ester in place of tetrahydrofuran in mobile phases for HPLC analysis of soybean phospholipids with ELSD. Balazs et al. (75) evaluated three HPLC methods for separation and quantitation of soybean phospholipids: a mixed-phase method, the AOCS Official Method (76), and a method proposed by the International Lecithin and Phospholipid Society (ILPS). The mixed-phase and AOCS methods are isocratic methods using silica columns with UV detection, whereas the ILPS method involves use of a gradient mobile phase with a diol–bonded silica column and evaporative light-scattering detection. The ILPS method showed the best precision and ruggedness and required the least amount of column stabilizing time.

HPLC–MS is a system with important applications and great potential for lipid analysis. Kusaka et al. (77) identified various saturated and unsaturated fatty acids as their anilides by HPLC–MS with atmospheric pressure ionization. Kuksis et al. (78) identified triacylglycerol species in fats by HPLC–MS with chloride-attachment negative chemical ionization. Huang et al. (79) used a combination of SFE and on-line particle beam HPLC–MS to identify individual triacylglycerols in fats. The particle beam interface can be used for a wide range of analytes, and the electron impact (EI) ion source generally gives typical EI spectra. Neff and Byrdwell (80) used coupled RPHPLC–MS with atmospheric pressure chemical ionization to identify and quantitate triacylglycerols in oils from new soybean lines.

Delgado-Zamarreno et al. (81,82) employed RPHPLC with electrochemical detection for on-line automatic determination of vitamins A, D_3, and E in milk and in butter and margarine. The on-line system comprised channels for alkaline hydrolysis and neutralization of test solutions and C_{18} cartridges for solid-phase extraction placed in the sample loop of a six-port injection valve of the LC system. An American Oil Chemists' Society (AOCS) Official Method (83) is applicable to HPLC separation, identification, and quantitative determination of individual triacylglycerols in edible fats and oils.

Supercritical Fluid Chromatography (SFC)

SFC is a relatively new separation technique that has features from both capillary GC (capillary columns, oven plus detectors) and HPLC (injection valve, pump plus detectors). SFC that utilizes $SC-CO_2$ as the mobile phase has also been interfaced to FTIR (84) and MS (85) systems. Most applications are carried out near or slightly above room temperature, thus making SFC an ideal technique for separating thermally labile compounds.

Capillary SFC with $SC-CO_2$ mobile phase and FID has been used for analysis of fish and vegetable oil triacylglycerols (86). A standard mixture of 13 triacylglycerols was used to optimize the procedure and aid in identifying peaks in real test samples. A method was also developed for quantitative analysis of triacylglycerols using silver ion–SFC (87). A series of vegetable, fish, and hydrogenated oils were separated using a combination of packed microcolumn silver ion–SFC and miniaturized ELSD. The application of silver ion–SFC, as well as other modes of silver ion chromatography of lipids, was recently reviewed (88).

Capillary SFC combined with on-line FTIR spectroscopy was used to determine the relative level of unsaturation and extent of isomerization in partially hydrogenated soybean oil (89) and to study triacylglycerols in cheddar cheese (a "natural" cheese) and in a recently introduced "unsaturated" cheese marketed in the United Kingdom under the "Flora" label (90). Capillary SFC was also used to analyze polyglycerol esters (91) and propoxylated glycerol esters (92), and to determine cholesterol in milk fat (93), fat-soluble vitamins as well as triacylglycerols in edible fats and oils (94,95), emulsifiers in a food product model system (96), and lipid classes in marine oils (97–99). Lou et al. (100) investigated pressure drop effects on selectivity and resolution in packed column SFC. Recent developments in SFC and SFE have been reviewed (101–104).

Gas Chromatography (GC)

GC is a powerful separation technique for resolving and quantifying lipids, including nonpolar, high-molecular-weight species such as triacylglycerols. Methyl ester derivatives of fatty acids are almost exclusively used by GC analysts. However, other derivatives can often provide improved peak resolution. An example is the enhanced separation of petroselenic (cis-6 18:1) and oleic (cis-9 18:1) acids found in carrot, parsley, and other *Umbelliferae* seed oils as their isopropyl esters (105). Complete separation of common fatty acids is usually achieved by using capillary columns with highly polar liquid phases (for example, cyanopropylsiloxane). These columns can readily separate fatty acids according to chain length and degree of unsaturation. Only certain geometric isomers, however, can be resolved, and positional isomers were separated in only a few cases. A preliminary separation with a nonpolar phase could rapidly yield useful information about the different chain lengths that are present in a given unknown mixture of FAMEs. Recent applications of capillary GC to separations of FAMEs were reviewed by Wolff (106).

A FAME can be identified by comparing its observed retention time to that of a standard on at least two columns with liquid phases having different polarities. However, commercial standards are not available for every compound of interest, particularly for unsaturated FAMEs and their isomers. In such a case, the identity of fatty acids (that are between 14 and 22 carbons long) can be deduced from the so-called equivalent chain length (ECL) value. This parameter has great significance, because it depends on the number of double bonds and their distances from both ends of the fatty acid chain. ECL values for unknowns are read directly from a straight line plot of the "logarithms of retention times" of a homologous series of straight-chain saturated FAME standards against "the number of carbon atoms" in the hydrocarbon chain of each compound. For an unknown FAME, ECL values are generated from the observed retention time relative to that of a standard (18:0). All retention times for standards and unknowns must be observed under identical isothermal experimental conditions. The flame ionization detector is the detector of choice for the qualitative and quantitative determination of lipids by GC. However, in order to obtain structural information about eluting components, infrared and mass spectrometers must be used as detectors for the gas chromatograph.

The content of milk fat and vegetable fats, particularly lauric fats (coconut and palm kernel oils) in chocolates, was determined based on capillary GC of the triacylglycerols and FAMEs (107). In the absence of lauric fats, the milk fat content was obtained from the sum of the C_{40} to C_{44} triacylglycerols. If lauric fats were present, the ratio of lauric acid to the sum of the minor fatty acids present between the peaks of myristic and palmitic acids was used to identify and estimate the content of lauric fat in the product. Formulas were proposed for calculation of the milk fat content in chocolates.

Wolff and Bayard (108) recommended use of commercially available 100-m CP-Sil 88 capillary columns for improved resolution of *trans*-18:1 fatty acid isomers following their fractionation by silver ion–TLC. A combination of silver ion–TLC and capillary column GC was used to determine the total *trans*-18:1 fatty acid content of milk fats as well as the isomeric patterns of *trans*-18:1 fatty acids in different fats and oils (109).

Duchateau et al. (110) developed a single, optimized capillary GC method for analysis of *cis* and *trans* fatty acid isomers in hydrogenated and refined vegetable oils. The authors noted that:

1. A single GC method should provide for maximum resolution between the *trans*- and *cis*-18:1 isomers, and the *trans*-13 18:1 isomer must be separated from the *cis*-9 18:1 isomer.

2. 20:1 should be sufficiently resolved from *cis*-, *cis*-, *cis*-18:3 (linolenic acid) to allow correct integration.

An optimization procedure was outlined based on careful selection of stationary phase and temperature program optimization using suitable software. *Trans* values obtained with the optimized GC method were in good agreement with values obtained with silver ion–HPLC.

A new polar capillary column (50% phenyl-, 50% methyl polysiloxane stationary phase) was used for analysis of the total unsaponifiable matter and the sterol fractions in vegetable oils (111). This allowed measurement of constituents not observed with the usual nonpolar or slightly polar columns. Rodriguez-Palmero et al. (112) described a rapid capillary column GC method for determination of the sterol content of prepared foods, avoiding the need for TLC fractionation of the unsaponifiable fraction. The procedure included lipid extraction with dichloromethane:methanol (2:1), saponification at 80°C, separation of the unsaponifiable matter with cyclohexane, and derivatization to form trimethylsilyl ethers, followed by GC and using 5α-cholestane as the internal standard.

Johnson et al. (113) developed an automated sample preparation system for determination of cholesterol in foods in which the saponification, pH adjustment, solid-phase extraction, and drying steps are carried out prior to analysis by capillary GC. Substitution of hexane-isopropanol for methanol-chloroform in the solid-phase extraction step significantly reduced solvent procurement and disposal costs and eliminated the health hazard associated with the use of chloroform. Toschi et al. (114) compared the standard capillary GC and an on-line HPLC–capillary GC method for evaluation of 3,5-stigmastadiene content of edible oils. Quantitation of 3,5-stigmastadiene and other steroidal hydrocarbons, which are formed during fat and oil refining (in particular bleaching), provides evidence of the presence of undeclared refined oil in virgin olive oil and other cold-pressed oils. Results with both methods were in agreement. However, the coupled (on-line) HPLC–GC method used only small amounts of solvents, did not require saponification and subsequent solvent extraction, and required a much shorter analysis time than the standard GC method. Recent reviews discuss the various techniques and applications of gas chromatography (115–117).

Fourier Transform Infrared (FTIR) Spectroscopy

Functional groups (isolated or conjugated double bonds, hydroxyl, epoxy, or ester functions) in fatty acids, FAMEs, or triacylglycerols give rise to unique absorption bands in the mid-infrared (IR) spectral region (wavenumbers 4000–600 cm^{-1}), making IR spectroscopy an important tool for confirmation of structure. A widely used application of dispersive IR or FTIR is the quantitative determination of total *trans* content in FAMEs (118). It is based on measuring the C-H out-of-plane deformation vibration at 966 cm^{-1} for analytes dissolved in the toxic and volatile solvent, carbon disulfide. Unfortunately, this band, which is uniquely characteristic of isolated *trans* unsaturation, overlaps with other broad features in the IR spectrum that lead to a strongly sloping background, converting the *trans* band into a shoulder at levels below 2% and reducing the accuracy of the determination. Many modifications to this method were proposed (119), including:

1. Use of attenuated total reflection (ATR) cells, which require neither weighing nor quantitative dilution of analytes in any solvent.
2. Taking the ratio of the single-beam spectrum of the *trans* fat or its corresponding FAME analyte to that of an appropriate reference material (such as unhydrogenated vegetable oil or its corresponding FAMEs) to obtain the *trans* band as a symmetric feature on a horizontal background.

Recently, a single-bounce-horizontal ATR (SB-HATR) cell requiring only 50 µL of neat test sample was tested for rapidly quantitating the *trans* content (as FAMEs) of 18 commercial food products (120). The SB-HATR procedure offers several advantages over capillary column GC procedures, including rapidity of analysis. However, issues including the identification of appropriate reference material(s) need to be resolved in order to determine the applicability of this novel and rapid techniques to routine analysis of *trans* fatty acids in foods, particularly foods of low *trans* fat content (120).

The use of FTIR spectroscopy in food analysis has been described by van de Voort and coworkers (121–124), who have recently developed FTIR spectroscopic methods for analysis of edible fats and oils, including iodine value and saponification number (125), free fatty acids (126), peroxide value (16), and *cis* and *trans* content (127). The method for *cis* and *trans* content was developed to determine percent *cis* and *trans* content simultaneously at a rate of less than 2 min per measurement. These workers also described the use of FTIR spectroscopy for monitoring the oxidation of edible oils in terms of hydroperoxide, alcohol, and total carbonyl content (128) and to determine the solid fat index of hydrogenated vegetable oil based on spectral data related to triacylglycerol weight/average molecular weight and degree of unsaturation of the fatty acids that make up the triacylglycerols (129). Lai et al. used FTIR spectroscopy to authenticate vegetable oils (130) and to carry out quantitative determinations of refined olive oil and other oils in extra virgin olive oil (131).

Luinge et al. (132) described the quantitative determination of fat, protein, and lactose content of milk using FTIR spectroscopy. IR spectra were recorded between 3000 and 1000 cm^{-1} using demineralized water as background to measure fat absorptions corresponding with asymmetric and symmetric CH_2 stretching at 2922 and 2852 cm^{-1}, respectively; C=O triacylglycerol stretching at 1746 cm^{-1}; the CH_2 deformation at 1466 cm^{-1}; and the ester C-O stretching at around 1160 cm^{-1}. Classical and inverse least squares, principal component, and partial least squares regression were applied. Fat and protein content were compared with Röse-Gottlieb and Kjeldahl reference values. Fat, protein, and lactose were also determined using an IR filter instrument. All methods performed comparably, and the FTIR method was deemed well suited for in-line analysis of fat, protein, and lactose.

Safar et al. (133) used ATR FTIR spectroscopy with principal component analysis to characterize vegetable oils, butters, and margarines as well as to precisely assign the mid-IR absorption bands of edible fats and oils. Villé et al. (134) developed an FTIR method for determination of phospholipid content in animal (pig) lipids based on use of L-α-phosphatidylcholine from egg yolk as a reference for phosphate band identification. The band between 1282 and 1020 cm^{-1} was recommended for determination of phospholipid content, and the band between 1785 and 1697 cm^{-1} was recommended for determination of total fat.

Near Infrared (NI) Spectroscopy

NI spectroscopy has become widely used in food analysis because it is rapid, requires little or no test sample preparation, and is nondestructive. NI spectra in the range of 700 to 2500 nm (14,300 to 4000 cm^{-1}) are due to overtone and combination bands of carbon-hydrogen, oxygen-hydrogen, and nitrogen-hydrogen. NI instruments are commonly used for analyzing grains, flour, meat, and oilseeds (135). The most common constituents determined are moisture, protein, and fat. NI calibration and data analysis are carried out by statistical treatments such as multiple linear regression and partial least squares regression.

Rodriguez-Otero et al. (136) used NI spectroscopy for neat measurement of fat, protein, and total solids in cheese. A set of 90 cow's milk cheeses (Tetilla, Arzua, and Edam) were used to calibrate the NI instrument by principal components analysis and modified partial least squares regression. The calibration was validated with an independent set of similar cheese samples. Lee et al. (137) used short-wavelength (700 to 1100 nm) NI reflectance spectroscopy with a bifurcated fiber optic probe to estimate the crude lipid content in the muscle of whole rainbow trout. Wold et al. (138) determined the average fat content of farmed Atlantic salmon fillets by NI transmittance spectroscopy, obtaining good correlations with fat content determined by chemical analysis.

NI spectroscopy has also been used to differentiate and classify vegetable oils (139,140), to identify undeclared vegetable oils in olive oils (141,142), to determine the fatty acid composition of sunflower seeds (143), to measure the oil and protein content of soybeans stored under actual elevator conditions (144), to evaluate the quality of soybeans stored at different moisture levels (145) or undergoing processing in an oil-milling plant (146), to determine the dimer and polymer triacylglycerols and acid value of used frying oils (147), and to determine the effects of frying time on the quality and stability of frying oil and potato chips (148).

Daun et al. (149) evaluated three whole-seed analyzers for measuring oil, protein, chlorophyll, and glucosinolates in intact canola seed. No significant differences were found between the three instruments for measurement of oil content. One of the instruments gave the best results for glucosinolates, chlorophyll, and protein. Cartagena et al. (150) prepared a general review of statistical treatments (chemometrics) applied to NI spectroscopy.

Gas Chromatography–Fourier Transform Infrared (GC–FTIR) Spectroscopy

Capillary GC–FTIR spectroscopy can be used to obtain on-line FTIR spectra of analytes eluting from the gas chromatograph (84). Aside from obtaining GC retention time data, double bond configuration can be confirmed for individual geometric isomers in complex mixtures of saturated and unsaturated FAMEs. The three available types of GC–FTIR interfaces have each been applied to lipid analysis. The light pipe (LP) interface (151) consists of a flow-through glass tube with alkali halide windows. FTIR spectra of FAMEs are measured in real time as the effluent exits the GC column and passes through the heated LP. A variation of the LP was adapted for the supercritical fluid chromatography (SFC)–FTIR analysis of intact triacylglycerols (89). The matrix isolation (MI) and the direct deposition (DD) interfaces (84) operate at cryogenic temperatures and provide an order of magnitude greater sensitivity (and cost several times more). With these systems the GC effluent is trapped under vacuum on a cryogenic surface for subsequent off-line signal averaging by FTIR. For example, FAME mixtures from hydrogenated vegetable oils (152) (Fig. 1), menhaden oil (153), and methyl ester (154) and DMOX (4,4-dimethyloxazoline) (155) derivatives of cyclic fatty acid monomer (CFAM) mixtures from heated oils have been analyzed by GC–MI–FTIR. The quantitative determination of FAMEs from margarines has also been reported with this technique (152). An advantage of the newer DD interface is that it also allows the on-the-fly measurement of FTIR spectra. Hence, during a GC run the analyst can monitor, at the same time, both the chromatographic peak and the FTIR spectrum of an eluting mixture component.

Jirovetz et al. (156) used GC–FTIR as well as GC–MS to examine the volatiles in the seed oil of *Hibiscus sabdariffa*, the calyx and leaves of which are used in India in the preparation of curries and pickles. More than 25 volatiles (mainly unsaturated hydrocarbons, alcohols, and aldehydes) were identified. Boosfeld and Vitzthum (157) employed GC–FTIR, GC–MS, and NMR spectroscopy to identify several unsaturated aldehydes (nona- and decadienals) in green coffee. Wahl et al. (158) used a combined GC–FTIR–MS system with a wide band (4000 to 500 cm^{-1}) mercury cadmium telluride (MCT) detector to identify *cis/trans* isomers in reference fatty acids and fish

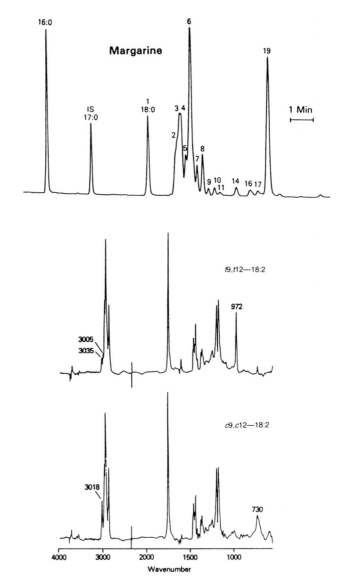

Fig. 1. *Top:* Partial capillary gas chromatogram of fatty acid methyl esters of a commercial margarine; column: CP-Sil 88 (Chrompack), 50 m × 0.22 mm i.d. Peak identification: 1, 18:0; 2-5, *trans* monoenes; 6–11, *cis* monoenes; 14, 16, 17, *trans* dienes; 19, *cis* diene. Various positional and geometric isomers overlap. *Middle:* Matrix isolation GC–FTIR spectrum of *t*9,*t*12 18:2 diene; the 972 cm^{-1} band is unique for the *trans* species. *Bottom:* Matrix isolation GC–FTIR spectrum of *c*9,*c*12 18:2 diene. The matrix isolation GC–FTIR spectra were acquired at 4 cm^{-1} resolution at a cryogenic temperature of ca 12 K (152).

oil. Both methyl ester and DMOX derivatives were examined. Position of the double bond was determined by MS of the DMOX derivatives. *Cis/trans* isomers were identified by examining the IR bands resulting from C–H out-of-plane bending near 720 cm^{-1} (*cis* isomers) and near 967 cm^{-1} (*trans* isomers).

Gas Chromatography–Mass Spectrometry (GC–MS)

GC–MS provides retention time and molecular weight data, as well as structural information (location of double bonds, hydroxyl groups, or alkyl branches) for lipid components (159). One way to determine the location of double bonds in FAMEs by GC–MS is to form addition compounds, such as dimethyldisulfide adducts or trimethylsilyl ether derivatives. Alternatively, unsaturated sites can be found with derivatives that can localize the charge of the molecular ion away from the double bonds. Pyrrolidine, picolinyl ester (160–162), DMOX (162–164), or nicotinate derivatives (160,165) localize the charge predominantly on the nitrogen atom of these functional groups and yield electron ionization (EI) mass spectra that exhibit distinctive fragmentation patterns from which double bond location can be easily determined. An important advantage of DMOX derivatives is that they readily chromatograph on the same polar phases used for FAMEs with no loss in resolution (166)—at times, under certain experimental conditions, with improved resolution (167) (Fig. 2). Garrido and Medina (168) described a procedure for direct conversion of either free or esterified fatty acids in oils or total lipid extracts into their DMOX derivatives. The procedure is rapid, eliminates saponification, and avoids formation of interfering dimethylacetals. The molecular weight, position of double bonds along the fatty acid chain, and ring size of fatty acid monomers in heated linseed oil were recently determined by GC-EIMS for a complex mixture of CFAM DMOX derivatives (166). DMOX derivatives were also used for GC–MS confirmation of the structure of a number of minor CFAM components in the oil (169).

Joh et al. (170) carried out a structural analysis of seed oil triacylglycerols by RP and silver ion HPLC of the triacylglycerols and GC–MS of the fatty acids in each HPLC fraction as the picolinyl esters. Laakso and Kallio (171) optimized conditions for MS analysis of triacylglycerols using a direct exposure probe and ammonia negative-ion chemical ionization. The reproducibility of the method was demonstrated with standards and a raspberry seed oil. GC–MS was also used to identify phenolic compounds present in virgin olive oil (172) and wax esters in the roe oil of amber fish (173).

Nuclear Magnetic Resonance (NMR) Spectroscopy

Wide-line NMR spectroscopy has been used by the edible oils and fats industry for process monitoring and quality control purposes. It has been applied to the rapid determination of the oil content of oil seeds and meals, solid fat content of food fats, and fat content of food products (174). AOCS methods are available for determination of oil content of oilseeds (175,176) and solid fat content of commercial fats

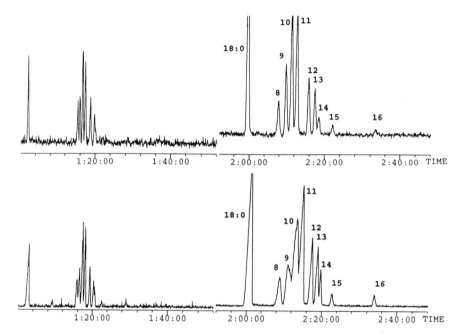

Fig. 2. GC separation of *trans* 18:1 positional isomers from partially hydrogenated soybean oil after isolation by silver ion HPLC. GC-EIMS chromatographic data are shown for two separate GC runs using the SP 2560, 100-m capillary column. The *top* and *bottom* traces were for two GC runs in which 0.4 and 1.0 µL portions of the test sample were injected, respectively. The M/Z 264 and 113 ion profiles were recorded for *trans* 18:1 positional isomer FAME and DMOX derivatives, respectively, present in the same test portion. The DMOX eluted about 1 h after the FAME derivatives under 140°C isothermal GC conditions. The numerical labels 8 through 16 next to GC peaks denote the identity of the Δ8 through Δ16 *trans* 18:1 positional isomers (167).

(177). AOCS Method Ak 3-94 (175), using a continuous-wave low-resolution spectrometer, requires that seed samples be dried prior to NMR measurement. AOCS Method Ak 4-95 (176), requiring a pulsed instrument, recommends that the moisture content of sunflower seeds and rapeseeds be between 6 and 9% and that the moisture content of soybeans be between 6 and 12%. Tiwari and Gambhir (178) used combined free induction decay and spin-echo NMR signals to measure the oil content of seeds without weighing or drying the seeds.

High-resolution proton and carbon-13 NMR are important tools available for confirming the identity and elucidating the structure of lipid molecules. Unfortunately, they are relatively insensitive and require large amounts of test samples (about 1 to 100 mg). The usefulness of high-resolution NMR is greatest for determining the structure of individual compounds, but it diminishes for analyzing components of complex mixtures. Carbon-13 NMR spectra exhibit highly characteristic signals such as those for olefinic, allylic, and ω1–3 carbon atoms (179).

Applications of this technique include studies of mixtures of mono-, di- and triacylglycerols (180); determination of identity and composition of butterfat, other fats, and mixtures of fats (181); and analysis of partially hydrogenated vegetable oils (182) as well as edible vegetable oils (183) and oils and fats containing double bonds close to the carboxyl group (fish oil) (184,185), oxygenated acids (epoxy oils) (186), hydroxy acids (187), branched-chain acids (188), and margarines (189). Henderson et al. (190) used proton, carbon-13, and two-dimensional proton–carbon-13 NMR to analyze complex triacylglycerol mixtures.

^{31}P-NMR is a preferred method to determine phospholipid composition (191,192). This approach was used to quantify phospholipids in canned peas (193). Hanna and Specchio (194) reviewed the application of NMR spectroscopy in food regulatory analysis.

Stable Carbon Isotope Ratio Analysis (SCIRA)

Measurement of the relative abundance of the naturally occurring stable carbon isotopes ^{12}C and ^{13}C has been applied to determination of the authenticity of a variety of foods (195,196). SCIRA is useful in food analysis and control because the content of stable and naturally occurring isotopes is often characteristic of the origin, purity, and processing of a food and its components. An important application is the differentiation of products prepared from corn or sugar cane (C_4 plants) and products prepared from other cultivated plants.

SCIRA is applicable to food analysis because the $^{13}C/^{12}C$ ratio of plant products depends on the photosynthetic pathway used by each plant (196,197). Most land plants use the C_3 pathway (normal or Calvin cycle) to incorporate CO_2 by carboxylation of ribulose diphosphate to phosphoglycerate (C_3 product). However, some plants, including corn, millet, and sugar cane, use the alternate C_4 (Hatch-Slack) pathway involving carboxylation of phosphoenolpyruvate to oxalacetate (C_4 product). A few plants, including pineapple and cacti, use the crassulacean acid metabolic pathway, capable of C_3 or C_4 CO_2 fixation (196).

SCIRA is carried out by combustion of a test portion to CO_2 and measurement of the $^{13}C/^{12}C$ ratio with an isotope ratio mass spectrometer or other instrument capable of determining the ^{13}C and ^{12}C content of CO_2 (195,198). Isotopic composition is expressed as $\delta^{13}C$ values in units permil (‰) and defined as the permil deviation of the respective isotope ratio of a sample vs. a standard. The $\delta^{13}C$ values are determined from comparisons with CO_2 generated from a reference standard (Pee Dee belemnite).

Gaffney et al. (199) determined that $\delta^{13}C$ values for beef (-13.1 ± 1.6) and pork (-12.5 ± 0.4) were consistent with the heavy use of corn and sorghum (C_4 plants) as livestock feeds in the United States. The $\delta^{13}C$ value for corn oil was determined to be -12.4 ± 0.6, vs. -27 to -28 for peanut, soybean, and other vegetable oils.

Since the fatty acid compositions of other vegetable oils are close to or overlap that of corn oil, and the large concentration of sterols in corn oil can swamp the sterols of blends, Rossell (200) evaluated the SCIRA technique for determining corn oil authenticity. Stable isotope ratios of corn oil from various regions in Asia, Africa, Australia, New Zealand, North America, and South America ranged from −13.87 to −16.36, an average (42 samples) of −14.95. SCIRA identified the presence of as little as 10% of a foreign oil in corn oil.

Magnetic Resonance Imaging (MRI)

The application of MRI to foods has grown rapidly in the past few years. The method is based on manipulation of magnetic field gradients oriented at right angles to each other to provide spacial encoding of signals from an object, which are converted by Fourier transform techniques to three-dimensional NMR images (201). The NMR imaging produces three-dimensional data by detecting two-dimensional cross sections in all directions. Most MR images of food are based on proton resonances, from either water or lipids. The real potential of MRI lies in following, in real time and noninvasively, the dynamic changes in foods as they are stored, processed, packaged, and distributed (202). Thus, various operations including drying, frying, freezing, heating, blanching, freeze drying, and rehydrating can be investigated under conditions identical to those used in industrial storage and processing.

Simoneau et al. (203) discussed the use of MRI for study of fat crystallization in bulk or in dispersed systems. Halloin et al. (204) applied two MRI techniques (spin-echo imaging (SEI) and chemical-shift imaging (CSI)) to study lipid distribution in pecan embryos. SEI of normal dry embryos showed a fairly uniform distribution of lipid in the tissues. Embryos damaged by insects or fungus gave images that were less intense than those of normal embryos, reflecting a lower oil content.

Hiel et al. (205) used MRI and NMR spectroscopy to measure oil content in French-style salad dressings. Imaging and NMR spectroscopy gave results within ±2% of expected values and were in good agreement with oil content determined by traditional methods. Pilhofer et al. (206) studied the formation and stability of foamed oil in water emulsions of vegetable oil, milk fat, or milk fat fractions. Tingley et al. (207) used MRI to study the distribution of muscle and fat in retail meat as a means of quality assessment. MRI can also be used to visualize oil and water concentration gradients during deep fat frying of food (208).

Data Analysis

Chemometric methods (multivariate data analyses) are used to analyze complex mixtures and validate food composition data. Multivariate classification methods based on fatty acid composition have been used to determine the geographical origin of olive oils (209–211). Various European virgin olive oils have been characterized by using volatile components determined by a dynamic headspace GC method (212). The GC data were subjected to discriminant and principal component analyses, and individual varieties were distinguished with a set of only six compounds. Pattern analysis of triglyceride data obtained by capillary GC has been used to identify individual vegetable oils and confirm adulteration (213). Principal component analysis of near IR spectroscopic data has been used to classify vegetable oils and detect adulteration (140). Principal component analysis has also been used to classify virgin olive oils produced in different geographical areas of Italy on the basis of phenolic and aroma compounds and fatty acid and triacylglycerol composition (214) and to characterize edible oils by their triacylglycerol composition determined by positive-ion atom bombardment–MS (215).

Aparicio and Alonso (216) described the use of an expert system for characterization of virgin olive oils to determine whether the oils are genuine and to classify the oils in relation to country and region where produced. The chemical parameters measured for the expert system database included fatty acids, alcohols, sterols, methylsterols, hydrocarbons, and triacylglycerols. Lipp (217) compared three methods (multiple linear regression, principal component regression, and partial least squares regression) for evaluation of GC triacylglycerol data applied to detection of foreign fats in butterfat. Partial least squares regression appeared to offer the lowest detection limit of about 2% foreign fat added to butterfat. Anklam et al. (218) used chemometric data obtained by HPLC triacylglycerol analysis (light-scattering detection) to detect undeclared vegetable fats in cocoa butter. Chemometric methods in analytical science have been reviewed recently (219).

References

1. Elmer-Frohlich, K., and Lachance, P.A. (1992) Faster and Easier Methods for Quantitative Lipid Extraction for Miniature Samples of Animal Tissues, *J. Am. Oil Chem. Soc. 69*, 243.
2. Lee, C.M., Trevino, B., and Chaiyawat, M. (1996) A Simple and Rapid Solvent Extraction Method for Determining Total Lipids in Fish Tissue, *J. AOAC Intern. 79*, 487.
3. Carpenter, D.E., and Ngeh-Ngwainbi, J. (1993) "Lipid Analysis" in *Methods of Analysis for Nutrition Labeling* (Sullivan, D.M., and Carpenter, D.E., eds.), AOAC International, Arlington, pp. 85–109.
4. Park, P.W., and Goins, R.E. (1994) *In situ* Preparation of Fatty Acid Methyl Esters for Analysis of Fatty Acid Composition in Foods, *J. Food Sci. 59*, 1262.
5. House, S.D., Larson, P.A., Johnson, R.R., DeVries, J.W., and Martin, D.L. (1994) Gas Chromatographic Determination of Total Fat Extracted From Food Samples Using Hydrolysis in the Presence of Antioxidant, *J. AOAC Intern. 77*, 960.
6. Rader, J.I., Angyal, G., O'Dell, R.G., Weaver, C.M., Sheppard, A.J., and Bueno, M.P. (1995) Determination of Total Fat and Saturated Fat in Foods by Packed Column Gas-Liquid Chromatography After Acid Hydrolysis, *Food Chem. 54*, 419.
7. *AOAC Official Methods of Analysis* (1997) 16th edn., 3rd suppl., Official Method 996.01, AOAC International, Gaithersburg, in press.
8. Lembke, P., and Engelhardt, H. (1993) Development of a New SFE Method for Rapid Determination of Total Fat Content of Food, *Chromatographia 35*, 509.
9. King, J.W., and List, G. (1996) *Supercritical Fluid Technology in Oil and Lipid Chemistry,* AOCS Press, Champaign, Illinois.
10. AOCS Recommended Practice Cd 8b-90 *Official Methods and Recommended Practices,* 4th edn., 1987–, American Oil Chemists' Society, Champaign, Illinois.

11. AOCS Official Method Cd 8-53 (1987) in *Official Methods and Recommended Practices*, 4th edn., 1987–. American Oil Chemists' Society, Champaign, Illinois.
12. Akaza, I., and Aota, N. (1990) Colorimetric Determination of Lipid Hydroperoxides in Oils and Fats with Microperoxidase, *Talanta 37*, 925.
13. Oishi, M., Onishi, K., Nishijima, M., Nakagomi, K., Nakazawa, H., Uchiyama, S., and Suzuki, S. (1992) Rapid and Simple Coulometric Measurements of Peroxide Value in Edible Oils and Fats, *J. AOAC Intern. 75*, 507.
14. Lovaas, E. (1992) A Sensitive Spectrophotometric Method for Lipid Hydroperoxide Determination, *J. Am. Oil Chem. Soc. 69*, 777.
15. Matthaus, B., Wiezorek, C., and Eichner, K. (1994) Fast Chemiluminescence Method for Detection of Oxidized Lipids, *Fat Sci. Technol. 96*, 95.
16. Van de Voort, F.R., Ismail, A.A., Sedman, J., Dubois, J., and Nicodemo, T. (1994) The Determination of Peroxide Value by Fourier Transform Infrared Spectroscopy, *J. Am. Oil Chem. Soc. 71*, 921.
17. Tur'yan, Ya.I., Berezin, O.Yu., Kuselman, I., and Shenhar, A. (1996) pH-metric Determination of Acid Values in Vegetable Oils without Titration, *J. Am. Oil Chem. Soc. 73*, 295.
18. Nomura, A., Yamada, J., Yarita, T., Sudo, Y., Kudo, S., and Nishizawa, Y. (1995) Determination of Free Fatty Acids by Supercritical Fluid Chromatography on an ODS Silica-Gel Column, *Anal. Sciences 11*, 385.
19. Buchberger, W., and Winna, K. (1996) Determination of Free Fatty Acids by Capillary Zone Electrophoresis, *Mikrochim. Acta 122*, 45.
20. Sebedio, J.-L. (1995) Classical Chemical Techniques for Fatty Acid Analysis, *New Trends in Lipid and Lipoprotein Analysis*, Sebedio J.-L., and Perkins, E.G. AOCS Press, Champaign, Illinois, p. 277.
21. Wolff, R.L., and Sebedio, J.-L. (1994) Characterization of Gamma-Linolenic Acid Geometrical Isomers in Borage Oil Subjected to Heat Treatments (Deodorization), *J. Am. Oil Chem. Soc. 71*, 117.
22. Wolff, R.L., and Sebedio, J.-L. (1991) Geometrical Isomers of Linolenic Acid in Low-Calorie Spreads Marketed in France, *J. Am. Oil Chem. Soc. 68*, 719.
23. Ratnayake, W.M.N., and Pelletier, G. (1992) Positional and Geometrical Isomers of Linoleic Acid in Partially Hydrogenated Oils, *J. Am. Oil Chem. Soc. 69*, 95.
24. Wolff, R.L., Nour, M., and Bayard, C.C. (1996) Participation of the *cis*-12 Ethylenic Bond to *cis-trans* Isomerization of the *cis*-9 and *cis*-15 Ethylenic Bonds in Heated α-Linolenic Acid, *J. Am. Oil Chem. Soc. 73*, 327.
25. Sebedio, J.-L., Septier, Ch., and Grandgirard, A. (1986) Fractionation of Commercial Frying Oil Samples Using Sep-Pak Cartridges, *J. Am. Oil Chem. Soc. 63*, 1541.
26. Kaluzny, M.A., Duncan, L.A., Merritt, M.V., and Epps, D.E. (1985) Rapid Separation of Lipid Classes in High Yield and Purity Using Bonded Phase Columns, *J. Lipid Res. 26*, 135.
27. Wilson, R., Henderson, R.J., Burkow, I.C., and Sargent, J.R. (1993) The Enrichment of n-3 Polyunsaturated Fatty Acids Using Aminopropyl Solid Phase Extraction Columns, *Lipids 2*, 51.
28. Prieto, J.A., Ebri, A., and Collar, C. (1992) Optimized Separation of Nonpolar and Polar Lipid Classes from Wheat Flour by Solid-Phase Extraction, *J. Am. Oil Chem. Soc. 69*, 387.
29. Vaghela, M.N., and Kilara, A., (1995) A Rapid Method for Extraction of Total Lipids from Whey Protein Concentrates and Separation of Lipid Classes with Solid Phase Extraction, *J. Am. Oil Chem. Soc. 72*, 1117.

30. Schmarr, H.-G., Gross, H.B., and Shibamoto, T. (1996) Analysis of Polar Cholesterol Oxidation Products. Evaluation of a New Method Involving Transesterification, Solid Phase Extraction, and Gas Chromatography, *J. Agric. Food Chem. 44,* 512.
31. Perez-Camino, M.C., Moreda, W., and Cert, A. (1996) Determination of Diacylglycerol Isomers in Vegetable Oils by Solid-Phase Extraction Followed by Gas Chromatography on a Polar Phase, *J. Chromatogr. A 721,* 305.
32. Christie, W.W. (1989) Silver Ion Chromatography Using Solid-Phase Extraction Columns Packed with a Bonded-Sulfonic Acid Phase, *J. Lipid Res. 30,* 1471.
33. Wachob, G.D. (1991) Solid Phase Extraction of Lipids, in *Analyses of Fats, Oils and Lipoproteins,* Perkins, E.G., American Oil Chemists' Society, Champaign, Illinois, pp. 122–137.
34. Ebeler, S.E., and Shibamoto, T. (1994) Overview and Recent Developments in Solid Phase Extraction for Separation of Lipid Classes, in *Lipid Chromatographic Analysis,* Shibamoto, T., Marcel Dekker Inc., New York, pp. 1–49.
35. Ebeler, S.E., and Ebeler, J.D. (1996) SPE Methodologies for the Separation of Lipids, *INFORM 7,* 1094.
36. Tanaka, M., Itoh, T., and Kaneko, H. (1980) Quantitative Determination of Isomeric Glycerides, Free Fatty Acids and Triglycerides by Thin Layer Chromatography–Flame Ionization Detector System, *Lipids 15,* 872.
37. Sebedio, J.L., Farquharson, T.E., and Ackman, R.G. (1985) Quantitative Analysis of Methyl Esters of Fatty acid Geometrical Isomers, and of Triglycerides Differing in Unsaturation, by the Iatroscan TLC/FID Technique Using $AgNO_3$ Impregnated Rods, *Lipids 20,* 555.
38. Przybylski, R., and Eskin, N.A.M. (1991) Phospholipid Composition of Canola Oils during the Early Stages of Processing as Measured by TLC with Flame Ionization Detector, *J. Am. Oil Chem. Soc. 68,* 241.
39. De Schrijver, R., and Vermeulen, D. (1991) Separation and Quantitation of Phospholipids in Animal Tissues by Iatroscan TLC/FID, *Lipids 26,* 74.
40. O'Keefe, S.F., Wiley, V.A., and Knauft, D.A. (1993) Comparison of Oxidative Stability of High- and Normal-Oleic Peanut Oils, *J. Am. Oil Chem. Soc. 70,* 489.
41. Molkentin, J., and Precht, D. (1995) Optimized Analysis of *trans*-Octadecenoic Acids in Edible Oils. *Chromatographia 41,* 267.
42. Wolff, R.L., Bayard, C.C., and Fabien, R.J. (1995) Evaluation of Sequential Methods for the Determination of Butterfat Fatty Acids. Application to the Study of Seasonal Variations in French Butters, *J. Am. Oil Chem. Soc. 72,* 1471.
43. Bayard, C.C., and Wolff, R.L. (1995) *Trans*-18:1 acids in French Tub Margarines and Shortenings:Recent Trends, *J. Am. Oil Chem. Soc. 72,* 1485.
44. Bayard, C.C., and Wolff, R.L. (1996) Analysis of *trans*-18:1 Isomer Content and Profile in Edible Refined Beef Tallow, *J. Am. Oil Chem. Soc. 73,* 531.
45. Peyrou, G., Rakotondrazafy, V., Mouloungui, Z., and Gaset, A. (1996) Separation and Quantitation of Mono-, Di- and Triglycerides and Free Oleic Acid Using Thin-Layer Chromatography with Flame-Ionization Detection, *Lipids 31,* 27.
46. Nakamura, T., Fukuda, M., and Tanaka, R. (1996) Estimation of Polyunsaturated Fatty Acid Content in Lipids of Aquatic Organisms Using Thin-Layer Chromatography on a Plain Silica Gel Plate, *Lipids 31,* 427.
47. Shukla, V.K.S. (1995) Thin-Layer Chromatography of Lipids, in *New Trends in Lipid and Lipoprotein Analyses,* Sebedio, J.-L., and Perkins, E.G., AOCS Press, Champaign, Illinois, pp. 17–23.

48. Sebedio, J.-L. (1995) Utilization of Thin-Layer Chromatography-Flame Ionization Detection for Lipid Analyses, in *New Trends in Lipid and Lipoprotein Analyses*, Sebedio, J.-L., and Perkins, E.G., AOCS Press, Champaign, Illinois, pp. 24–37.
49. Aitzetmuller, K. (1972) Investigation of Artifacts Formed in Frying Fats, *Fette Seifen Anstrichm. 74*, 598.
50. Perrin, J.-L., Redero, F., and Prevot, A. (1984) Dosage Rapide des Polymeres de Triglycerides par Chromatographie d'Exclusion, *Rev. Franç. Corps Gras 31*, 131.
51. Christopoulou, C.N., and Perkins, E.G. (1989) High Performance Size Exclusion Chromatography of Monomer, Dimer and Trimer Mixtures, *J. Am. Oil Chem. Soc. 66*, 1338.
52. Romero, A., Cuesta, C., and Sanchez-Muniz, F.J. (1995) Quantitation and Distribution of Polar Compounds in Extra Virgin Olive Oil Used in Fryings with Turnover of Fresh Oil, *Fat Sci. Technol. 97*, 403.
53. Romero, A., Sanchez-Muniz, F.J., Tulasne, C., and Cuesta, C. (1995) High-Performance Size-Exclusion Chromatographic Studies on a High-Oleic Sunflower Oil during Potato Frying, *J. Am. Oil Chem. Soc. 72*, 1513.
54. Marquez-Ruiz, G., Perez-Camino, M.C., and Dobarganes, M.C. (1990) Combination of Adsorption and Size-Exclusion Chromatography for the Determination of Fatty Acid Monomers, Dimers and Polymers, *J. Chromatogr. 514*, 37.
55. Marquez-Ruiz, G., Tasioula, M., and Dobarganes, M.C. (1995) Quantitation and Distribution of Altered Fatty Acids in Frying Fats, *J. Am. Oil Chem. Soc. 72*, 1171.
56. Firestone, D. (1994) Gel-Permeation Liquid Chromatographic Method for Determination of Polymerized Triglycerides in Oils and Fats: Summary of Collaborative Study, *J. AOAC Intern. 77*, 957.
57. Biedermann, M., Grob, K., and Mariani, C. (1995) On-Line LC-UV-GC-FID for the Determination of Δ^7- and $\Delta^{8(14)}$-Sterols and its Application for the Detection of Adulterated Olive Oils, *Riv. Ital. Sostanze Grasse 72*, 339.
58. Neff, W.E., Adlof, R.O., List, G.R., and El-Agaimy, M. (1994) Analysis of Vegetable Oil Triacylglycerols by Silver Ion High Performance Liquid Chromatography with Flame Ionization Detection, *J. Liq. Chromatogr. 17*, 3951.
59. Nikolova-Damyanova, B., Herslof, B.G., and Christie, W.W. (1992) Silver Ion High-Performance Liquid Chromatography of Derivatives of Isomeric Fatty Acids, *J. Chromatogr. A 609*, 133.
60. Adlof, R.O., Copes, L.C., and Emken, E.A. (1995) Analysis of Monoenoic Fatty Acid Distribution in Hydrogenated Vegetable Oils by Silver-Ion High-Performance Liquid Chromatography, *J. Am. Oil Chem. Soc. 72*, 571.
61. Sempore, G., and Bezard, J. (1991) Determination of the Molecular Species of Oil Triacylglycerols by Reverse-Phase and Chiral-Phase High-Performance Liquid Chromatography, *J. Am. Oil Chem. Soc. 68*, 702.
62. Myher, J.J., Kuksis, A., Geher, K., Park, P.W., and Diersen-Schade, D.A. (1996) Stereospecific Analysis of Triacylglycerols Rich in Long-Chain Polyunsaturated Fatty Acids, *Lipids 31*, 207.
63. Christie, W.W. (1994) Silver ion and Chiral Chromatography in the Analysis of Triacylglycerols, *Progress in Lipid Research 33*, 9.
64. Svensson, L., Sisfontes, L., Nyborg, G., and Blomstrand, R. (1982) High Performance Liquid Chromatography and Glass Capillary Gas Chromatography of Geometric and Positional Isomers of Long Chain Monounsaturated Fatty Acids, *Lipids 17*, 50.

65. Deffense, E. (1993) Nouvelle Méthode d'Analyse pour Séparer, via HPLC, les Isomères de Position 1-2 et 1-3 des Triglycerides Mono-Insaturés des Graisses Vegetales, *Rev. Franç. Corps Gras 40*, 33.
66. Dionisi, F., Prodolliet, J., and Tagliaferri, E. (1995) Assessment of Olive Oil Adulteration by Reversed-Phase High-Performance Liquid Chromatography/Amperometric Detection of Tocopherols and Tocotrienols, *J. Am. Oil Chem. Soc. 72*, 1505.
67. Sebedio, J.-L., Prevost, J., Ribot, E., and Grandgirard, A. (1994) Utilization of High-Performance Liquid Chromatography as an Enrichment Step for the Determination of Cyclic Fatty Acid Monomers in Heated Fats and Biological Samples, *J. Chromatogr. A 659*, 101.
68. Antonopoulou, S., Andrikopoulos, N.K., and Demopoulos, C.A. (1994) Separation of the Main Neutral Lipids into Classes and Species by RPHPLC and UV Detection, *J. Liq. Chromatogr. 17*, 633.
69. Christie, W.W. (1987) *High-Performance Liquid Chromatography and Lipids*, Pergamon Press, New York.
70. Miwa, H., and Yamamoto, M. (1996) Rapid Liquid Chromatographic Determination of Fatty Acids as 2-Nitrophenylhydrazine Derivatives, *J. AOAC Intern. 79*, 493.
71. Van der Meeren, P., Vanderdeelen, J., Huys, M., and Baert, L. (1988) Simple and Tapid Method for high-Performance Liquid Chromatographic Separation and Quantitation of Soybean Phospholipids, *J. Chromatogr. 447*, 436.
72. Vaghela, M.N., and Kilara, A. (1995) Quantitative Analysis of Phospholipids from Whey Protein Concentrates by High-Performance Liquid Chromatography with a Narrow-Bore Column and an Evaporative Light-Scattering Detector, *J. Am. Oil Chem. Soc. 72*, 729.
73. Picchioni, G.A., Watada, A.E., and Whitaker, B.D. (1996) Quantitative High-Performance Liquid Chromatography Analysis of Plant Phospholipids and Glycolipids Using Light-Scattering Detection, *Lipids 31*, 217.
74. Abidi, S.L., Mounts, T.L., and Finn, T. (1996) A Preferred Solvent System for High-Performance Liquid Chromatographic Analysis of Soybean Phospholipids with Evaporative Light-Scattering Detection, *J. Am. Oil Chem. Soc. 73*, 535.
75. Balazs, P.E., Schmit, P.L., and Szuhaj, B.F. (1996) High-Performance Liquid Chromatographic Separations of Soy Phospholipids, *J. Am. Oil Chem. Soc. 73*, 193.
76. AOCS Official Method Ja 7b-91 in *Official Methods and Recommended Practices*, 4th edn., 1987–, American Oil Chemists' Society, Champaign, Illinois.
77. Kusaka, T., Ikeda, M., Nakano, H., and Numajiri, Y. (1988) Liquid Chromatography/Mass Spectrometry of Fatty Acids as their Anilides, *J. Biochem. 104*, 495.
78. Kuksis, A., Marai, L., and Myher, J.J. (1991) Reversed-Phase Liquid Chromatography-Mass Spectrometry of Complex Mixtures of Natural Triacylglycerols with Chloride-Attachment Negative Chemical Ionization, *J. Chromatogr. 588*, 73.
79. Huang, A.S., Robinson, L.R., Gursky, L.G., Profits, R., and Sabidong, C.G. (1994) Identification and Quantification of SALATRIM 23CA in Foods by the Combination of Supercritical Fluid Extraction, Particle Beam LC–Mass Spectrometry, and HPLC with Light-Scattering Detector, *J. Agric. Food Chem. 42*, 468.
80. Neff, W.E., and Byrdwell, W.C. (1995) Soybean Oil Triacylglycerol Analysis by Reversed-Phase High-Performance Liquid Chromatography Coupled with Atmospheric Pressure Chemical Ionization Mass Spectrometry, *J. Am. Oil Chem. Soc. 72*, 1185.

81. Delgado-Zamarreno, M.M., Sanchez-Perez, A., Gomez-Perez, M.C., and Hernandez-Mendez, J. (1995) Directly Coupled Sample Treatment–High-Performance Liquid Chromatography for On-Line Automatic Detemination of Liposoluble Vitamins in Milk, *J. Chromatogr. A 694*, 399.
82. Delgado-Zamarreno, M.M., Sanchez-Perez, A., Gomez-Perez, M.C., and Hernandez-Mendez, J. (1995) Automatic Determination of Liposoluble Vitamins in Butter and Margarine Using Triton X-100 Aqueous Micellar Solution by Liquid Chromatography with Electrochemical Detection, *Anal. Chim. Acta 315*, 201.
83. AOCS Official Method Ce 5c-93 (1993) in *Official Methods and Recommended Practices*, 4th edn., 1987–, American Oil Chemists' Society, Champaign, Illinois.
84. Mossoba, M.M. (1993) Applications of Capillary GC–FTIR, *INFORM 4*, 854–859.
85. Snyder, J.M., Taylor, S.L., and King, J.W. (1993) Analysis of Tocopherols by Capillary Supercritical Fluid Chromatography and Mass Spectrometry, *J. Am. Oil Chem. Soc. 70*, 349.
86. Baiocchi, C., Saini, G., Cocito, C., Giacosa, D., Roggero, M.A., Marengo, E., and Favale, M. (1993) Analysis of Vegetable and Fish Oils by Capillary Supercritical Fluid Chromatography with Flame Ionization Detection, *Chromatographia 37*, 525.
87. Blomberg, L.G., Demirbüker, M., and Andersson, P.E. (1993) Argentation Supercritical Fluid Chromatography for Quantitative Analysis of Triacylglycerols, *J. Am. Oil Chem. Soc. 70*, 939.
88. Dobson, G., Christie, W.W., and Nikolova-Damyanova, B. (1995) Silver Ion Chromatography of Lipids and Fatty Acids, *J. Chromatogr. B 671*, 197.
89. Calvey, E.M., McDonald, R.E., Page, S.W., Mossoba, M.M., and Taylor, L.T. (1991) Evaluation of SFC/FTIR for Examination of Hydrogenated Soybean Oil, *J. Agric. Food Chem. 39*, 542.
90. Kaplan, M., Davidson, G., and Poliakoff, M. (1994) Capillary Supercritical Fluid Chromatography–Fourier Transform Infrared Spectroscopy of Triglycerides and the Qualitative Analysis of Normal and "Unsaturated" Cheeses, *J. Chromatogr. A 673*, 231.
91. Chester, T.L., and Innis, D.P. (1986) Effect of Free Hydroxy Groups in the Separation of Polyglycerol Esters by Capillary Supercritical Fluid Chromatography, *J. High Res. Chromatogr. & Chromatogr. Comm. 9*, 178.
92. Lu, X.J., Myers, M.R., and Artz, W.E. (1993) Supercritical Fluid Chromatographic Analysis of the Propoxylated Glycerol Esters of Oleic Acid, *J. Am. Oil Chem. Soc. 70*, 355.
93. Huber, W., Molero, A., Pereyra, E., and Martinez de la Ossa, E. (1995) Determination of Cholesterol in Milk Fat by Supercritical Fluid Chromatography, *J. Chromatogr. A 715*, 333.
94. Manninen, P., Laakso, P., and Kallio, H. (1995) Method for Characterization of Triacylglycerols and Fat-Soluble Vitamins in Edible Oils and Fats by Supercritical Fluid Chromatography, *J. Am. Oil Chem. Soc. 72*, 1001.
95. Ibanez, E., Tabera, J., Regero, G., and Herraiz, M. (1995) Optimization of Separation of Fat- Soluble Vitamins by Supercritical Fluid Chromatography Using Serial Micropacked Columns, *J. Agric. Food Chem. 43*, 2667.
96. Artz, W.E., and Myers, M.R. (1995) Supercritical Fluid Extraction and Chromatography of Emulsifiers, *J. Amer. Oil Chem. Soc. 72*, 219.
97. Staby, A., Borch-Jensen, C., Balchen, S., and Mollerup, J. (1994) Supercritical Fluid Chromatographic Analysis of Fish Oils, *J. Am. Oil Chem. Soc. 71*, 355.

98. Staby, A., Borch-Jensen, C., Balchen, S., and Mollerup, J. (1994) Quantitative Analysis of Marine Oils by Capillary Supercritical Fluid Chromatography, *Chromatographia 39,* 697.
99. Borch-Jensen, C., and Mollerup, J. (1996) Supercritical Fluid Chromatography of Fish, Shark and Seal Oils, *Chromatographia 42,* 252.
100. Lou, X., Janssen, H.-G., Snijders, H., and Cramers, C.A. (1996) Pressure Drop Effects on Selectivity and Resolution in Packed Column Supercritical Fluid Chromatography, *J. High Resol. Chromatogr. 19,* 449.
101. Laakso, P. (1992) Supercritical Fluid Chromatography of Lipids, in *Advances in Lipid Methodology—One,* Christie, W.W., The Oily Press, Ayr, Scotland, pp. 81–119.
102. Levy, J.M., and Houck, R.K. (1993) Developments in Off-Line Collection for Supercritical Fluid Extraction, *American Laboratory 25(7),* 36R.
103. King, J.W. (1993) Analysis of Fats and Oils by SFE and SFC, *INFORM 4,* 1089.
104. Chester, T.L., Pinkston, J.D., and Raynie, D.E. (1994) Supercritical Fluid Chromatography and Extraction, *Anal. Chem. 66,* 106R.
105. Wolff, R.L., and Vandamme, F.F. (1992) Separation of Petroselenic (*cis*-6 18:1) and Oleic (*cis*-9 18:1) Acids by Gas-Liquid Chromatography of Their Isopropyl Esters, *J. Am. Oil Chem. Soc. 69,* 1228.
106. Wolff, R.L. (1995) Recent Applications of Capillary Gas-Liquid Chromatography to Some Difficult Separations of Positional or Geometrical Isomers of Unsaturated Fatty Acids, in *New Trends in Lipid and Lipoprotein Analyses,* Sebedio, J.-L., and Perkins, E.G. AOCS Press, Champaign, Illinois, pp. 147–180.
107. Pontillon, J. (1995) Determination of Milk Fat in Chocolates by Gas-Liquid Chromatography of Triglycerides and Fatty Acids, *J. Am. Oil Chem. Soc. 72,* 861.
108. Wolff, R.L., and Bayard, C.C. (1995) Improvement in the Resolution of Individual *trans*-18:1 Isomers by Capillary Gas-Liquid Chromatography: Use of a 100 m. CP-Sil 88 Column, *J. Am. Oil Chem. Soc. 72,* 1197.
109. Molkentin, J., and Precht, D. (1995) Optimized Analysis of *trans*-Octadecenoic Acids in Edible Fats, *Chromatographia 41,* 267.
110. Duchateau, G.S.M.J.E., van Oosten, H.J., and Vasconcellos, M.A. (1996) Analysis of *cis*- and *trans*-Fatty Acid Isomers in Hydrogenated and Refined Vegetable Oils by Capillary Gas-Liquid Chromatography, *J. Am. Oil Chem. Soc. 73,* 275.
111. Frega, N., Bocci, F., Giovannoni, G., and Lercker, G. (1993) High Resolution GC of Unsaponifiable Matter and Sterol Fraction in Vegetable Oils, *Chromatographia 36,* 215.
112. Rodriguez-Palmero, M., de la Presa-Owens, S., Castellote-Bargallo, A.I., Lopez Sabater, M.C., Rivero-Urgell, M., and de la Torre–Boronat, M.C. (1994) Determination of Sterol Content in Different Food Samples by Capillary Gas Chromatography, *J. Chromatogr. A 672,* 267.
113. Johnson, J.H., McIntyre, P., and Zdunek, J. (1995) Automated Sample Preparation for Cholesterol Determination in Foods, *J. Chromatogr. A 718,* 371.
114. Toschi, T.G., Bendini, A., and Lercker, G. (1996) Evaluation of 3,5-Stigmastadiene Content of Edible Oils: Comparison Between the Traditional Capillary Gas Chromatographic Method and the On-Line High Performance Liquid Chromatography–Capillary Gas Chromatographic Analysis, *Chromatographia 43,* 195.
115. Eiceman, G.A., Hill, H.H., and Davani, B. (1994) Gas Chromatography, *Anal. Chem. 66,* 621R.

116. Mordret, F.X., and Coustille, J.L. (1995) Gas-Liquid Chromatography: Choice and Optimization of Operating Conditions, in *New Trends in Lipid and Lipoprotein Analyses*, Sebedio, J.-L., and Perkins, E.G. AOCS Press, Champaign, Illinois, pp. 133–146.
117. Wolff, R.L. (1995) Recent Applications of Capillary Gas-Liquid Chromatography to Some Difficult Separations of Positional or Geometrical Isomers of Unsaturated Fatty Acids, in *New Trends in Lipid and Lipoprotein Analyses*, Sebedio, J.-L., and Perkins, E.G. AOCS Press, Champaign, Illinois, pp. 147–180.
118. Firestone, D., and Sheppard, A. (1992) Determination of *trans* Fatty Acids, in *Lipid Methodology—One*, Christie, W.W. The Oily Press, Ayr, Scotland, pp. 273–322.
119. Mossoba, M.M., and McDonald, R.E. (1995) Quantitation of the *trans* Content of Hydrogenated Oils by Infrared Spectroscopy, *INFORM 6*, 461.
120. Ali, L.H., Angyal, G., Weaver, C.M., Rader, J.I., and Mossoba, M.M. (1996) Determination of Total *trans* Fatty Acids in Foods: Comparison of Capillary Column Gas Chromatography and Single Bounce Horizontal Attenuated Total Reflection Infrared Spectroscopy, *J.Am. Oil Chem. Soc. 73*, 1688.
121. van de Voort, F.R., and Ismail, A.A. (1991) Proximate Analysis of Foods by Mid-FTIR Spectroscopy, *Trends in Food Science and Technology 2*, 13–17.
122. van de Voort, F.R. (1992) Fourier Transform Infrared Spectroscopy Applied to Food Analysis, *Food Research International 25*, 397.
123. van de Voort, F.R., Sedman, J., and Ismail, A.A. (1993) A Rapid FTIR Quality-Control Method for Determining Fat and Moisture in High-Fat Products, *Food Chem. 48*, 213.
124. van de Voort, F.R. (1994) FTIR Spectroscopy in Edible Oil Analysis, *INFORM 5*, 1038.
125. van de Voort, F.R., Sedman, J., Emo, G., and Ismail, A.A. (1992) Rapid and Direct Iodine Value and Saponification Number Determination of Fats and Oils by Attenuated Total Reflectance/Fourier Transform Infrared Spectroscopy, *J. Am. Oil Chem. Soc. 69*, 1118.
126. Ismail, A.A., van de Voort, F.R., Emo, G., and Sedman, J. (1993) Rapid Quantitative Determination of Free Fatty Acids in Fats and Oils by Fourier Transform Infrared Spectroscopy, *J. Am. Oil Chem. Soc. 70*, 335.
127. van de Voort, F.R., Ismail, A.A., and Sedman, J. (1995) A Rapid Automated Method for the Determination of *cis* and *trans* Content of Fats and Oils by Fourier Transform Infrared Spectroscopy, *J. Am. Oil Chem. Soc. 72*, 873.
128. van de Voort, F.R., Ismail, A.A., Sedman, J., and Emo, G. (1994) Monitoring the Oxidation of Edible Oils by Fourier Transform Infrared Spectroscopy, *J. Am. Oil Chem. Soc. 71*, 243.
129. van de Voort, F.R., Memon, K.P., Sedman, T., and Ismail, A.A. (1996) Determination of Solid Fat Index by Fourier Transform Infrared Spectroscopy, *J. Am. Oil Chem. Soc. 73*, 411.
130. Lai, Y.W., Kemsley, E.K., and Wilson, R.H. (1994) Potential of Fourier Transform Infrared Spectroscopy for the Authentication of Vegetable Oils, *J. Agric. Food Chem. 42*, 1154.
131. Lai, Y.W., Kemsley, E.K., and Wilson, R.H. (1995) Quantitative Analysis of Potential Adulterants of Extra Virgin Olive Oil Using Infrared Spectroscopy, *Food Chem. 53*, 95.
132. Luinge, H.J., Hop, E., Lutz, E.T.G., van Hemert, J.A., and de Jong, E.A.M. (1993) Determination of the Fat, Protein and Lactose Content of Milk Using Fourier Trans-

form Infrared Spectrometry, *Anal. Chim. Acta 284,* 419.
133. Safar, M., Bertrand, D., Robert, P., Devaux, M.F., and Genot, C. (1994) Characterization of Edible Oils, Butters and Margarines by Fourier Transform Infrared Spectroscopy with Attenuated Total Reflectance, *J. Am. Oil Chem. Soc. 71,* 371.
134. Villé, H., Maes, G., De Schrijver, R., Spincemaille, G., Rombouts, G., and Geers, R. (1995) Determination of Phospholipid Content of Intramuscular Fat by Fourier Transform Infrared Spectroscopy, *Meat Sci. 41,* 283.
135. Osborne, B.G., and Fearn, T. (1996) *Near Infrared Spectroscopy in Food Analysis,* John Wiley and Sons, New York.
136. Rodriguez-Otero, J.L., Hermida, M., and Cepeda, A. (1995) Determination of Fat, Protein, and Total Solids in Cheese by Near-Infrared Reflectance Spectroscopy, *J. AOAC Intern. 78,* 802.
137. Lee, M.H., Cavinato, A.G., Mayes, D.M., and Rasco, B.A. (1992) Noninvasive Short-Wavelength Near-Infrared Spectroscopic Method to Estimate the Crude Lipid Content in the Muscle of Intact Rainbow Trout, *J. Agric. Food Chem. 40,* 2176.
138. Wold, J.P., Jakobsen, T., and Krane, L. (1996) Atlantic Salmon Average Fat Content Estimated by Near-Infrared Transmittance Spectroscopy, *J. Food Sci. 61,* 74.
139. Bewig, K.M., Clarke, A.D., Roberts, C., and Unklesbay, N. (1994) Discriminant Analysis of Vegetable Oils by Near-Infrared Reflectance Spectroscopy, *J. Am. Oil Chem. Soc. 71,* 195.
140. Sato, T. (1994) Application of Principal-Component Analysis on Near-Infrared Spectroscopic Data of Vegetable Oils for their Classification, *J. Am. Oil Chem. Soc. 71,* 293.
141. Wesley, I.J., Barnes, R.J., and McGill, A.E.J. (1995) Measurement of Adulteration of Olive Oils by Near-Infrared Spectroscopy, *J. Am. Oil Chem. Soc. 72,* 289.
142. Wesley, I.J., Pacheco, F., and McGill, A.E.J. (1996) Identification of Adulterants in Olive Oils. *J. Am. Oil Chem. Soc. 73,* 515.
143. Sato, T., Takahata, Y., Noda, T., Yanagisawa, T., Morishita, T., and Sakai, S. (1995) Nondestructive Determination of Fatty Acid Composition of Husked Sunflower *(Helianthus annua* L.) Seeds by Near-Infrared Spectroscopy, *J. Am. Oil Chem. Soc. 72,* 1177.
144. Hurburgh, C.R., Jr. (1994) Identification and Segregation of High-Value Soybeans at a County Elevator, *J. Am. Oil Chem. Soc. 71,* 1073.
145. Frankel, E.N., Nash, A.M., and Snyder, J.M. (1987) A Methodology Study to Evaluate Quality of Soybeans Stored at Different Moisture Levels, *J. Am. Oil Chem. Soc. 64,* 987.
146. Sato, T., Abe, H., Kawano, S., Ueno, G., Suzuki, K., and Iwamoto, M. (1994) Near-Infrared Spectroscopic Analysis of Deterioration Indices of Soybeans for Process Control in Oil Milling Plant, *J. Am. Oil Chem. Soc. 71,* 1049.
147. Boot, A.J., and Speek, A.J. (1994) Determination of the Sum of Dimer and Polymer Triglycerides and of Acid Value of Used Frying Fats and Oils by Near-Infrared Reflectance Spectroscopy, *J. AOAC Intern. 77,* 1184.
148. Warner, K., Orr, P., Parrott, L., and Glynn, M. (1994) Effects of Frying Oil Composition on Potato Chip Stability, *J. Am. Oil Chem. Soc. 71,* 1117.
149. Daun, J.K., Clear, K.M., and Williams, P. (1994) Comparison of Three Whole Seed Near-Infrared Analyzers for Measuring Quality Components of Canola Seed, *J. Am. Oil Chem. Soc. 71,* 1063.
150. Cartagena, M., Mozayeni, F., and Szajer, G. (1994) Chemometric Applications of NIR

Spectroscopy, *INFORM 5,* 1146.
151. Sebedio, J.-L., Le Quere, J.L., Semon, E., Morin, O., Prevost, J., and Grandgirard, A. (1987) Heat Treatment of Vegetable Oils. II. GC–MS and GC–FTIR Spectra of Some Isolated Cyclic Fatty Acid Monomers, *J. Am. Oil Chem. Soc. 64,* 1324.
152. Mossoba, M.M., McDonald, R.E., Chen, J.-Y.T., Armstrong, -Y. D.J., and Page, S.W. (1990) Identification and Quantitation of *trans*-9, *trans*-12-Octadecadienoic Acid Methyl Ester and Related Compounds in Hydrogenated Soybean Oil and Margarines by Capillary Gas Chromatography/Matrix Isolation/Fourier Transform Infrared Spectroscopy, *J. Agric. Food Chem. 38,* 86.
153. Mossoba, M.M., McDonald, R.E., and Prosser, A. R. (1993) GC-MI–FTIR Determination of *trans* Monounsaturated and Saturated FAMEs in Partially Hydrogenated Menhaden Oil, *J. Agric. Food Chem. 41,* 1998.
154. Mossoba, M.M., Yurawecz, M.P., Lin, H.S., McDonald, R.E., Flickinger, B.D., and Perkins, E.G. (1995) Application of GC-MI–FTIR Spectroscopy to the Structural Elucidation of Cyclic Fatty acid Monomers, *American Laboratory 27(14),* 16K.
155. Mossoba, M.M., Yurawecz, M.P., Roach, J.A.G., McDonald, R.E., Flickinger, B.D., and Perkins, E.G. (1996) Analysis of CFAM DMOX Derivatives by GC-MI–FTIR Spectroscopy, *J. Agric. Food Chem. 44,* 3193.
156. Jirovetz, L., Jäger, W., Remberg, G., Espinosa-Gonzalez, J., Morales, R., Woidich, A., and Nikiforov, A. (1992) Analysis of the Volatiles in the Seed Oil of *Hibiscus sabdariffa (Malvaceae)* by Means of GC-MS and GC–FTIR, *J. Agric. Food Chem. 40,* 1186.
157. Boosfeld, J., and Vitzthum, O.G. (1995) Unsaturated Aldehydes Identification from Green Coffee, *J. Food Sci. 60,* 1092.
158. Wahl, H.G., Habel, S.-Y., Schmieder, N., and Liebich, H. M. (1994) Identification of *cis/trans* Isomers of Methyl Ester and Oxazoline Derivatives of Unsaturated Fatty Acids Using GC-FTIR–MS, *J. High Res. Chromatogr. 17,* 543.
159. Christie, W.W. (1989) *Gas Chromatography and Lipids.* The Oily Press, Ayr, Scotland, pp. 161–184.
160. Harvey, D.J. (1992) Mass Spectrometry of Picolinyl and Other Nitrogen-Containing Derivatives of Lipids, in *Advances in Lipid Methodology—One,* Christie, W.W., The Oily Press, Ayr, Scotland, pp. 19–80.
161. Christie, W.W. (1993) Determination of Fatty Acid Structure, *INFORM 4,* 85.
162. Dobson, G., and Christie, W.W. (1996) Structural Analysis of Fatty Acids by Mass Spectroscopy of Picolinyl Esters and Dimethyloxazoline Derivatives, *Trends in Anal. Chem. 15,* 130.
163. Zhang, J.Y., Wang, H.Y., Yu, Q.T., Yu, X.J., Liu, B.N., and Huang, Z.H. (1989) The Structures of Cyclopentenyl Fatty Acids in the Seed Oils of *Flacourtiaceae* Species by GC–MS of their 4,4-Dimethyloxazoline Derivatives, *J. Am. Oil Chem. Soc. 66,* 242.
164. Mossoba, M.M., Yurawecz, M.P., Roach, J.A.G., Lin, H.S., McDonald, R.E., Flickinger, B.D., and Perkins, E.G. (1994) Rapid Determination of Double Bond Configuration and Position Along the Hydrocarbon Chain in Cyclic Fatty Acid Monomers, *Lipids 29,* 893.
165. Keusgen, M., Curtis, J.M., and Ayer, S.W. (1996) The Use of Nicotinates and Sulfoquinovosyl Monoacylglycerols in the Analysis of Monounsaturated n-3 Fatty Acids by Mass Spectrometry, *Lipids 31,* 231.
166. Mossoba, M.M., Yurawecz, M.P., Roach, J.A.G., Lin, H.S., McDonald, R.F., Flickinger, B.D., and Perkins, E.G. (1995) Elucidation of Cyclic Fatty Acid and Monomer Structures, Cyclic and bicyclic ring sizes and double bond position and

configuration. *J. Am. Oil Chem. Soc. 72,* 721.
167. Mossoba, M.M., McDonald, R.E., Roach, J.A.G., Fingerhut, D.D., Yurawecz, M.P., and Sehat, N. (1997) Spectral Confirmation of *trans* Monounsaturated C_{18} Fatty Acid Positional Isomers, *J. Am. Oil Chem. Soc.,* in press.
168. Garrido, J.L., and Medina, I. (1994) One-Step Conversion of Fatty Acids into their 2-Alkenyl-4,4-Dimethyloxazoline Derivatives Directly from Total Lipids, *J. Chromatogr. A 673,* 101.
169. Mossoba, M.M., Yurawecz, P.M., Roach, J.A.G., McDonald, R.E., and Perkins, E.G. (1996) Confirmatory Mass-Spectral Data for Cyclic Fatty Acid Monomers, *J. Am. Oil Chem. Soc. 73,* 1317.
170. Joh, Y.-G., Kim, S.J., and Christie, W.W. (1995) The Structure of the Triacylglycerols, Containing Punicic Acid, in the Seed Oil of *Trichosanthes kirilowii, J. Am. Oil Chem. Soc. 72,* 1037.
171. Laakso, P., and Kallio, H. (1996) Optimization of the Mass Spectrometric Analysis of Triacylglycerols Using Negative-Ion Chemical Ionization with Ammonia, *Lipids 31,* 33.
172. Angerosa, F., d'Alessandro, N., Konstantinoua, P., and Di Giacinto, L. (1995) GC–MS Evaluation of Phenolic Compounds in Virgin Olive Oil, *J. Agric. Food Chem. 43,* 1802.
173. Joh, Y.-G., Brechany, E.Y., and Christie, W.W. (1995) Characterization of Wax Esters in the Roe Oil of Amber Fish, *Seriola aureovittata, J. Am. Oil Chem. Soc. 72,* 707.
174. Horman, I. (1984) NMR Spectroscopy, in *Analysis of Foods and Beverages,* Charalambous, G., Academic Press, Orlando, Florida, pp. 205–264.
175. AOCS Recommended Practice Ak 3-94 (1994) in *Official Methods and Recommended Practices,* 4th edn., 1987–, American Oil Chemists' Society, Champaign, Illinois.
176. AOCS Recommended Practice Ak 4-95 (1995) in *Official Methods and Recommended Practices,* 4th edn., 1987–, American Oil Chemists' Society, Champaign, Illinois.
177. AOCS Official Method Cd 16b-93 (1993) in *Official Methods and Recommended Practices,* 4th edn., 1987–, American Oil Chemists' Society, Champaign, Illinois.
178. Tiwari, P.N., and Gambhir, P.N. (1995) Seed Oil Determination without Weighing and Drying the Seeds by Combined Free Induction Decay and Spin-Echo Nuclear Magnetic Resonance Signals, *J. Am. Oil Chem. Soc. 72,* 1017.
179. Gunstone, F.D. (1994) High Resolution ^{13}C NMR. A Technique for the Study of Lipid Structure and Composition, *Prog. Lipid Res. 33,* 19.
180. Gunstone, F.D. (1991) ^{13}C-NMR Studies of Mono-, Di- and Tri-Acylglycerols Leading to Qualitative and Semiquantitative Information about Mixtures of these Glycerol Esters, *Chem. Phys. Lipids 58,* 219.
181. Gunstone, F.D. (1993) Information on the Composition of Fats from their High-Resolution ^{13}C Nuclear Magnetic Resonance Spectra, *J. Am. Oil Chem. Soc. 70,* 361.
182. Gunstone, F.D. (1993) The Composition of Hydrogenated Fats by High-Resolution ^{13}C Nuclear Magnetic Resonance Spectroscopy, *J. Am. Oil Chem. Soc. 70,* 965.
183. Wollenberg, K.F. (1990) Quantitative High Resolution ^{13}C Nuclear Magnetic Resonance of the Olefinic and Carbonyl Carbons of Edible Vegetable Oils, *J. Am. Oil Chem. Soc. 67,* 487.
184. Aursand, M., Rainuzzo, J.R., and Grasdalen, H. (1993) Quantitative High-Resolution ^{13}C and ^{1}H Nuclear Magnetic Resonance of ω3 Fatty Acids from White Muscle of

Atlantic Salmon (*Salmo salar*), *J. Am. Oil Chem. Soc. 70,* 971.
185. Aursand, M., Jorgensen, L., and Grasdalen, H. (1995) Positional Distribution of ω3 Fatty Acids in Marine Lipid Triacylglycerols by High Resolution ^{13}C Nuclear Magnetic Resonance Spectroscopy, *J. Am. Oil Chem. Soc. 72,* 293.
186. Gunstone, F.D. (1993) The Study of Natural Epoxy Oils and Epoxidized Vegetable Oils by ^{13}C Nuclear Magnetic Resonance Spectroscopy, *J. Am. Oil Chem. Soc. 70,* 1139.
187. Pfeffer, P.E., Sonnet, P.E., Schwartz, D.P., Osman, S.F., and Weisleder, D. (1992) Effects of Bis-Homoallylic and Homoallylic Hydroxyl Substitution on the Olefinic ^{13}C Resonance Shifts in Fatty Acid Methyl Esters, *Lipids 27,* 285.
188. Gunstone, F.D. (1993) High Resolution ^{13}C NMR Study of Synthetic Branched-Chain Acids and of Wool Wax Acids and Isostearic Acid, *Chem. Phys. Lipids 65,* 155.
189. Lie Ken Jie, M.S.F., Lam, C.C., Pasha, M.K., Stefenov, K.L., and Marekov, I. (1996) ^{13}C Nuclear Magnetic Resonance Spectroscopic Analysis of the Triacylglycerol Composition of Some Margarines, *J. Am. Oil Chem. Soc. 73,* 1011.
190. Henderson, J.M., Petersheim, M., Templeman, G.J., and Softly, B.J. (1994) Quantitation and Structure Elucidation of the Positional Isomers in a Triacylglycerol Mixture Using Protein and Carbon One- and Two-Dimensional NMR, *J. Agric. Food Chem. 42,* 435.
191. De Kock, J. (1993) The European Analytical Subgroup of I.L.P.S.—A Joint Effort to Clarify Lecithin and Phospholipid Analysis, *Fat Sci. Technol. 95,* 352.
192. Gunstone, F.D. (1995) Information About Fatty Acids and Lipids Derived by ^{13}C Nuclear Magnetic Resonance Spectroscopy, in *New Trends in Lipid and Lipoprotein Analyses,* Sebedio, J.-L., and Perkins, E.G., AOCS Press, Champaign, Illinois, pp. 250–264.
193. Murcia, M.A., and Villalain, J. (1993) Phospholipid Composition of Canned Peas by ^{31}P-NMR, *J. Sci. Food Agric. 61,* 345.
194. Hanna, G.M., and Specchio, J.J. (1995) Selected NMR Applications in Food Regulatory Analysis, *Food Testing and Anal. 1(2),* 43.
195. Winkler, F.J., and Schmidt, H.-L. (1980) Scope of the Application of ^{13}C Isotope Mass Spectrometry in Food Analysis, *Z. Lebensm. Unters. Forsch. 171,* 85.
196. Winkler, F.J. (1984) Application of Natural Abundance Stable Isotope Mass Spectrometry in Food Control, in *Chromatography and Mass Spectrometry in Nutrition Science and Food Safety,* Frigerio, A., and Milon, H., Elsevier Science Publishers, Amsterdam, pp. 173–190.
197. Pollard, M. (1993) Tales Told by Dry Bones, *Chem. Ind.,* No. 10, pp. 359–362.
198. Doner, L.W., and Phillips, J.G. (1981) Detection of High Fructose Corn Syrup in Apple Juice by Mass Spectrometric ^{13}C/^{12}C Analysis: Collaborative Study, *J. Assoc. Off. Anal. Chem. 64,* 85.
199. Gaffney, J., Irsa, A., Friedman, L., and Emken, E. (1979) ^{13}C-^{12}C Analysis of Vegetable Oils, Starches, Proteins and Soy-Meat Mixtures, *J. Agric. Food Chem. 27,* 475.
200. Rossell, J.B. (1994) Stable Carbon Isotope Ratios in Establishing Maize Oil Purity, *Fat Sci. Technol. 96,* 304.
201. Haase, A. (1992) Introduction to NMR Imaging, *Trends in Food Sci. Technol. 3,* 206.
202. Hills, B. (1995) Food Processing: An MRI Perspective, *Trends in Food Sci. Technol. 6,* 111.
203. Simoneau, C., McCarthy, M.J., Reid, D.S., and German, J.B. (1992) Measurement of Fat Crystallization Using NMR Imaging and Spectroscopy, *Trends in Food Sci.*

Technol. 3, 208.
204. Halloin, J.M., Cooper, T.G., Potchen, E.J., and Thompson, T.E. (1993) Proton Magnetic Resonance Imaging of Lipid in Pecan Embryos, *J. Am. Oil Chem. Soc. 70,* 1259.
205. Heil, J.R., Perkins, W.E., and McCarthy, M.J. (1990) Use of Magnetic Resonance Procedures for Measurement of Oil in French-Style Dressings, *J. Food Sci. 55,* 763.
206. Pilhofer, G.M., Lee, H.-C., McCarthy, M.J., Tong, P.S., and German, J.B. (1994) Functionality of Milk Fat in Foam Formation and Stability, *J. Dairy Sci. 77,* 55.
207. Tingley, J.M., Pope, J.M., Baumgartner, P.A., and Sarafis, V. (1995) Magnetic Resonance Imaging of Fat and Muscle Distribution in Meat, *Intern. J. Food Sci. Technol. 30,* 437.
208. Singh, R.P. (1995) Heat and Mass Transfer in Foods During Deep-Fat Frying, *Food Technol. 49,* 134.
209. Forina, M., Armanino, C., Lanteri, S., and Tiscornia, E. (1983) Classification of Olive Oils from their Fatty Acid Composition, in *Food Research and Data Analysis,* Martins, H., and Russwurm, H., Applied Science Publishers, London, pp. 189–214.
210. Alonso, M.V., and Aparicio, R. (1993) Characterization of European Virgin Olive Oils Using Fatty Acids, *Grasas y Aceites 44,* 18.
211. Tsimidou, M., and Karakostas, K.X. (1993) Geographical Classification of Greek Virgin Olive Oils by Nonparametric Multivariate Evaluation of Fatty Acid Composition, *J. Sci. Food Agric. 62,* 253.
212. Morales, M.T., and Aparicio, R. (1993) Characterizing Some European Olive Oil Varieties by Volatiles Using Statistical Tools, *Grasas y Aceites 44,* 113.
213. Chaves Das Neves, H.J., and Vasconcelos, A.M.P. (1989) Characterization of Fatty Oils by Pattern Recognition of Triglyceride Profiles, *J. High Res. Chromatogr. 12,* 226.
214. Montedoro, G.F., Servili, M., Baldioli, M., Selvaggini, R., Perretti, G., Magnarini, C., Cossignani, L., and Damiani, P. (1995) Characterization of Some Italian Virgin Olive Oils in Relation to Origin Area, *Riv. Ital. Sostanze Grasse 72,* 403.
215. Lamberto, M., and Saitta, M. (1995) Principal Component Analysis in Fast Atom Bombardment–Mass Spectrometry of Triacylglycerols in Edible Oils, *J. Am. Oil Chem. Soc. 72,* 867.
216. Aparicio, R., and Alonso, V. (1994) Characterization of Virgin Olive Oils by SEXIA Expert System, *Prog. Lipid Res. 33,* 29.
217. Lipp, M. (1996) Comparison of PLS, PCR, and MLR for the Quantitative Determination of Foreign Oils and Fats in Butter Fats of Several European Countries by their Triglyceride Composition, *Z. Lebensm. Unters. Forsch. 202,* 193.
218. Anklam, E., Lipp, M., and Wagner, B. (1996) HPLC with Light Scatter Detector and Chemometric Data Evaluation for the Analysis of Cocoa Butter and Vegetable Fats, *Fett/Lipid 98,* 55.
219. Brown, S.D., Blank, T.B., Sum, S.T., and Weyer, L.G. (1994) Chemometrics. *Anal. Chem. 66,* 315R.

Chapter 2
Use of Stable Isotopes to Study Incorporation of Dietary Fat into Blood Lipids

J.P. DeLany and T.-S. Lu

Pennington Biomedical Research Center, 6400 Perkins Road, Baton Rouge, LA 70808

Introduction

A considerable amount of research on human fatty acid metabolism has been conducted using deuterium-labeled fatty acids, particularly by Emken and coworkers (1–7). Our study made use of individually ^{13}C-labeled fatty acids enabling us to investigate differences in the oxidized amounts of ingested ^{13}C-labeled fatty acids (the primary aim of these studies); the incorporation of each fatty acid into the various blood lipid classes; and interconversion of fatty acids, all at the same time (Fig. 1). Interconversion comprises chain elongation and desaturation, which can be examined with carboxyl- or methyl-labeled fatty acids, and chain shortening, which can be probed with methyl-labeled fatty acids. Subjects consume a ^{13}C-labeled fatty acid blended in a hot liquid meal, and blood samples are collected over 8 h. Lipid classes of plasma total lipid extracts are separated by thin-layer chromatography (TLC). Fatty acids are esterified, separated by gas chromatography (GC), and analyzed by a new combustion system interfaced to an isotope ratio mass spectrometer. This technique allows for the on-line measurement of ^{13}C enrichment of each fatty acid in a single run.

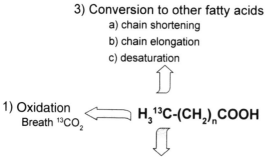

FIG. 1. Study of fatty acid metabolism with ^{13}C-labeled fatty acids. Oxidation of each fatty acid to breath $^{13}CO_2$, incorporation into blood lipids, and conversion into other fatty acids can be studied simultaneously.

Much of our methodology using ^{13}C-labeled fatty acids is adapted directly from the work of Emken and coworkers (1–7). Several studies in animals have examined the oxidation of different fatty acids. The most complete series of fatty acid oxidation measurements was recently carried out in rats (8). In these studies, it was shown that saturated fatty acid oxidation decreased with increasing carbon length (lauric > myristic > palmitic > stearic). With the improvement in mass spectrometers and the increased availability of ^{13}C-labeled substrates, $^{13}CO_2$ breath tests have become a feasible method to study oxidation in humans (9,10). In one such report, various ^{13}C-labeled lipids were examined for use in diagnosis of fat malabsorption (11). It was observed that octanoic acid was oxidized much faster than palmitic acid. One of the first studies to examine differences in oxidation of fatty acids systematically was conducted in males consuming a test diet of normal foods with added ^{13}C-labeled fatty acids (12). Three 18-carbon fatty acids were examined that differed in degree of unsaturation. Oxidation of oleic acid (18:1n-9) was greater than that of linoleic (18:2n-6), which was greater in turn than that of stearic (18:0).

In the studies using ^{13}C-labeled fatty acids, the label was at the carboxyl end (the first carbon to be cleaved in oxidation). However, there is some evidence, in humans and rats, that long-chain saturated fatty acids are only partially chain-shortened (13,14), possibly in peroxisomes, and diverted to other pathways, such as incorporation of a shortened fatty acid into blood lipid. This is one reason we began examining chain shortening of the labeled fatty acids.

The current studies were greatly facilitated by isotope ratio-monitoring gas chromatography–mass spectrometry technology (also known as GC-combustion IRMS). Described in the late 1970s (15,16), this technology has recently become available commercially, and a study in which [U-^{13}C]-labeled stearic acid was studied has been reported (17).

Methods

Protocol

For 7 days prior to initiation of fatty acid metabolism tests, subjects were fed a standard diet, of defined composition, to meet energy requirements. After 5 days of the standard diet, breath and blood samples were collected to determine background ^{13}C contribution from the test diet alone. The published studies of ^{13}C-labeled fatty acids involved subjects taking the fatty acid in a capsule with breakfast. We doubted that this was the best protocol, especially for the long-chain saturated fatty acids, even if a correction was made for differential absorption, because the high melting point of the long-chain fatty acids (62°C for palmitic, 69°C for stearic—both well above body temperature) could lead to very low absorption. Therefore, we administered the fatty acids in a hot blended mixture, similar to that used by Emken et al. (1) in their studies of deuterium-labeled fatty acids. The reported absorption, even of tristearin, was better than 93% when using this protocol. In our studies, each ^{13}C-labeled fatty acid was blended in a hot liquid meal (85°C, Ensure®). The ^{13}C-labeled fatty acid doses to obtain adequate label in breath samples were based on previous data (11,12). The dose of the long-chain saturated ^{13}C-fatty acids was initially 15 mg/kg but was subsequently reduced to 10 mg/kg body weight, which was the dose used for each ^{13}C-labeled fatty acid. Plasma samples were taken every 2 h for 8 h and again 24 h after the dose. Two fatty acids were tested per week, while the subject was maintained on the standard diet. Subjects signed a consent form, approved by the LSU Institutional Review Board, before undergoing these studies.

Diets

The metabolism of the labeled fatty acids would be affected by the endogenous pool size and the amount of the fatty acid in the diet. Thus, precise composition and amounts of the fatty acids in the diet were defined. Menus were formulated to provide 15% protein, 45% carbohydrate, and 40% fat, with a P/S ratio of 0.51. A lunch consisting of the same composition was consumed during the test, and dinner was served following the test. The same menu cycle was served every day a fatty acid metabolism test was conducted.

Lipid Class Extraction and Separation

Total plasma lipids were extracted with 2:1 chloroform:methanol, with known amounts of triheptadecanoin, cholesteryl heptadecanoate, and diheptadecanyl-L-α-phosphatidylcholine added as internal standards. Preparative TLC was used to isolate triglycerides (TG), cholesteryl esters (CE), and phospholipids (PL) using a mixture of hexane, ethyl ether, and acetic acid (79:20:1, v/v/v), and bands were visualized with iodine. Methyl esters of the separated lipids were prepared by heating with a solution of BF_3-methanol, methanol, and benzene (34:30:36, v/v/v) at 100°C for 45 min, for triglycerides and cholesteryl esters, and for 90 min, for phospholipids. The reaction mixtures were extracted after addition of saturated NaCl solution. Samples were dried under nitrogen and stored in hexane under nitrogen at −20°C until GC-combustion IRMS analysis was performed.

GC-Combustion Isotope Ratio Mass Spectrometry

Quantitation of ^{13}C-labeled fatty acids incorporated into plasma lipid classes was achieved by GC-combustion IRMS analysis of their methyl esters. The basic concept of this system is shown in Fig. 2. Fatty acid methyl esters are separated by conventional GC technology. However, longer columns are generally required, because of the loss of resolution in the relatively large dead spaces (compared to the GC column) associated with the combustion reactor and water trap. The individual fatty acids then pass through a combustion reactor and are combusted to CO_2 and water. The water is removed from the stream, and each bolus of CO_2 enters the isotope ratio mass spectrometer for separation and quantitation of $^{13}CO_2$ and $^{12}CO_2$.

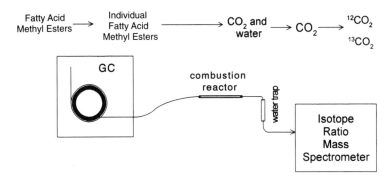

FIG. 2. Basic concepts of GC-combustion isotope ratio mass spectrometry. Fatty acid methyl esters are separated by GC, and each peak is oxidized to CO_2 and water. The water is removed from the stream, and the CO_2 enters the mass spectrometer, where the $^{12}CO_2$ and $^{13}CO_2$ are separated and quantitated.

The system utilized a Varian 3400 gas chromatograph (Sugarland, TX), equipped with a 100-m × 0.32-mm SP2380 fused-silica capillary column (Supelco, Bellefonte, PA). Operating conditions were splitless injection with helium as the carrier gas at an inlet pressure of 15 psi and injection temperature of 235°C. The column oven temperature started at 50°C for 2 min and was then programmed from 50°C to 170°C at 25°C/min and held for 27 min. The second ramp increased the temperature from 170°C to 180°C at 5°C/min and was held for 15 min, followed by a 5°C/min increase to 220°C, with a hold time of 2 min. The final ramp increased temperature at 5°C/min to 240°C with a hold of 25 min. The exit of the chromatographic column was connected via a three-way connector (Valco Instruments, Houston, TX) to the combustion interface (Finnigan, Bremen, Germany). Outputs were connected to a vent line controlled by an air-actuated T-valve and a combustion reactor consisting of a nonporous ceramic tube (32 cm length, 0.5 mm i.d.) packed with copper and platinum wires. The copper wire is oxidized to copper oxide by a flow of oxygen through the reactor before analyses are begun, to serve as a source of oxygen for combustion. The platinum wire serves as a catalyst. The combustion reactor was operated at 840°C, and water generated from the oxidation of organic materials was removed as the effluent stream passed through a semipermeable Nafion (DuPont) membrane tube. The mass spectrometer used for the analysis was a Finnigan MAT delta-S isotope ratio mass spectrometer (Bremen, Germany) with an accelerating potential of 3 kV. The ion source and analyzer regions were separately pumped. Ion currents were measured continuously for m/z 44, 45, and 46 using triple Faraday cups connected to high-speed amplifiers. The data acquired were processed by ISODAT software (18) from Finnigan MAT as the ratio of mass 45/44 against the standard. The isotopic abundance is expressed as $\delta^{13}C$ ‰ (parts per thousand) vs. standard, using the following equation:

$$\delta^{13}C = \frac{(^{13}C/^{12}C)_{sa} - (^{13}C/^{12}C)_{st}}{(^{13}C/^{12}C)_{st}} \times 1000 \qquad [1]$$

The $\delta^{13}C$ values were transformed to ^{13}C atom percent (AP) using the following formula:

$$AP = \frac{100R(\delta/1000 + 1)}{1 + R(\delta/1000 + 1)} \qquad [2]$$

where R is the $^{13}C/^{12}C$ of international Pee Dee Belemnite (PDB) standard ($R = 0.0112372$) and δ is the value of the sample. The ^{13}C enrichment (APE, for AP excess) can be calculated by using the following formula:

$$APE = AP_{sample} - AP_{baseline} \qquad [3]$$

Since measured compounds are fatty acid methyl esters, the $\delta^{13}C$ values were corrected to those for the respective underivatized molecules by using the following formula:

$$APE_c = APE(n + m)/n \qquad [4]$$

where APE_c is the corrected APE for the underivatized compound, n is the number of carbon atoms of underivatized compound, and m is the number of carbon atoms added by derivatization (which for methylation is one). If the number of ^{13}C-labeled atoms per molecule (n') is known, the results can be expressed as the molecule percentage excess (MPE):

$$MPE_c = APE_c(n/n') \qquad [5]$$

For the single ^{13}C-labeled fatty acids used, $n' = 1$, $MPE_c = (APE_c)n$.

The quantitation and absolute amount data of fatty acid methyl esters in the plasma lipid samples were based on the known weights of heptadecanoic acid, as the internal standard, added to the total lipid extract prior to the separation and conversion of the lipid classes to their methyl esters. Response factors were determined by analysis of standard mixtures containing weighed amounts of pure fatty acid methyl esters, purchased from Nu-Chek Prep, Inc. (Elysian, MN).

Results

A printout of results from our standard fatty acid mixture (approximately 2 µg each fatty acid) is presented in Fig. 3. The bottom panel is a tracing of the mass 44 (normal CO_2) signal in volts. This tracing looks nearly identical to a flame ionization tracing from a GC. Peak numbers 6, 13, and 24 represent inputs of standard CO_2 gas with a known ^{13}C abundance, used to calculate the enrichments of the fatty acid peaks. The identity of the peaks in Fig. 3 are as follows: (3) 12:0, (4) 14:0, (5) ?, (6) Std CO_2 gas, (7) 16:0, (8) 16:1, (9) 17:0, (10) 18:0, (11) *trans*-18:1n-9, (12)

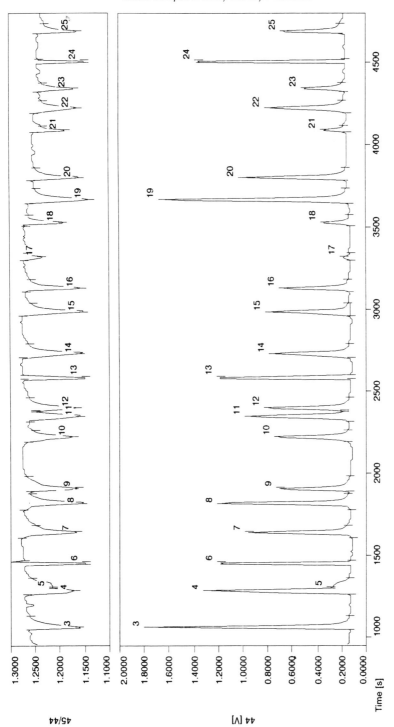

FIG. 3. Tracings of fatty acid methyl ester standards. The bottom tracing is the signal for the mass 44 detector (normal CO_2); the ratio of mass 45 ($^{13}CO_2$)/ mass 44 is shown in the upper panel.

cis-18:1n-9, (13) Std CO_2 gas, (14) 18:2n-6, (15) 18:3n-6, (16) 18:3n-3, (17) ?, (18) ?, (19) 20:3n-6, (20) 20:4n-6, (21) 20:5n-3, (22) 24:0, (23) 24:1, (24) Std CO_2 gas, (25) 22:6n-3. The upper panel of Fig. 3 is a tracing of the mass 45 ($^{13}CO_2$)/ mass 44 ratio, which is used to calculate the enrichment of ^{13}C. Since these were our unlabeled fatty acid methyl ester standards, all peaks have low enrichment. Although there is an automated peak detection algorithm (which was used for each peak in Fig. 3), it is often necessary to define some peaks manually because of the complexity of the 45/44 calculations (18).

The reproducibility of the measurement of enrichment of some of the standard fatty acids is depicted in Fig. 4. Most of the standards had a ^{13}C abundance of approximately −34 ‰, with the exception of myristic and arachidonic, which had a somewhat higher abundance of ^{13}C. The standard deviations are also relatively low for this series of 14 injections of our standard mixture.

The reproducibility of the measurement of quantity of the standard fatty acids (6–10 nmol injected) is depicted in Fig. 5. These values were calculated from injections of our standard mixture of fatty acid methyl esters using the 17:0 internal standard and the calculated response factors. As with the ^{13}C enrichments, the standard deviations are also relatively low for the amount of fatty acid injected, although the error increases with the later-eluting peaks.

Figure 6 depicts the tracings for a triglyceride sample 6 h after administration of ^{13}C-labeled stearic acid. The large enrichment is apparent for stearate (the peak

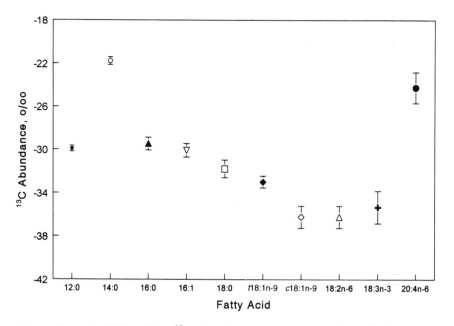

FIG. 4. Reproducibility of the ^{13}C abundance measurements of standard fatty acid methyl ester mixture. Symbols indicate means ± standard deviations of 14 replicates.

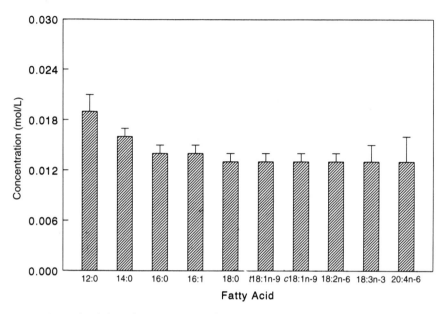

FIG. 5. Reproducibility of measurement of quantity of fatty acid injected using 17:0 as internal standard and individual response factors (mean ± standard deviation).

labeled 13) in the mass 45/44 tracing. The enrichments in phospholipids and cholesterol esters are generally not as high. This same injection is used to calculate any chain shortening of stearate to palmitate (peak at approximately 1600 seconds) or desaturation to oleic acid (the large peak immediately after stearate). This sample was injected 5 times to determine the reproducibility of the measurement of enrichments in an actual sample. The high enrichment in the stearate peak of 839.3 ‰ was quite reproducible, with a standard deviation of 14.1 ‰. Although, because of scaling for the large enrichment of stearate, it does not appear that there was enrichment in other fatty acids, there is measurable enrichment in both palmitic and oleic acids. The enrichment of palmitic acid in this sample was 20.5 ± 1.3 ‰, and that of oleic acid was 14.5 ± 2.4 ‰ above baseline. To put these enrichments in perspective of the more widely used APE, an enrichment of 1 ‰ is equivalent to 0.0012 APE, and an enrichment of 1 APE is equivalent to 850 ‰. Therefore, the stearate peak enrichment of 839.3 ‰ is nearly 1 APE above the baseline sample.

Summary

In summary, we have developed a methodology for studying several aspects of fatty acid metabolism following administration of ^{13}C-labeled fatty acids. In addition to

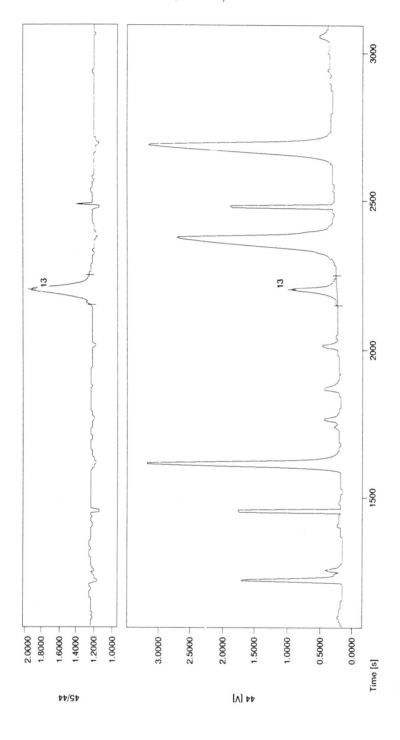

FIG. 6. Tracing for a triglyceride sample 6 hours after administration of ^{13}C-labled stearate. The high enrichment of stearate (peak # 13) is apparent in the 45/44 tracing in the upper panel.

examining fatty acid oxidation, we can study incorporation of fatty acids into the various blood lipid classes and fatty acid interconversion at the same time. The reproducibility of the ^{13}C measures ranges from approximately 1 to 2 ‰ at baseline and low-enrichment samples to 14.5 ‰ for stearate in an enriched (839.3 ‰) triglyceride sample. However, for some samples the reproducibility is worse. The measurements of quantity of fatty acid are also quite reproducible. We can reliably detect enrichment of administered ^{13}C-labeled fatty acid in triglycerides, phospholipids, and cholesterol esters as well as interconversion to other fatty acids.

References

1. Emken, E.A., Adlof, R.O., Rohwedder, W.K., and Gulley, R.M. (1993) Influence of Linoleic Acid on Desaturation and Uptake of Deuterium-Labeled Palmitic and Stearic Acids in Humans, *Biochim. Biophys. Acta. 1170*, 173–181.
2. Emken, E.A., Adlof, R.O., and Abraham, S. (1991) Metabolism of Meadowfoam Oil Fatty Acids in Mice, *Lipids, 26(6)*, 736–742.
3. Rohwedder, W.K., Duval, S.M., Wolf, D.J., and Emken, E.A. (1990) Measurement of the Metabolic Interconversion of Deuterium-Labeled Fatty Acids by Gas Chromatography/Mass Spectrometry, *Lipids 25*, 401–405.
4. Emken, E.A., Adlof, R.O., Hachey, D.L., Garza, C., Thomas, M.R., and Brown-Booth, L. (1989) Incorporation of Deuterium-Labeled Fatty Acids into Human Milk, Plasma, and Lipoprotein Phospholipids and Cholesteryl Esters, *J. Lipid Res. 30*, 395–402.
5. Beyers, E.C., and Emken, E.A. (1991) Metabolites of *cis, trans,* and *trans,cis* Isomers of Linoleic Acid in Mice and Incorporation into Tissue Lipids, *Biochim. Biophys. Acta. 1082*, 275–284.
6. Emken, E.A., Rohwedder, W.K., Adlof, R.O., Rakoff, H., and Gulley, R.M. (1987) Metabolism in Humans of *cis*-12,*trans*-15-Octadecadienoic Acid Relative to Palmitic, Stearic, Oleic and Linoleic Acids, *Lipids, 22(7)*, 495–504.
7. Emken, E.A., Adlof, R.O., Rakoff, H., Rohwedder, W.K., and Gulley, R.M. (1990) Metabolism *in vivo* of Deuterium-Labeled Linolenic and Linoleic Acids in Humans, *Biochem. Soc. Transact. 18*, 766–769.
8. Leyton, J., Drury, P.J., and Crawford, M.A. (1987) Differential Oxidation of Saturated and Unsaturated Fatty Acids *in vivo* in the Rat, *Brit. J. Nutr. 57*, 383–393.
9. Schoeller, D.A., Schneider, J.F., Solomons, N.W., Watkins, J.B., and Klein, P.D. (1977) Clinical Diagnosis with the Stable Isotope ^{13}C in CO_2 Breath Tests: Methodology And Fundamental Considerations, *J. Lab. Clin. Med. 90*, 412–421.
10. Jones, P.J.H., Pencharz, P.B., and Clandinin, M.T. (1985) Absorption of ^{13}C-Labeled Stearic, Oleic, and Linoleic Acids in Humans: Application to Breath Tests, *J. Lab. Clin. Med. 105*, 647–652.
11. Watkins, J.B., Klein, P.D., Schoeller, D.A., Kirschner, B.S., Park, R., and Perman, J.A. (1982) Diagnosis and Differentiation of Fat Malabsorption in Children Using ^{13}C-Labeled Lipids: Trioctanoin, Triolein, and Palmitic Acid Breath Tests, *Gastroenterol. 82*, 911–917.
12. Jones, P.J.H., Pencharz, P.B., Bell, L., and Clandinin, M.T. (1985) Model for Determination of ^{13}C Substrate Oxidation Rates in Humans in the Fed State, *Am. J. Clin. Nutr. 41*, 1277–1282.

13. Weinman, E.O., Chaikoff, I.L., Stevens, B.P., and Dauben, W.G. (1951) Conversion of First and Sixth Carbons of Stearic Acid to Carbon Dioxide by Rats, *J. Biol. Chem. 191*, 523–529.
14. Clandinin, M.T., Khetarpal, S., Kielo, E. S., French, M.A., Tokarska, B., and Goh, Y.K. (1988) Chain Shortening of Palmitic Acid in Human Subjects, *Am. J. Clin. Nutr. 48*, 587–591.
15. Sano, M., Yotsui, Y., Abe, H., and Hayes, J.M. (1976) A New Technique for the Detection of Metabolites Labeled by the Isotope ^{13}C Using Mass Fragmentography, *Biomed. Mass Spectrom. 3*, 1–3.
16. Matthews, D.E., and Hayes, J.M. (1978) Isotope-Ratio-Monitoring Gas Chromatography–Mass Spectrometry, *Anal. Chem. 50(11)*, 1465–1473.
17. Goodman, K.J., and Brenna, J.T. (1992) High Sensitivity Tracer Detection Using High-Precision Gas Chromatography–Combustion Isotope Ratio Mass Spectrometry and Highly Enriched [U-^{13}C]-Labeled Precursors, *Anal. Chem. 64*, 1088–1095.
18. Ricci, M.P., Merritt, D.A., Freeman, K.H., and Hayes, J.M. (1994) Acquisition and Processing of Data for Isotope-Ratio-Monitoring Mass Spectrometry, *Org. Geochem. 21(6/7)*, 561–571.

Chapter 3
Qualitative and Quantitative Analysis of Triacylglycerols Using Atmospheric-Pressure Chemical Ionization Mass Spectrometry

W.C. Byrdwell and W.E. Neff

FQS, NCAUR, ARS, USDA, 1815 N. University St., Peoria, IL 61604

Introduction

Analytical techniques applied to triacylglycerols (TAGs) have steadily improved over the last decade. Improved column technology and instrumentation for high-performance liquid chromatographic (HPLC) separations have led to improved resolution of molecular species of TAG. Nevertheless, complete characterization of TAG mixtures must still involve the combination of several analytical approaches, because no single analytical method has been demonstrated that can elucidate all of the structural information contained within the numerous geometric, positional, and stereochemical isomers of a typical TAG mixture. Early (1–4) and recent (5–8) reviews have outlined the multiple-step approach necessary for complete characterization of the different structural aspects of TAG species. The advent of liquid chromatography/mass spectrometry (LC/MS) soft-ionization techniques has allowed mass spectrometry to be applied to structural characterization of TAGs. The effectiveness of LC/MS to identify overlapped or underresolved TAGs in chromatographic peaks by mass has made it the most effective method for qualitative detection of TAGs in use. More TAG molecular species may be identified using LC/MS than using LC with any other detector.

Several soft-ionization mass spectrometric methods have been applied to triacylglycerol analysis. These have included LC with direct inlet introduction (1,9–17), desorption chemical ionization (DCI) (18–26), direct insertion probe introduction (DIP) (25,27–40), fast atom bombardment (FAB) (41–43), electrospray ionization (ESI) (44), and other techniques (45–50). These ionization techniques have allowed analysis of TAGs from a wide variety of lipid sources. Of these techniques, those that incorporated online HPLC proved to be the most effective for analysis of complex lipid mixtures. Electrospray ionization has been shown to be most effective for analysis of inherently or easily charged molecules such as phospholipids, whereas direct inlet introduction has been shown to be perhaps the most simple yet effective method for introduction of neutral TAG species into a mass spectrometer.

Our recent publications (51–55) as well as others (56) have demonstrated that atmospheric-pressure chemical ionization mass spectrometry (APCI–MS) is similarly a simple yet effective ionization interface. Reversed-phase HPLC with

APCI–MS (RPHPLC/APCI–MS) has been applied to synthetic standards (51), to normal and genetically modified seed oils (52,54), and to TAGs containing unique functional groups (53). We demonstrated that the mass spectra of triacylglycerols obtained using APCI–MS showed very little fragmentation, resulting primarily in diacylglycerol [M-RCOO]$^+$ (or [DG]$^+$) ions and protonated molecular ions, [M + H]$^+$. This fragmentation behavior is very similar to that obtained using direct inlet ionization mass spectrometry (9–17). Using both ionization methods, the degree of unsaturation was found to be the primary factor in determining the amounts of diacylglycerol versus protonated triacylglycerol ions formed (15,55). Using LC/APCI–MS, the mass spectra of triacylglycerols containing more than four double bonds showed protonated TAG ions as base peaks. Spectra of TAGs containing less than three sites of unsaturation had diacylglycerol fragments as base peaks. Spectra of TAGs having three or four sites of unsaturation had either a diacylglycerol base peak or a triacylglycerol base peak, depending on the specific structure. No TAG [M + H]$^+$ ion was observable in spectra of trisaturated TAG.

Reported herein is the use of APCI–MS for qualitative and quantitative analysis of a synthetic mixture containing 35 TAGs, and to analysis of normal and randomized lard, a canola oil blend, and an interesterified blend. From the qualitative analysis of TAG mixtures we defined a simple mathematical relationship, called the "triacylglycerol quotient," from which we developed a mathematical model that allowed approximation of the relative amounts of [DG]$^+$ and [M + H]$^+$ ions expected for a TAG component. Also presented is a comparison of quantitative results based on several approaches:

1. Linear interpolation of response factors obtained for a simple, six-component mixture of homogeneous TAG using d_{12}-PPP as internal standard.
2. Calculation of response factors based on the mathematical model of the triacylglycerol quotient.
3. Calculation of response factors obtained using a synthetic mixture of model TAG.
4. Response factors calculated from comparison of randomized samples to their statistically expected compositions.
5. Response factors calculated from comparison of the fatty acid (FA) composition calculated from the TAG compositions to the FA composition obtained by calibrated GC with a flame ionization detector (FID).
6. Quantitation by RPHPLC with FID.

Materials And Methods

Triacylglycerols: Mono-Acid TAG Mixture for Calibration

Triacylglycerol standard mixture "HPLC #G-2" was obtained from Nu-Check Prep (Elysian, MN). The TAG standard mixture contained equal weights of the following monoacid triacylglycerols: tripalmitin (tri-16:0, PPP), tripalmitolein (tri-16:1, PoPoPo), tristearin (tri-18:0, SSS), triolein (tri-18:1, OOO), trilinolein (tri-18:2, LLL), and trilinolenin (tri-18:3, LnLnLn). Stock solutions of the homogeneous triacylglycerol standards were prepared by dissolving 75 mg of the triacylglycerol standard mixture in 1 mL of chloroform. From this stock solution, the following concentrations (total TAG) were prepared: 30 µg/µL, 15 µg/µL, 6 µg/µL, 3 µg/µL, 0.6 µg/µL, 0.3 µg/µL, and 0.06 µg/µL. This resulted in solutions containing the following amounts of individual TAG standards: 5 µg/µL, 2.5 µg/µL, 1.0 µg/µL, 0.5 µg/µL, 0.1 µg/µL, 0.05 µg/µL, 0.01 µg/µL. D_{12}-Deuterated tripalmitin (tri d_4-PPP), prepared in our laboratory in a similar manner as described in (57), was added to each standard solution as internal standard at a concentration of 1.01 µg/µL.

Synthetic Mixture of TAG from FA

Fatty acids (all >99% pure) were purchased from Nu-Chek Prep (Elysian, MN). Five fatty acids were used: palmitic acid (P), stearic acid (S), oleic acid (O), linoleic acid (L), and linolenic acid (Ln). The mixed fatty acid TAG standard was prepared by the *p*-toluenesulfonic acid-catalyzed esterification of glycerol with the appropriate fatty acids (58). Glycerol and *p*-toluenesulfonic acid (Fisher Scientific Co., Fair Lawn, NJ) were used as received. All solvents were HPLC quality and were used without further purification.

Normal and Randomized Lard Samples

Refined and deodorized lard was obtained from Colfax (Pawtucket, RI). The lard was randomized by Nabisco Foods (Indianapolis, IN) using a sodium methoxide-catalyzed method similar to that described below.

Normal and Interesterified Canola Oil Blend

Canola seeds were obtained from Calgene, Inc. (Davis, CA). The crude canola oil (CNO) was obtained by extraction, in duplicate, of 15 g of canola seeds according to a previously described sonification-hexane extraction procedure, (59) to yield 4.5–6.0 g of oil. Soybean oil saturates (SBOS) flakes were a commercial preparation obtained from Riceland Foods (Stuttgart, AR) or Bunge Foods (Bradley, IL).

Liquid canola oil (80% by weight) and SBOS (20% by weight) were blended and dried by heating them under a water aspirator vacuum to produce normal canola oil–SBOS. This blend was interesterified by the following procedure: The dried blend (1600 g) was charged into a stirred stainless steel autoclave and brought to 70°C. Dry sodium methoxide (catalyst grade) from Amspec Co. (Gloucester City, NJ), 8 g (or 0.5% by weight), was added, and vigorous stirring was continued for 30 min at a temperature of 70°C. The catalyst was deactivated by the addition of 2% aqueous citric acid. The interesterification reaction was stirred for an additional 15 min; filter aid was stirred into the product, which was filtered under vacuum to remove soaps.

Gas Chromatography

Fatty acid methyl esters (FAMEs) were analyzed using gas chromatography with a Supelco SP2380 0.25 mm × 30 m capillary column (Supelco, Bellefonte Park, IL) in a Varian 3400 GC (Palo Alto, CA) under the following conditions: inlet temperature = 240°C; detector temperature = 280°C; initial temperature = 150°C; initial time = 35 min; ramp to 210°C at 3°C/min. Methyl esters were made by sodium methoxide esterification according to the method of Glass (60).

Liquid Chromatography

Solvents were purchased from Aldrich Chemical Co. (Milwaukee, WI) or EM Science (Gibbstown, NJ). Solvents were HPLC grade or the highest available quality and were used without further purification. The HPLC pump was an LDC 4100 MS (Thermo Separation Products, Schaumburg, IL) quaternary pump system with membrane degasser. Several column systems and solvent programs were used for the various samples studied. A gradient solvent program, described previously (54), with propionitrile (PrCN), dichloromethane (DCM), and acetonitrile (ACN) was used to separate the mixture of homogeneous TAGs containing d_{12}-tripalmitin as internal standard.

Separation of the 35-component synthetic mixture of TAGs was accomplished using gradient elution as follows: initial 65% ACN, 35% DCM, held for 20 min; linear from 20 to 25 min to 60% ACN, 40% DCM, held until 35 min; linear from 35 to 40 min to 55% ACN, 45% DCM, held until 50 min; linear from 50 to 60 min to 55% DCM, 45% ACN, held until 85 min. The columns used were an Adsorbosphere C18 (Alltech Associates, Deerfield, IL), 25 cm × 4.6 mm, 5 µm (12% carbon load) in series with an Adsorbosphere UHS C18 25 cm × 4.6 mm, 10 µm (30% carbon load). The flow rate was 1 mL/min. The column effluent was split so that ~850 µL/min went to an evaporative light scattering detector (ELSD) and ~150 µL/min went to the APCI interface. 5 µL of sample solution was injected.

Normal and randomized lard samples and normal and interesterified canola oil blend samples were separated using a method that provided improved resolution compared to the separation described above. The columns used were two Inertsil ODS-80A (GL Sciences, Keystone Scientific, Bellefonte Park, PA), 25 cm × 4.6 mm, 5 µm in series. The gradient elution was as follows: initial 70% ACN, 30% DCM, held for 40 min; linear from 40 to 45 min to 65% ACN, 35% DCM, held until 55 min; linear from 55 to 60 min to 60% ACN, 40% DCM, held until 70 min; linear from 70 to 80 min to 55% ACN, 45% DCM; linear from 80 to 85 min to 70% ACN, 30% DCM. The flow rate throughout was 0.85 µL/min. The column effluent was split so that ~720 µL/min went to an evaporative light scattering detector (ELSD) and ~130 µL/min went to the APCI interface. 10 µL of each sample solution was injected.

The ELSD was an ELSD MKIII (Varex, Burtonsville, MD). The drift tube was set at 140°C, and the gas flow was 2.0 standard liters per minute. High-purity N_2 was used as the nebulizer gas.

Mass Spectrometry

A Finnigan MAT (San Jose, CA) SSQ 710C mass spectrometer fitted with an atmospheric-pressure chemical ionization source was used to acquire mass spectral data. The vaporizer was operated at 400°C, and the capillary heater was operated at 265°C. The corona current was set to 6.0 µA throughout. High-purity nitrogen was used for the sheath and auxiliary gases, which were set to 35 psi and 5 mL/min, respectively. Spectra were obtained from 400 amu to 1100 amu, with a scan time of 2.67 s.

Lipase Hydrolysis

Positional isomers were determined using lipase hydrolysis as described in (61). Briefly, lipase (E.C. 3.1.1.3) was used to cleave the fatty acids at glycerol 1,3 carbons, and the lipolysis products were separated using solid-phase extraction. The FAMEs of the lipolysis products were analyzed using GC–FID as described above.

Calculations

The expected composition of the synthetic mixture was calculated from the fatty acid composition. If FAs are esterified with glycerol, the number of possible TAG molecular species, excluding isomers, can be calculated from the following equation (62):

$$\text{\#TAG molecular species} = (n^3 + 3n^2 + 2n)/6 \quad [1]$$

where n is the number of FAs. When 5 FAs are used, 35 TAG molecular species are possible. The 5 FAs would be randomly distributed among all possible molecular species. Formulas for calculating the composition of a randomly distributed mixture of TAGs were derived from probability theory (63). The following equations were used to calculate the percentage of TAGs containing one, two, or three different FA:

One FA:

$$\text{TAG\%} = \frac{(\text{FA\%} \times \text{FA\%} \times \text{FA\%})}{10{,}000} \quad [2]$$

Two FA:

$$\text{TAG\%} = \frac{(\text{FA1\%} \times \text{FA1\%} \times \text{FA2\%}) \times 3}{10{,}000} \quad [3]$$

Three FA:

$$\text{TAG\%} = \frac{(\text{FA1\%} \times \text{FA2\%} \times \text{FA3\%}) \times 6}{10{,}000} \quad [4]$$

These equations were also used to calculate the theoretical compositions of the randomized lard and interesterified CNO samples from their FA compositions determined by calibrated GC–FID.

The percent relative errors for TAGs were calculated by subtracting the statistically expected amount from the determined amount, multiplying by 100, and then dividing by the statistically expected amount. The percent relative errors for TAGs in a sample were averaged to give the average relative error. TAGs present at low levels (< 0.1%) had higher percent relative errors, so they were not included in calculation of the average relative error for a sample. Similarly, the percent relative errors for the FAs were calculated by comparing the FA composition calculated from the TAG composition to the FA composition determined by calibrated GC–FID. The individual FA relative errors were averaged to give the average relative error for the FA composition.

Results

35-Component Synthetic Mixture

The APCI–MS spectra obtained for all triacylglycerols in the samples studied exhibited behavior similar to that reported previously (51–55). The only fragments observed in most spectra were those for the diacylglycerol, $[M-RCOO]^+$ or $[DG]^+$, and the protonated molecular ion, $[M + H]^+$. The sum of all fragment ion abundances over time gave the total ion chromatogram (TIC) or reconstructed ion chromatogram (RIC). Fig. 1 shows a conventional LC–FID chromatogram and an LC/APCI–MS reconstructed ion chromatogram for the RPHPLC separation of a synthetic mixture of 35 TAGs produced by interesterification of five fatty acids with glycerol.

Qualitative analysis of TAG species in simple mixtures (those that are completely chromatographically resolved) may be accomplished by identification of fragment peaks in mass spectra arising from given chromatographic peaks. In spectra obtained for a complex mixture of TAGs all of the fragment peaks in a mass spectrum do not arise from a single TAG species, so identification of overlapped TAG species can appear complicated. In complex mixtures of TAGs, qualitative identification of TAG species was accomplished by the use of extracted ion chromatograms (EICs) to identify elution times for each of the members of the homologous set of TAG species containing an individual $[DG]^+$ fragment.

EICs for some $[DG]^+$ fragments of TAG in the synthetic mixture are shown in Fig. 2. Since the synthetic mixture was made from five FAs, $[DG]^+$ fragments arose from every combination of these FA. Also, every $[DG]^+$ fragment was combined with each of the five FAs, to make five TAGs containing each $[DG]^+$ fragment. Therefore, each EIC contained five peaks, corresponding to the five TAGs containing the $[DG]^+$ fragment. Each TAG species had up to three $[DG]^+$ fragments and therefore had peaks at corresponding retention times in up to three different EICs. TAGs having any sites of unsaturation also produced an $[M + H]^+$ ion at the same retention time. The appearance of diagnostic fragments in multiple EICs allowed facile, unambiguous qualitative identification of TAG species present even at levels below 0.1%.

An example of the use of EICs for TAG identification is demonstrated by identifying the fragments attributable to the TAG "LLP" (dilinoleoylpalmitoylglycerol)

Analysis of Triacylglycerols by APCI-MS 51

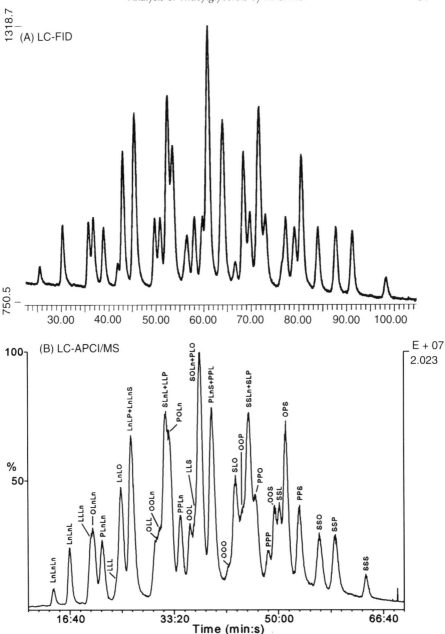

FIG. 1. Separation of synthetic mixture of 35 triacylglycerols by (A) liquid chromatography with flame ionization detection (LC–FID) and (B) liquid chromatography with atmospheric-pressure chemical ionization mass spectrometric detection (LC–APCI/MS). Fatty acyl chain abbreviations: Ln, linolenate (18:3); L, linoleate (18:2); O, oleate (18:1); S, stearate (18:0); P, palmitate (16:0).

in Fig. 2: the L peak in the EIC for the [PP] + diacylglycerol fragment (m/z: 551.5) occurred at the same time as the P peak in the EIC for the [PL]$^+$ diacylglycerol fragment (m/z: 575.5). Along with the [M + H]$^+$ peak in the EIC for m/z 831.7 (not shown), these fragments provide several pieces of confirmatory evidence that LLP was present.

Quantitative analysis was performed by summing together the areas under [DG]$^+$ and [M + H]$^+$ peaks in EICs attributable to each TAG species. The sum of the areas under peaks for all fragments for a particular TAG was divided by the total area under all EIC peaks for all masses to obtain a percentage composition for each TAG component. Qualitative (and quantitative) analysis was complicated by the presence of TAG species having the same masses and therefore the same equivalent carbon numbers (ECN = number of carbons in chain − 2 × number of double bonds). These species were either completely resolved chromatographically, partially resolved, or completely overlapped, depending on the nature of the FAs in the carbon chains. The more disparate the FAs making up the TAGs, the more species with the same ECN

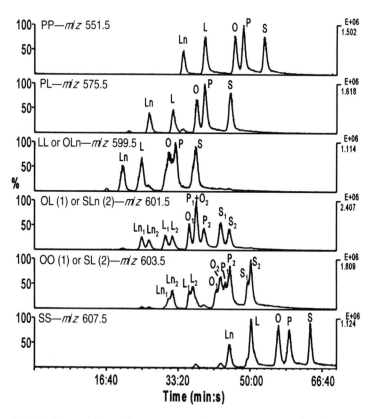

FIG. 2. Extracted ion chromatograms from LC–APCI/MS data for synthetic mixture of 35 triacylglycerols. Abbreviations are as in Fig. 1.

were resolved. Among overlapping species, TAGs having the least difference between the FAs eluted first, whereas TAGs having very different FAs eluted last. Examples of species having the same ECN showing complete resolution, partial resolution, and complete overlap are seen in the EICs for m/z 601.5, m/z 603.5, and m/z 599.5, respectively, shown in Fig. 2. Peaks in the EIC of m/z 601.5 were almost completely resolved, and their areas were integrated easily. However, the P_1 peak of [OL]$^+$ did overlap with the O_2 peak of [SLn]$^+$ and require apportionment. Since the Ln, L, and S peaks of this EIC were resolved, the $P_1 + O_2$ peak was apportioned so that O_1/O_2 and P_1/P_2 were in the same proportion as the ratio of [(Ln$_1$ + L$_1$ + S$_1$)/(Ln$_2$ + L$_2$ + S$_2$)]. Partially resolved species such as those shown in the EIC for m/z 603.5 were quantitated by manual integration of the overlapped peaks. Unresolved species, such as those of [LL]$^+$ and [OLn]$^+$ in the EIC of m/z 599.5, were apportioned according to their statistically expected proportions, calculated using the average uncorrected amounts of Ln, L O, P, and S obtained from all EICs. In the results presented here, Ln, L O, P, and S peak areas from more EICs were used for the apportionment than were previously used, resulting in small differences in composition from that already reported (55).

Comparison of the uncorrected TAG percentage composition obtained from raw integrated peak areas, listed in Table 1, to the composition expected from the statistical distribution of the five fatty acids revealed that response factors were necessary to account for differences in TAG response. Several methods were devised and tested for production of response factors. Since five FAs were interesterified with glycerol to produce the 35-component synthetic mixture, the FAs should be randomly distributed between all possible TAG molecular species. The composition should be that given from the equations, given in the preceding section, for calculation of the statistical amount of each TAG species expected from a given FA composition. The statistically expected composition was calculated from the fatty acid composition determined by calibrated GC–FID, listed in Table 2. Response factors were calculated for each TAG molecular species by dividing its statistically expected percentage by its empirically obtained percentage. When multiplied by these factors, the uncorrected percentage composition was adjusted to equal the statistical composition. Response factors thus calculated were further applied to the same TAG species present in the interesterified canola oil/SBOS blend. However, other methods of response factor calculation were found that provided less average relative error, so this method was not applied to all samples.

The differences in TAG response resulted primarily from the different amounts of [DG]$^+$ versus [M + H]$^+$ ions formed during ionization. Since [DG]$^+$ ions had lower masses than [M + H]$^+$ ions, they were propagated through the interface and mass detection system more efficiently and gave more signal. Therefore, highly saturated species, which gave predominantly [DG]$^+$ ions, produced more signal than highly unsaturated species, which produced mostly [M + H]$^+$ ions. To begin to model the fragmentation behavior of TAG, we defined a term called the triacylglycerol quotient (TAG quotient) as follows:

TABLE 1
35-Component Synthetic Mixture TAG Percentage Composition

TAG	Average Raw MS % (n = 3)	Statistical (%)	GC–FID RF-Adjusted % (n = 3)	Average LC–FID % (n = 3)
POS	6.3 ± 0.5	5.2	5.4 ± 0.5	5.7 ± 0.2
PLS	5.8 ± 0.3	5.1	5.1 ± 0.2	6.3 ± 0.2
PLO	4.5 ± 0.2	5.0	4.2 ± 0.2	7.0 ± 0.3
LOS	4.4 ± 0.7	5.0	4.5 ± 0.6	4.9 ± 0.2
PLnS	4.7 ± 0.3	4.8	4.6 ± 0.1	4.1 ± 0.1
PLnO	4.3 ± 0.1	4.7	4.4 ± 0.1	5.5 ± 0.2
LnOS	3.9 ± 0.1	4.6	4.5 ± 0.1	4.2 ± 0.2
PLnL	4.5 ± 0.2	4.6	4.7 ± 0.1	5.6 ± 0.2
LnLS	3.5 ± 0.1	4.5	4.1 ± 0.1	3.8 ± 0.1
LnLO	3.7 ± 0.4	4.4	4.5 ± 0.3	4.8 ± 0.3
PPS	3.6 ± 0.5	2.7	2.6 ± 0.4	2.3 ± 0.2
SSP	3.0 ± 0.3	2.7	2.4 ± 0.2	2.3 ± 0.3
PPO	3.7 ± 0.3	2.6	2.9 ± 0.2	2.9 ± 0.1
PPL	3.1 ± 0.2	2.6	2.5 ± 0.2	3.4 ± 0.1
SSO	2.9 ± 0.3	2.6	2.8 ± 0.0	2.6 ± 0.1
OOP	2.9 ± 0.3	2.6	2.7 ± 0.1	2.7 ± 0.0
SSL	2.8 ± 0.2	2.5	2.7 ± 0.2	2.3 ± 0.1
OOS	2.5 ± 0.0	2.5	2.6 ± 0.1	2.6 ± 0.1
LLP	2.6 ± 0.1	2.5	2.4 ± 0.1	3.2 ± 0.1
LLS	2.3 ± 0.1	2.4	2.4 ± 0.0	1.7 ± 0.1
OOL	2.6 ± 0.2	2.4	2.9 ± 0.1	2.0 ± 0.2
PPLn	3.1 ± 0.2	2.4	2.7 ± 0.1	1.8 ± 0.1
LLO	2.6 ± 0.1	2.4	2.9 ± 0.0	1.9 ± 0.0
SSLn	2.3 ± 0.2	2.4	2.5 ± 0.2	1.9 ± 0.1
OOLn	1.6 ± 0.0	2.3	2.0 ± 0.0	2.1 ± 0.1
LLLn	2.0 ± 0.1	2.2	2.5 ± 0.0	1.8 ± 0.1
LnLnP	1.9 ± 0.1	2.1	2.2 ± 0.1	2.1 ± 0.1
LnLnS	1.8 ± 0.1	2.1	2.3 ± 0.1	1.7 ± 0.1
LnLnO	1.4 ± 0.1	2.1	1.9 ± 0.0	2.4 ± 0.1
LnLnL	1.4 ± 0.2	2.0	2.0 ± 0.0	1.7 ± 0.2
PPP	1.5 ± 0.1	0.9	1.0 ± 0.0	0.5 ± 0.1
SSS	1.0 ± 0.2	0.9	0.9 ± 0.1	0.8 ± 0.1
OOO	0.7 ± 0.3	0.8	0.7 ± 0.2	0.5 ± 0.0
LLL	0.6 ± 0.0	0.8	0.7 ± 0.0	0.5 ± 0.1
LnLnLn	0.4 ± 0.0	0.6	0.6 ± 0.0	0.5 ± 0.0
Sum	100.0	100.0	100.0	100.0
Average Relative Error (%)	18.2 (n = 35)	—	8.8 (n = 35)	18.0 (n = 35)

$$\text{TAG quotient} = \frac{[\text{TAG} + \text{H}]^+}{[\text{TAG} + \text{H}]^+ + \Sigma[\text{DG}]^+} \qquad [5]$$

where $[\text{TAG} + \text{H}]^+$ is the peak height of the $[\text{M} + \text{H}]^+$ peak in a mass spectrum, or the area under the $[\text{M} + \text{H}]^+$ peak in an EIC, and $\Sigma[\text{DG}]^+$ is the sum of the peak heights of all $[\text{DG}]^+$ fragments in a mass spectrum arising from a TAG species or the area

TABLE 2
35-Component Synthetic Mixture Fatty Acid Percentage Composition

FA	From raw MS TAG	Calibrated GC–FID	From GC–FID RF-adj. TAG	From LC–FID TAG
P	24.0	20.9	20.8	22.2
Ln	16.0	18.5	18.5	17.5
L	19.0	19.8	19.9	20.1
O	19.6	20.2	20.2	20.8
S	21.3	20.6	20.6	19.3
Sum	100.0	100.0	100.0	100.0
Average Relative Error (%)	7.9	—	0.2	3.7

under all [DG]$^+$ peaks in EICs for a TAG. Using this equation, all saturated TAGs, which gave no [M + H]$^+$, had TAG quotients of 0. Highly unsaturated TAG, such as LnLnLn, had TAG quotients as high at 0.9 [where 1.0 = 100% [TAG + H]$^+$ ion]. We observed that the TAG quotient behaved in a manner which was approximately proportional to the number of sites of unsaturation in a TAG, to the one-third power:

$$\text{TAG Quotient} \propto (\text{\# sites of unsaturation})^{1/3}$$

Empirical constants were added to set the maximum value (~0.86) and the inflection point (~3.45 sites) to yield the final equation, called the *triacylglycerol quotient 1/3 power fit*:

$$\text{TAG quotient 1/3 power fit} = C_1 \times \left\{ \frac{(\text{sites} - C_2)^{1/3} + C_2^{1/3}}{(9 - C_2)^{1/3} + C_2^{1/3}} \right\}$$

where $C_1 = 0.8569$ and $C_2 = 3.445$, and "sites" is the number of sites of unsaturation in the TAG. This equation allowed approximation of the relative amounts of [M + H]$^+$ and [DG]$^+$ ions expected for a TAG with a given number of sites of unsaturation. Since it was observed that TAG MS signal response was dependent on the relative amounts of these ions formed, response factors were calculated from the inverse of the triacylglycerol quotient 1/3 power fit. Since 16-carbon saturates and 18-carbon saturates both gave TAG quotients of 0, but the 16-carbon chain had a lower mass and therefore gave more signal, another factor called the *18/16 factor* was included in the calculation of response factors from the TAG quotient. The response factors calculated from the combination of these empirically developed equations produced a TAG composition that had only 10.8% average relative error (data not shown) compared to the statistically expected composition. The FA composition calculated from the TAG quotient-normalized data showed 3.3% average relative error compared to the FA composition determined by GC–FID. Nevertheless, other methods for production of response factors were developed that were more effective and provided less average error. Therefore, no additional details regarding the TAG quotient calculations will be given.

Quantitation based on calibration curves was also performed. Calibration curves were constructed for a set of six monoacid standards containing deuterated tripalmitin (d_{12}-PPP) as internal standard. Equations and correlation coefficients of the calibration curves for six monoacid TAG species have been presented elsewhere (55). These data demonstrated that calibration curves with a high degree of linearity could be obtained for a few TAG species for which standards are available, but that even simple natural TAG mixtures contained enough species that producing calibration standards for each was impractical. The response factors for the six monoacid TAG species were interpolated to provide response factors for all TAG species. The TAG composition obtained using response factors calculated from the interpolated calibration curve slopes (data not shown) had 17.4% average relative error compared to the statistically predicted composition. The FA composition resulting from this adjusted TAG composition had an average relative error of 7.9% compared to the FA composition obtained by calibrated GC–FID. These values indicated that this method of normalization provided no improvement in average error compared to the uncorrected data. Other more effective methods for calculation of response factors were developed, so the use of calibration curves will not be discussed further.

Another approach to development of response factors was to calculate response factors for individual fatty acids, which were then used to calculate response factors for TAGs. First, a fatty acid composition was calculated from the TAG composition determined by LC–APCI/MS, as given in Table 2. The FA composition determined by calibrated GC–FID was divided by the FA composition calculated from the TAG composition, to produce a number that represents the underresponse or overresponse of each FA. When normalized to the smallest value, these numbers produced FA response factors, which were multiplied together to produce TAG response factors. The FA response factors calculated from the data in Table 2 were

P: 1.0000

Ln: 1.2853

L: 1.1827

O: 1.2020

S: 1.1140

These were multiplied together to yield response factors for each TAG containing these FAs. Application of the TAG response factors thus derived from FA response factors to the uncorrected LC/MS data resulted in the GC–FID–adjusted composition given in Table 1. The TAG composition thus determined had an average relative error of 8.8% compared to the statistically predicted composition. This average error was the least error produced by any quantitation method applied to this synthetic mixture. The fatty acid composition resulting from this TAG composition was in excellent agreement with the FA composition obtained by GC–FID, exhibiting only 0.2% average relative error.

Randomized and Normal Lard

Reconstructed ion chromatograms for normal and randomized lard are given in Fig. 3. Elimination of the guard column and use of improved columns and solvent gradient resulted in separations showing better resolution than previously reported for these samples (55). This improved resolution allowed more manual integration of partially overlapped peaks, with less need for statistical apportionment. The EICs for m/z 599.5 and m/z 603.5, shown in Figs. 4 and 5, exemplify the improved resolution, compared to the EICs shown in Fig. 2, which were obtained using the previous chromatographic method. Mass spectra obtained for several TAG from natural lard are shown in Fig. 6. These spectra demonstrate the tendency toward less $[M + H]^+$ ion with more saturation in the acyl chains.

The RIC of randomized lard, Fig. 3B, showed the presence of significantly more early-eluting diacylglycerols than were present in the normal lard. These diacylglycerols were a common byproduct of randomization or interesterification reactions. Quantitation of the diacylglycerols was not performed. The uncorrected TAG composition of the randomized lard, obtained from LC/MS data, is given in Table 3. Since this sample was randomized, its composition could be compared to the statistically expected composition calculated from the FA composition determined by calibrated GC–FID, given in Table 4. To ensure complete randomization, lipase hydrolysis was performed on the randomized lard. The FA composition at TAG position 2 is given in Table 4. This agreed well with the overall FA composition obtained by GC–FID, indicating a high degree of randomization. The uncorrected TAG composition exhibited 29.3% average relative error, compared to the statistically expected composition. The FA composition calculated from the raw TAG composition, given in Table 4, exhibited 21.8% average relative error compared to the FA composition determined by calibrated GC–FID.

Response factors to correct the raw LC/MS data were derived from the calibrated GC–FID data. Response factors for each FA were calculated, and these were multiplied together to produce TAG response factors as previously described. The average FA response factors obtained by dividing the FA composition determined by calibrated GC–FID by the FA composition calculated from the 'raw' LC/MS data, for each of six runs, were

M: 1.1854
Po: 1.0000
P: 1.5329
Ln: 1.0552
L: 1.4669
O: 1.7007
S: 1.4464

These FA responses were multiplied together to produce response factors for each of the TAG species containing these FAs. The adjusted TAG composition obtained by

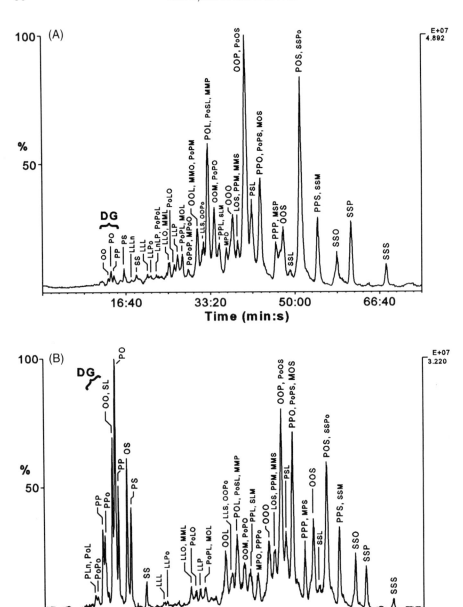

FIG. 3. Reconstructed ion chromatograms by atmospheric-pressure chemical ionization mass spectrometry of (A) normal lard and (B) randomized lard. Abbreviations: DG, diacylglycerols; Po, palmitoleate (16:1); M, myristate (14:0); other abbreviations as in Fig. 1.

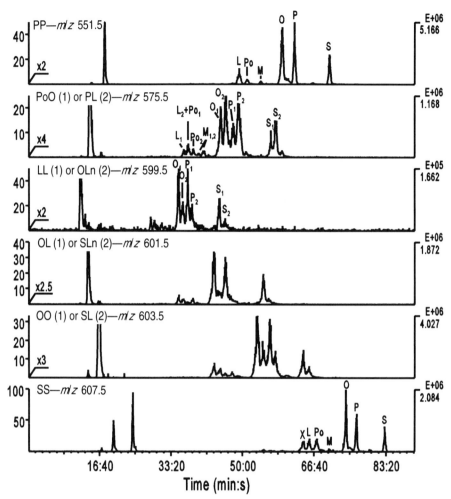

FIG. 4. Extracted ion chromatograms from LC–APCI/MS data for randomized lard. Abbreviations: DG, diacylglycerols; Po, palmitoleate (16:1); M, myristate (14:0); X, isotope contribution from (m/z −2); other abbreviations as in Fig. 1.

application of these response factors is given in Table 3. Application of the response factors thus derived resulted in average relative error of 16.7% per TAG, compared to the statistically expected amount. The TAG composition obtained by LC–FID showed 62.8% average relative error. TAG that were not identified by LC–FID, and therefore had a relative error of 100%, were not included in the calculation of the average relative error. Exclusion of LLS (dilinoleoylstearoylglycerol) and PPM (dipalmitoylmyristoylglycerol) from the LC–FID error calculation reduced the average error to 28.9% average relative error, as previously reported (55). Thus, for

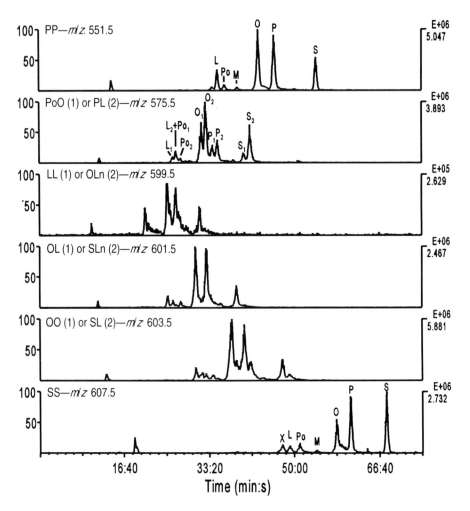

FIG. 5. Extracted ion chromatograms from LC–APCI/MS data for normal (nonrandomized) lard. Abbreviations as in Fig. 4.

most TAGs the LC–FID TAG composition had nearly the same average relative error as the raw LC/MS data.

The FA composition resultant from the GC–FID–adjusted TAG composition is given in Table 4. This FA composition showed good agreement with the FA composition determined by calibrated GC–FID, having only 1.4% average relative error per FA. The LC–FID TAG composition produced a FA composition that had nearly the same average relative error, 21.6%, as the FA composition obtained from the uncorrected LC/MS data.

The uncorrected TAG composition of the normal lard, obtained from LC/MS data, is given in Table 5. Typical EICs from the separation of normal lard are shown

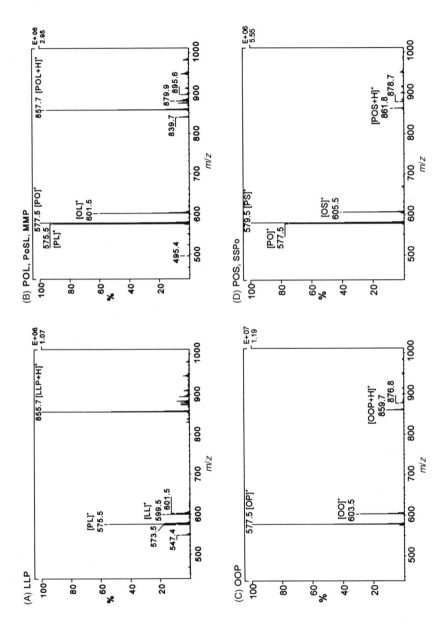

FIG. 6. LC/APCI-MS mass spectra from normal lard: (A) LLP, (B) POL, (C) OOP, (D) POS. Abbreviations as in Fig. 1.

TABLE 3
Randomized Lard TAG Percentage Composition

TAG	Raw MS % (n = 6)	Statistical (%)	GC–FID RF-adjusted (%)	LC–FID (%)
OOP	12.6 ± 1.2	14.8	15.3 ± 1.5	17.5
POS	11.4 ± 1.2	12.1	11.8 ± 1.1	12.4
PPO	10.8 ± 1.1	10.9	11.7 ± 1.0	12.5
OOS	7.3 ± 0.6	8.3	8.3 ± 0.6	12.4
OOO	5.2 ± 0.4	6.7	7.0 ± 0.5	7.5
PLO	4.8 ± 0.2	5.7	5.0 ± 0.3	5.8
PPS	4.5 ± 0.3	4.5	4.2 ± 0.2	3.9
OOL	3.2 ± 0.1	3.9	3.7 ± 0.1	3.4
SSO	3.2 ± 0.3	3.4	3.1 ± 0.3	2.9
LOS	3.5 ± 0.6	3.2	3.5 ± 0.5	4.0
PPP	2.6 ± 0.3	2.7	2.6 ± 0.3	1.4
SSP	2.4 ± 0.2	2.5	2.1 ± 0.2	1.6
PLS	3.0 ± 0.2	2.4	2.6 ± 0.2	3.4
PPL	2.2 ± 0.2	2.1	2.1 ± 0.2	1.0
PoPO	2.9 ± 0.3	2.0	2.1 ± 0.2	2.9
MPO	1.8 ± 0.3	1.4	1.5 ± 0.1	—
OOPo	1.2 ± 0.1	1.3	1.0 ± 0.0	—
PoOS	2.0 ± 0.2	1.1	1.3 ± 0.2	—
OOM	0.7 ± 0.1	1.0	0.6 ± 0.1	—
MOS	0.8 ± 0.1	0.8	0.6 ± 0.1	—
PoPS	1.8 ± 0.2	0.8	1.1 ± 0.1	—
LLO	0.7 ± 0.0	0.8	0.7 ± 0.0	0.5
PPPo	0.9 ± 0.1	0.7	0.6 ± 0.1	—
SSL	1.1 ± 0.1	0.7	0.9 ± 0.1	—
MPS	1.1 ± 0.2	0.6	0.8 ± 0.1	—
LLP	0.6 ± 0.1	0.6	0.5 ± 0.1	0.5
PPM	0.7 ± 0.1	0.5	0.5 ± 0.1	2.4
PoLO	0.5 ± 0.0	0.5	0.3 ± 0.0	0.3
SSS	0.5 ± 0.0	0.5	0.4 ± 0.0	0.8
MLO	0.4 ± 0.0	0.4	0.3 ± 0.0	—
PoPL	0.5 ± 0.1	0.4	0.3 ± 0.0	—
LLS	0.7 ± 0.1	0.3	0.6 ± 0.1	1.6
MPL	0.4 ± 0.0	0.3	0.3 ± 0.0	—
SSPo	0.4 ± 0.0	0.2	0.2 ± 0.0	—
PLnO	0.2 ± 0.0	0.2	0.2 ± 0.0	—
PoLS	0.4 ± 0.0	0.2	0.2 ± 0.0	—
SSM	0.1 ± 0.0	0.2	0.1 ± 0.0	—
MLS	0.2 ± 0.0	0.2	0.1 ± 0.0	—
Other	2.8	1.4	1.6	1.3
Sum	100.0	100.0	100.0	100.0
Average Relative Error (%)	29.3 (n = 38)	—	16.7 (n = 38)	62.8 (n = 21)

TABLE 4
Randomized Lard Fatty Acid Percentage Composition

FA	From raw MS TAGs	Calibrated GC–FID	From GC–FID RF-adj. TAGs	From LC–FID TAGs	TAG position 2
M	2.6	2.0	1.9	0.8	1.0
Po	4.2	2.7	2.7	1.1	2.5
P	30.0	29.9	29.9	29.7	29.5
Ln	0.4	0.3	0.3	0.0	0.3
L	8.3	7.9	7.9	7.8	8.9
O	36.8	40.6	40.7	44.0	40.7
S	17.8	16.7	16.6	16.6	17.1
Sum	100.0	100.0	100.0	100.0	100.0
Average Relative Error (%)	21.8	—	1.4	21.6	—

in Figure 5. The FA composition resulting from this TAG composition is given in Table 6. Table 6 also lists the composition of the FA at position 2 of the TAG backbone. The FA composition at position 2 differed significantly from the overall composition obtained by GC–FID, and indicated nonrandom distribution of FA through the TAG molecular species. Therefore, the TAG compositions obtained by LC/MS and LC–FID could not be compared to the statistically predicted TAG composition. However, the FA compositions resultant from different quantitation methods did provide a measure of the errors associated with the different quantitation methods. As with the previous samples, the FA composition obtained from the LC–FID data had average relative error similar to that of the uncorrected LC/MS data.

TAG response factors for the normal lard samples were calculated from FA response factors, as described above. The FA composition obtained by calibrated GC–FID was divided by the FA composition resultant from the TAG composition obtained by LC/MS, and normalized the smallest value to obtain these FA response factors:

M: 1.3419

Po: 1.0000

P: 1.6568

Ln: 1.1757

L: 1.4217

O: 1.7095

S: 1.3327

These FA response factors were multiplied together to obtain response factors for each of the TAG molecular species. The TAG composition calculated by application of these response factors is given in Table 5. The FA composition that resulted from this TAG composition is given in Table 6. This FA composition showed excellent

TABLE 5
Normal Lard TAG Percentage Composition

TAG	Raw MS % (n = 6)	GC–FID RF-adjusted (%)	LC–FID (%)	Randomized lard RF-adjusted (%)
OOP	16.1 ± 1.2	20.7 ± 0.9	25.7	18.7 ± 1.4
POS	16.1 ± 1.1	16.2 ± 0.9	18.2	17.1 ± 1.1
PPO	6.0 ± 0.8	7.5 ± 0.7	7.6	6.0 ± 0.8
OOS	4.3 ± 0.5	4.4 ± 0.5	7.2	4.8 ± 0.6
OOO	4.3 ± 0.3	5.7 ± 0.4	5.2	5.4 ± 0.4
PLO	6.5 ± 0.7	6.9 ± 0.7	7.5	7.6 ± 0.9
PPS	4.2 ± 0.4	4.0 ± 0.3	3.5	4.1 ± 0.4
OOL	2.3 ± 0.2	2.5 ± 0.1	2.3	2.7 ± 0.3
SSO	2.7 ± 0.2	2.2 ± 0.1	1.5	2.8 ± 0.2
LOS	2.2 ± 0.4	1.9 ± 0.4	1.0	2.0 ± 0.3
PPP	1.6 ± 0.2	1.9 ± 0.2	1.2	1.6 ± 0.2
SSP	3.8 ± 0.2	3.0 ± 0.2	2.9	4.0 ± 0.2
PLS	4.4 ± 0.3	3.7 ± 0.3	5.0	3.5 ± 0.2
PPL	1.3 ± 0.1	1.3 ± 0.2	1.0	1.2 ± 0.1
PoPO	3.4 ± 0.2	2.5 ± 0.1	3.6	2.2 ± 0.1
MPO	1.2 ± 0.2	1.2 ± 0.2	—	1.0 ± 0.2
OOPo	1.9 ± 0.2	1.5 ± 0.1	—	2.0 ± 0.2
PoOS	2.3 ± 0.2	1.4 ± 0.0	—	1.3 ± 0.1
OOM	1.1 ± 0.1	1.2 ± 0.1	—	1.5 ± 0.1
MOS	0.9 ± 0.1	0.7 ± 0.1	—	0.9 ± 0.1
PoPS	1.3 ± 0.2	0.8 ± 0.1	—	0.6 ± 0.1
LLO	0.8 ± 0.1	0.7 ± 0.1	0.5	0.9 ± 0.1
PPPo	0.4 ± 0.1	0.3 ± 0.1	—	0.3 ± 0.1
SSL	0.6 ± 0.1	0.4 ± 0.0	—	0.3 ± 0.0
MPS	1.0 ± 0.1	0.7 ± 0.1	—	0.5 ± 0.1
LLP	1.0 ± 0.1	0.9 ± 0.1	0.8	1.0 ± 0.1
PPM	0.4 ± 0.1	0.4 ± 0.0	1.5	0.3 ± 0.0
PoLO	0.7 ± 0.1	0.4 ± 0.0	0.4	0.7 ± 0.1
SSS	1.1 ± 0.1	0.7 ± 0.1	0.9	1.1 ± 0.1
MLO	0.5 ± 0.1	0.4 ± 0.0	—	0.5 ± 0.1
PoPL	0.5 ± 0.0	0.3 ± 0.0	—	0.4 ± 0.0
LLS	0.6 ± 0.1	0.4 ± 0.1	1.5	0.3 ± 0.1
MPL	0.2 ± 0.0	0.2 ± 0.0	—	0.2 ± 0.0
SSPo	0.3 ± 0.0	0.2 ± 0.0	—	0.2 ± 0.0
PLnO	0.3 ± 0.0	0.2 ± 0.0	—	0.2 ± 0.0
PoLS	0.5 ± 0.1	0.2 ± 0.0	—	0.3 ± 0.0
SSM	0.1 ± 0.0	0.1 ± 0.0	—	0.2 ± 0.0
MLS	0.2 ± 0.0	0.2 ± 0.0	—	0.2 ± 0.0
Other	2.8	1.8	1.0	1.4
Sum	100.0	100.0	100.0	100.0

agreement with the FA composition obtained by calibrated GC–FID, having only 0.8% average relative error per FA.

Because the randomized lard was similar in composition to the normal lard, and because the composition of the randomized lard was known from the statistical

TABLE 6
Normal Lard Fatty Acid Percentage Composition

FA	From raw MS TAG	Calibrated GC–FID	From GC–FID RF-adjusted TAG	From LC–FID TAG	From randomized lard RF-adjusted TAG	TAG position 2
M	2.4	2.1	2.0	0.5	2.0	2.3
Po	4.4	2.8	2.9	1.4	2.9	3.9
P	28.5	30.4	30.2	31.8	28.6	57.3
Ln	0.4	0.3	0.3	0.0	0.3	0.4
L	8.7	7.9	7.8	7.7	8.2	5.1
O	36.5	40.1	40.4	42.6	39.9	22.9
S	19.2	16.4	16.4	16.1	18.1	8.1
Sum	100.0	100.0	100.0	100.1	100.0	100.0
Average Relative Error (%)	20.8	—	0.8	23.7	4.9	—

distribution of its FA composition, response factors were calculated from randomized lard and applied to normal lard. Response factors for each TAG were obtained by dividing the statistical amount of each TAG of randomized lard by the amount determined by uncorrected LC/MS. The uncorrected TAG composition of normal lard was multiplied by these response factors to obtain the "Randomized lard RF-adjusted%" composition column in Table 5. The FA composition calculated from this adjusted TAG composition is given in Table 6. This FA composition produced 4.9% average relative error per FA.

Interesterified and Normal Canola Oil/CSOS Blend

Reconstructed ion chromatograms of the canola oil/CSOS (80:20) blend and the interesterified blend are shown in Fig. 7. As with the lard samples, the interesterified blend contained a significantly larger amount of diacylglycerols, as a byproduct of the interesterification process, than did the normal (noninteresterified) blend. Comparison of the EICs from these two samples dramatically showed the compositional differences between these samples. The EIC of [SS]$^+$ (m/z 607.5) from the normal blend, Fig. 8, revealed that most of the [SS]$^+$ came from two species: SSP and SSS. These two components were present in noticeably larger amounts in the RIC from the normal blend, shown in Fig. 7A, than in the RIC for the interesterified blend, shown in Fig. 7B. These two components came directly from the cottonseed oil stearin flakes used in the blend. The EICs of the other [DG]$^+$ contained only small amounts of saturates (P and S). Conversely, the EICs from the interesterified blend, Fig. 9, showed that palmitic acid and stearic acid were distributed throughout all possible TAG species. Also, the EIC for [SS]$^+$ (m/z 607.5) from the interesterified blend exhibited peaks from all FAs in similar proportions as all other EICs, rather than being concentrated in the two trisaturated species.

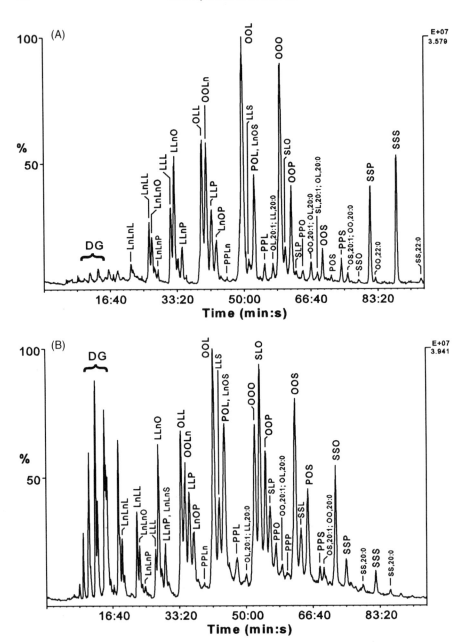

FIG. 7. Reconstructed ion chromatograms by atmospheric-pressure chemical ionization mass spectrometry of (A) canola oil/soybean oil saturates blend (80:20) and (B) interesterified canola oil/soybean oil saturates blend (80:20). Abbreviations: DG, diacylglycerols; other abbreviations as in Fig. 1.

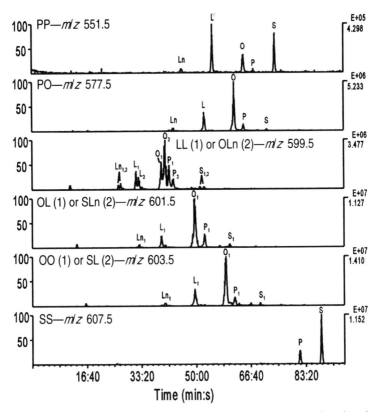

FIG. 8. Extracted ion chromatograms from LC–APCI/MS data for canola oil/soybean oil saturates blend (80:20). Abbreviations: X, C^{13} isotope contribution from $(m/z - 2)$; other abbreviations as in Fig. 1.

The uncorrected TAG composition of the interesterified blend is given in Table 7. Because this sample was interesterified, its composition was determined by the statistical distribution of the FAs as determined by calibrated GC–FID, given in Table 8. The uncorrected TAG composition obtained by LC/MS exhibited 29.2% average relative error compared to the statistically predicted composition. The FA composition calculated from the uncorrected TAG composition had an average relative error of 11.8% per FA.

As demonstrated above, the FA composition determined by calibrated GC–FID was divided by the FA composition calculated from the raw LC/MS data and normalized to the smallest value (excluding Po, since it was present at <1.0%) to produce FA response factors. These FA response factors were then multiplied together to produce TAG response factors. The FA response factors thus obtained were

Po: 0.4716

P: 1.0000

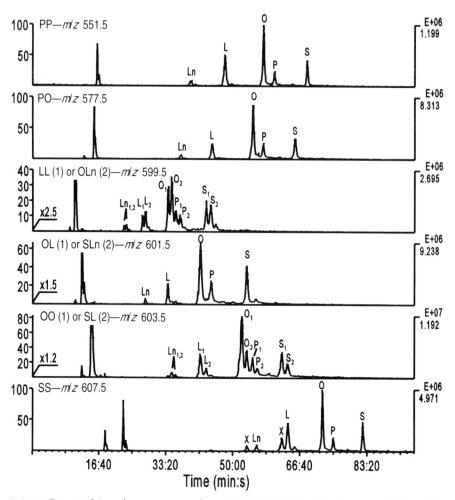

FIG. 9. Extracted ion chromatograms from LC–APCI/MS data for interesterified canola oil/soybean oil saturates blend (80:20). Abbreviations: X, C^{13} isotope contribution from $(m/z - 2)$; other abbreviations as in Fig. 1.

Ln: 1.2242
L: 1.3923
O: 1.4924
S: 1.3633

The TAG composition obtained by application of these response factors is given in Table 7 as the "GC–FID RF-Adjusted%." This TAG composition agreed well with the statistically predicted composition, having 9.9% average relative error. The FA composition calculated from this TAG composition was in exceptional agreement

TABLE 7
Interesterified Canola Oil/CSOS Blend TAG Percentage Composition

TAG	Raw MS (%)	Statistical (%)	GC–FID RF-adjusted (%)	Mix 5 adjusted (%)	LC–FID (%)
OOL	9.7	12.7	11.7	8.8	10.3
OOS	9.4	11.4	11.1	9.2	11.6
LOS	9.6	11.0	10.6	10.6	11.5
OOO	7.7	8.8	10.0	9.2	8.0
LLO	5.1	6.1	5.7	4.6	4.5
SSO	4.6	4.9	5.0	4.0	4.2
OOP	5.6	4.5	4.9	4.8	4.8
OOLn	3.8	4.3	4.1	5.2	4.3
PLO	4.9	4.3	4.0	5.4	8.3
LnLO	3.9	4.1	3.9	4.6	4.4
POS	4.7	3.9	3.7	3.8	3.9
LnOS	3.8	3.7	3.7	4.3	0.0
LLS	3.2	2.6	3.3	3.2	1.9
SSL	2.5	2.4	2.5	2.2	2.4
PLS	2.4	1.9	1.8	2.1	2.9
LnLS	2.3	1.8	2.0	2.8	0.0
PLnO	1.7	1.5	1.2	1.8	0.7
LLP	1.6	1.0	1.2	1.5	5.7
LLLn	1.0	1.0	0.9	1.1	0.9
LLL	1.0	1.0	1.0	1.2	0.5
SSP	1.1	0.8	0.8	1.0	0.8
SSLn	0.7	0.8	0.6	0.6	0.0
PPO	1.5	0.8	0.9	1.0	1.3
SSS	0.9	0.7	0.9	0.7	2.1
LnLnO	0.8	0.7	0.7	1.2	0.9
PLnL	1.0	0.7	0.7	1.0	1.4
PLnS	1.0	0.6	0.7	1.0	0.0
PPL	0.6	0.4	0.4	0.5	0.7
LnLnL	0.5	0.3	0.4	0.7	0.9
PPS	0.6	0.3	0.3	0.4	0.7
LnLnS	0.5	0.3	0.4	0.6	0.0
PPLn	0.2	0.1	0.1	0.2	0.1
LnLnP	0.3	0.1	0.2	0.3	0.2
OOPo	0.1	0.1	0.0	0.0	0.0
PoLO	0.7	0.1	0.3	0.0	0.0
PoOS	0.1	0.1	0.0	0.0	0.0
Other	0.8	0.3	0.3	0.2	0.1
Sum	100.0	100.0	100.0	100.0	100.0
Average Relative Error (%)	29.2 ($n = 31$)	—	9.9 ($n = 31$)	29.9 ($n = 31$)	62.7 ($n = 26$)

with the FA composition determined by calibrated GC–FID, having only 0.4% average relative error per FA. The average relative errors for the adjusted TAG composition and the resultant FA composition were the smallest average relative errors obtained using any of the quantitation methods applied to this sample.

TABLE 8
Interesterified Canola Oil/CSOS Blend Fatty Acid Percentage Composition

FA	From raw MS TAG	Calibrated GC–FID	From GC–FID RF-adjusted TAG	From Mix 5 adjusted TAG	From LC–FID TAG
Po	0.6	0.2	0.2	0.0	0.0
P	10.3	7.5	7.5	9.0	11.5
Ln	8.1	7.2	7.2	9.6	5.4
L	21.0	21.4	21.2	21.1	23.5
O	40.8	44.4	44.5	41.6	41.8
S	19.3	19.2	19.3	18.6	17.9
Sum	100.0	100.0	100.0	100.0	100.0
Average Relative Error (%)	11.8	—	0.4	12.8	20.2

To determine whether response factors from one sample could successfully be applied to a dissimilar sample, response factors calculated from the 35-component synthetic mixture were applied to quantitation of the CNO/CSOS interesterified blend TAGs. The statistical TAG composition of the synthetic mixture was divided by the TAG composition of the mixture obtained by LC/MS to produce a response factor for each TAG. These response factors were then applied to the interesterified blend uncorrected TAG composition to produce the "Mix 5 Adjusted%" composition given in Table 7. This adjusted composition had 29.9% average relative error, compared to the statistically predicted composition. This error was similar to that of the uncorrected data, indicating that no improvement was achieved by application of this quantitation method. The FA composition resultant from the normalized TAG composition is given in Table 8. The average relative error of the adjusted FA composition was 12.8%. This was again similar to the uncorrected results, indicating that no improvement was obtained by this quantitation method. This quantitation method was not further applied to other samples.

The LC–FID data yielded a TAG composition that exhibited 62.7% average relative error. This high amount of error resulted from high relative errors of several components such as LLP and SSS. The FA composition resultant from the TAG composition determined by LC–FID had an average relative error of 20.2% per FA.

The uncorrected TAG composition of the normal canola oil/CSOS blend is given in Table 9. Since this sample was not interesterified, the composition was not given by the statistical distribution of FAs. The FA composition calculated from the uncorrected TAG composition gave an average relative error of 20.8%, compared to the composition obtained by calibrated GC–FID, as given in Table 10.

Response factors based on the FA composition determined by GC–FID were calculated, as described above. The FA response factors used to calculate the TAG response factors for this sample were

Po: 0.4716

P: 1.0000

TABLE 9
Canola Oil/CSOS Blend TAG Percentage Composition

TAG	Raw MS (%)	GC–FID RF-adjusted (%)	Randomized CNO adjusted (%)	LC–FID (%)
OOL	18.0	15.4	23.4	15.7
OOS	1.5	2.2	1.8	0.9
LOS	1.8	2.4	2.1	2.0
OOO	14.1	13.5	15.8	16.1
LLO	7.4	5.5	8.7	5.4
SSO	0.1	0.2	0.1	0.3
OOP	5.3	4.2	4.2	4.1
OOLn	6.9	4.8	7.7	7.6
PLO	5.6	3.9	4.8	5.4
LnLO	4.7	2.9	4.9	6.1
POS	0.4	0.4	0.3	0.1
LnOS	1.7	1.8	1.7	0.0
LLS	0.9	1.0	0.7	0.8
SSL	0.2	0.3	0.2	0.1
PLS	0.6	0.7	0.5	0.5
LnLS	1.2	1.1	0.9	0.0
PLnO	2.2	1.3	1.8	2.5
LLP	3.1	1.9	1.9	2.9
LLLn	1.8	1.0	1.7	0.9
LLL	2.8	1.8	2.8	1.3
SSP	5.2	9.2	3.8	5.9
SSLn	0.1	0.1	0.1	0.0
PPO	0.6	0.4	0.3	0.6
SSS	6.1	19.5	5.0	15.5
LnLnO	1.5	0.8	1.2	1.5
PLnL	1.4	0.7	0.9	1.7
PLnS	0.3	0.3	0.2	0.0
PPL	0.8	0.4	0.4	0.2
LnLnL	0.6	0.2	0.4	0.5
PPS	0.8	0.8	0.5	0.7
LnLnS	0.6	0.4	0.3	0.0
PPLn	0.2	0.1	0.1	0.1
LnLnP	0.4	0.2	0.2	0.4
OOPo	0.3	0.1	0.3	0.0
PoLO	0.1	0.0	0.0	0.0
PoOS	0.1	0.0	0.1	0.0
Other	0.8	0.2	0.3	0.3
Sum	100.0	100.0	100.0	100.0

Ln: 1.2242
L: 1.3923
O: 1.4924
S: 1.3633

The TAG composition obtained using these response factors is listed in Table 9. The FA composition that results from this normalized TAG composition is given in

TABLE 10
Canola Oil/CSOS Blend Fatty Acid Percentage Composition

FA	From raw MS TAG	Calibrated GC–FID	From GC–FID RF-adjusted TAG	From randomized CNO adjusted TAG	From LC–FID TAG
Po	0.4	0.2	0.1	0.2	0.0
P	9.9	8.2	8.8	7.1	9.0
Ln	9.1	6.7	5.9	8.2	8.1
L	23.2	20.7	17.5	24.4	18.7
O	44.3	44.6	37.9	49.4	42.9
S	13.1	19.7	29.8	10.8	21.3
Sum	100.0	100.0	100.0	100.0	100.0
Average Relative Error (%)	20.8	—	20.2	22.0	10.7

Table 10. The FA composition obtained using this normalization method had 20.2% average relative error.

Quantitation was also performed on the CNO/CSOS blend by application of empirical response factors obtained from the interesterified blend. The statistical amount of each TAG in the interesterified blend was divided by the amount of each TAG in the raw LC/MS data for the blend. The response factors thus obtained were then applied to the uncorrected LC/MS data for the noninteresterified sample. The resultant TAG composition is shown in Table 9 as "Randomized CNO RF-adjusted %. The fatty acid composition obtained from this TAG composition is given in Table 10. The FA composition had an average relative error of 22.0% per FA.

The TAG composition obtained by LC–FID is listed in Table 9. The FA composition that was calculated based on that TAG composition is given in Table 10. This FA composition had an average relative error of 10.7% per FA.

Discussion

The utility of atmospheric-pressure chemical ionization mass spectrometry for qualitative analysis of TAG mixtures has been well demonstrated for numerous natural TAG mixtures, including TAGs having unique functional groups (53), TAG from normal and genetically modified seed oils (52,54), and TAG from hydroperoxidized samples (56). In mass spectra of all samples, similar fragmentation behavior was observed. Two primary types of ions were consistently formed: protonated molecular ions and fragments from which one acyl chain was lost to form a diacylglycerol ion. This fragmentation is similar to that observed using LC/direct inlet MS. Also, as with LC/direct inlet MS, the amount of unsaturation was the main factor in determining the relative amounts of diglyceride vs. protonated molecular ion formed. As with direct inlet MS, the qualitative appearance of the spectra has a direct effect on the response factors necessary for quantitation using this ionization technique.

Comparison of APCI–MS spectra reported elsewhere (56) to those reported previously by us and other authors show apparent contradictions in the masses obtained for diacylglycerol fragments and protonated molecular ions. A difference of 2 mass units was observed for peaks in all APCI–MS spectra reported elsewhere (56) compared to results presented by us (51–55) and others (1,6,9,12,16,48). This discrepancy most likely arose from differences in mass axis calibration.

Quantitation

Quantitation of compounds is most often performed by running standards of known concentrations, constructing calibration curves, and then matching a compound's detector response to a concentration on its calibration curve. We demonstrated previously that calibration curves for individual TAG species could be constructed that showed a high degree of linearity and so were suitable for quantitiation of the species for which curves were constructed. However, natural TAG mixtures contain a sufficiently large number of TAG species that it is impractical to construct calibration curves for every TAG in a mixture. Also, calibration curves may be constructed only when standards of known concentration are available. For many samples, such as genetically modified seed oils, appropriate standards are not generally available. Therefore, alternative methods of quantitation were needed. One approach used was to construct calibration curves for several representative species and then interpolate response factors for all TAG species from these. This approach was tested using a synthetic 35-component TAG mixture and was found to produce average relative error (17.4%) that was very similar to the error obtained by the uncorrected data.

TAG Quotient

In an attempt to characterize the qualitative behavior of TAGs during APCI–MS analysis, a factor called the triacylglycerol quotient was defined. The triacylglycerol quotient for a number of TAG species was plotted versus the number of sites of unsaturation, and it was seen that the median values of the triglyceride quotient gave a curve with a shape that could be approximated by an equation of the type $y = x^{1/3}$, where x was the number of sites of unsaturation. This equation allowed approximation of the relative amounts of $[M + H]^+$ and $[DG]^+$ ions expected for a TAG with a given number of sites of unsaturation. Since it was observed that TAG MS signal response was dependent on the relative amounts of these ions formed, response factors were calculated from the inverse of the triacylglycerol quotient 1/3 power fit. Similar dependence of correction factors on the amount of unsaturation was reported using LC/direct inlet MS (15). The average relative errors obtained by application of response factors calculated from the triacylglycerol quotient 1/3 power fit were less than the errors given by some other methods of quantitation, but this approach nevertheless had drawbacks that necessitated the development of other quantitation methods. This method offered no means of compensation for contributions of carbon-13 isotope abundances to peak areas. In our chromatographic system the L peak of an EIC often occurred at the nearly the same retention time as the Ln peak of the EIC having a m/z 2 mass units higher. For instance, in Fig. 9, the L peak of m/z 601.5 occurs at nearly the same time as the Ln peak of m/z 603.5. In cases where the amount of L greatly exceeded the amount of Ln, this resulted in an isotopic contribution of the L peak to the area of the Ln peak, giving a larger-than-actual area for the Ln. Other quantitation methods, discussed below, compensate for carbon-13 isotope contributions.

TAG Standard Method

It was hoped that a multiple-component calibration mixture could provide response factors for TAG species in a wide variety of samples. Toward that end, a 35-component TAG mixture was synthesized. The composition of the mixture was determined by the statistical distribution of the FA used in the synthesis. Response factors were calculated by dividing the "known" amounts of TAG by the amounts determined by LC/MS. When the response factors thus obtained were applied to samples of disparate composition, such as the interesterified canola oil/CSOS blend, the average relative error was similar to that of the uncorrected results. This was a result of concentration dependence of signal response for TAG species, so that response factors obtained for TAGs present in high concentrations were not applicable to samples that contained only small amounts of the TAG.

Randomized Fat Method

A similar, yet more effective method was devised. When most samples are randomized (or interesterified), a mixture of TAGs is produced that is not greatly different in composition from the nonrandomized sample, and its composition can be calculated from the random distribution of its FAs. Furthermore, the randomized and nonrandomized samples should have the same FA composition (compare Table 4 to Table 6, and Table 8 to Table 10). Thus, the randomized sample represented a mixture with known composition similar to that of the nonrandomized sample. Response factors obtained by dividing the statistically predicted composition of the randomized sample by the actual composition obtained by LC/MS were applied to the nonrandomized sample to produce an adjusted composition. Since the normal lard sample was not randomized, its TAG composition may not be compared to the statistical composition, but the FA composition obtained from the TAG composition may be compared to the FA composition obtained by GC–FID for error comparison. Results presented here and previously (55) indicated that the FA composition obtained from the TAG composition normalized using this method had less average relative error than LC–FID or any other method, except the one discussed in the following section. Thus, this method represents a useful approach for reducing the amount of error associated with quantitation using LC/APCI–MS.

GC–FID FA Composition Method

The final method for TAG composition quantitation developed by us employed the FA composition obtained by calibrated GC–FID. Response factors were first calculated for each fatty acid by dividing the FA composition obtained by calibrated GC–FID by the FA composition calculated from the uncorrected TAG composition. FA response factors were multiplied together to give TAG response factors. It was seen, in the results presented here and those presented previously (55), that this method of normalization produced the lowest average relative error in the TAG composition, and also in the resultant FA composition, of all the quantitation methods tested. In the results presented here, exceptional agreement was achieved between the calculated FA composition and the FA composition obtained by calibrated GC–FID. This quantitation method is effective for two reasons:

1. By calculating FA response factors, the overall underresponse or overresponse of particular FA is inherently incorporated into the normalization scheme. Thus, factors such as the contribution from carbon-13 isotopic abundances to the TAG having 2 mass units lower (mentioned above), which could add to integrated peak areas, are compensated for by the FA response factor.

2. This quantitation method is equally applicable to normal and randomized samples. The quantitation method discussed in the preceding section used response factors calculated from randomized samples for normalization of nonrandomized samples, so the randomized sample was simply set to equal the statistical composition. Using the method based on the FA composition determined by GC–FID, response factors are calculated and applied to the randomized samples just as they were calculated and applied to nonrandomized samples.

Furthermore, the ready availability and low cost of GC–FID make it an attractive technique for routine use.

This quantitation method has one drawback, exemplified by the canola oil/CSOS blend. By multiplying together the FA response factors to produce TAG response factors, the inherent assumption was made that the FAs were distributed uniformly throughout the TAG species (not necessarily in statistical proportions, just uniformly distributed). This assumption created no problems for most natural samples, as seen by the excellent agreement of the FA compositions calculated from the adjusted TAG composition for normal lard. However, when the FAs were concentrated in only a few TAG species, such as palmitic and stearic acid present in the CNO/CSOS blend, then significant errors occurred. Fortunately, the average relative error of the FA composition acted as an indicator of the uniformity of the distribution. Thus, for all samples except the noninteresterified blend, the average relative error of the FA composition obtained from the normalized results showed excellent agreement with the FA composition obtained by GC–FID. Only in the case of the noninteresterified blend did the FA composition of the normalized data show significant error compared to the GC–FID data. Thus, this quantitation method provided a self-check of its effectiveness. In the case of the noninteresterified blend, the error of the FA composition for the normalized data was similar to that of the uncorrected data, so the use of the FA-derived response factors provided no benefit. No other quantitation method has been found that provides better agreement than this approach, so in such cases the uncorrected composition should be used for comparison to the uncorrected composition of other samples.

Summary

Overall, then, we have demonstrated that RPHPLC/APCI–MS was used very effectively for conclusive identification of molecular species of TAGs present in mixtures, even at levels below 0.1%. The simple appearance of spectra and the use of extracted ion chromatograms made qualitative analysis straightforward and facile. It was also shown that the qualitative appearance of spectra had a distinct effect on the quantitation of TAG species. Characterization of the fragmentation behavior of TAG allowed us to develop the triacylglycerol quotient and to develop a model equation to produce response factors that yielded quantitation that had less average relative error than uncorrected results.

Two of the methods for quantitation proved superior. Quantitation using response factors calculated from randomized samples applied to nonrandomized samples of similar composition gave significantly less average relative error in the FA compositions than uncorrected results. The quantitation method that produced the least average relative error in TAG compositions and resultant FA compositions was based on response factors calculated from the FA composition determined by GC–FID. The FA composition of the normalized data also provided a check of the quality of the normalization. Response factor–adjusted natural and randomized samples gave excellent agreement with results obtained by GC–FID. Only samples in which FAs were concentrated into very few TAG species produced large average relative errors using this quantitation method. The TAG compositions of samples obtained by LC–FID were shown to exhibit average relative error similar to the uncorrected LC/APCI–MS results. The inability of LC–FID to differentiate overlapped peaks resulted in large relative errors for some TAG species.

References

1. Kuksis, A., Marai, L., and Myher, J.J. (1983) Strategy of Glycerolipid Separation and Quantitation by Complementary Analytical Techniques, *J. Chromatogr. 273*, 43–66.
2. Kuksis, A., and Myher, J.J. (1986) Lipids and Their Constituents, *J. Chromatogr. 379*, 57–90.
3. Shukla, V.K.S. (1988) Recent Advances in the High-Performance Liquid Chromatography of Lipids, *Prog. Lipid Res. 27*, 5–38.
4. Wojtusik, M.J., Brown, P.R., and Turcotte, J.G. (1989) Separation and Detection of Triacylglycerols by High-Performance Liquid Chromatography, *Chem. Rev. 89*, 397–406.
5. Myher, J.J., and Kuksis, A. (1995) General Strategies in Chromatographic Analysis of Lipids, *J. Chromatogr. B671*, 3–33.
6. Kuksis, A., and Myher, J.J. (1995) Application of Tandem Mass Spectrometry for the Analysis of Long-Chain Carboxylic Acids, *J. Chromatogr. B 671*, 35–70.

7. Ruiz-Gutierrez, V., and Barron, L.J.R. (1995) Methods for the Analysis of Triacylglycerols, *J. Chromatogr. B671*, 133–168.
8. Laakso, P. (1996) Analysis of Triacylglycerols—Approaching the Molecular Composition of Natural Mixtures, *Food Rev. Int. 12*, 199–250.
9. Marai, L., Kuksis, A., and Myher, J.J. (1994) Reversed-Phase Liquid Chromatograph–Mass Spectrometry of the Uncommon Triacylglycerol Structures Generated by Randomization of Butteroil, *J. Chromatogr. A672*, 87–99.
10. Kim, H.Y., and Salem Jr., N. (1993) Liquid Chromatography–Mass Spectrometry of Lipids, *Prog. Lipid Res. 32*, 221–245.
11. Myher, J.J., Kuksis, A., and Marai, L. (1993) Identification of the Less Common Isologous Short-Chain Triacylglycerols in the Most Volatile 2.5% Molecular Distillate of Butter Oil, *J. Am. Oil Chem. Soc. 70*, 1183–1191.
12. Kuksis, A., Marai, L., and Myher, J.J. (1991) Reversed-Phase Liquid Chromatography–Mass Spectrometry of Complex Mixtures of Natural Triacylglycerols with Chloride-Attachment Negative Chemical Ionization, *J. Chromatogr. 588*, 73–87.
13. Kuksis, A., Marai, L., and Myher, J.J. (1991) Plasma Lipid Profiling by Liquid Chromatography with Chloride-Attachment Mass Spectrometry, *Lipids 26*, 240–246.
14. Kuksis, A., Marai, L., Myher, J.J., Cerbulis, J., and Farrell Jr., H.M. (1986) Comparative Study of the Molecular Species of Chloro-Propanediol Diesters and Triacylglycerols in Milk Fat, *Lipids 21*, 183–190.
15. Myher, J.J., Kuksis, A., Marai, L., and Manganaro, F. (1984) Quantitation of Natural Triacylglycerols by Reversed-Phase Liquid Chromatography with Direct Liquid Inlet Mass Spectrometry, *J. Chromatogr. 283*, 289–301.
16. Kuksis, A., Myher, J.J., and Marai, L. (1984) Lipid Methodology—Chromatography and Beyond. Part I. GC/MS and LC/MS of Glycerolipids, *J. Am. Oil Chem. Soc. 61*, 1582–1589.
17. Kuksis, A., Myher, J.J., and Marai, L. (1985) Lipid Methodology—Chromatography and Beyond. Part II. GC/MS, LC/MS and Specific Enzymic Hydrolysis of Glycerolipids, *J. Am. Oil Chem. Soc. 62*, 762–767.
18. Spanos, G.A., Schwartz, S.J., van Breemen, R.B., and Huang, C.-H. (1995) High-Performance Liquid Chromatography with Light-Scattering Detection and Desorption Chemical Ionization Tandem Mass Spectrometry of Milk Fat Triacylglycerols, *Lipids 30*, 85–90.
19. Kallio, H., and Rua, P. (1994) Distribution of the Major Fatty Acids of Human Milk Between *sn*-2 and *sn*-1,3 Positions of Triacylglycerols, *J. Am. Oil Chem. Soc. 71*, 985–992.
20. Anderson, M.A., Collier, L., Dilliplane, R., and Ayorinde, F.O. (1993) Mass Spectrometric Characterization of *Vernonia galamensis* Oil, *J. Am. Oil Chem. Soc. 70*, 905–908.
21. Laakso, P., and Kallio, H. (1993) Triacylglycerols of Winter Butterfat Containing Configurational Isomers of Monoenoic Fatty Acyl Residues. I. Disaturated Monoenoic Triacylglycerols, *J. Am. Oil Chem. Soc. 70*, 1161–1171.
22. Laakso, P., and Kallio, H. (1993) Triacylglycerols of Winter Butterfat Containing Configurational Isomers of Monoenoic Fatty Acyl Residues. II. Saturated Dimonoenoic Triacylglycerols, *J. Am. Oil Chem. Soc. 70*, 1173–1176.
23. Mares, P., Rezanka, T., and Novak, M. (1991) Analysis of Human Blood Plasma Triacylglycerols Using Capillary Gas Chromatography, Silver Ion Thin-Layer

Chromatographic Fractionation and Desorption Chemical Ionization Mass Spectrometry, *J. Chromatogr. 568*, 1–10.
24. Rezanka, T., and Mares, P. (1991) Determination of Plant Triacylglycerols Using Capillary Gas Chromatography, High-Performance Liquid Chromatography and Mass Spectrometry, *J. Chromatogr. 542*, 145–159.
25. Merritt Jr., C., Vajdi, M., Kayser, S.G., Halliday, J.W., and Bazinet, M.L. (1982) Validation of Computational Methods for Triglyceride Composition of Fats and Oils by Liquid Chromatography and Mass Spectrometry, *J. Am. Oil Chem. Soc. 59*, 422–432.
26. Rezanka, T., Mares, P., Husek, P., and Podojil, M. (1986) Gas Chromatography–Mass Spectrometry and Desorption Chemical Ionization Mass Spectrometry of Triacylglycerols from the Green Alga *Chlorella kessleri*, *J. Chromatogr. 355*, 265–271.
27. Taylor, D.C., Giblin, M., Reed, D.W., Hogge, L.R., Olson, D.J., and MacKenzie, S.L. (1995) Stereospecific Analysis and Mass Spectrometry of Triacylglycerols from *Arabidopsis thaliana* (L.) Heynh. Columbia Seed, *J. Am. Oil Chem. Soc. 72*, 305–308.
28. Manninen, P., Laakso, P., and Kallio, H. (1995) Method for Characterization of Triacylglycerols and Fat-Soluble Vitamins in Edible Oils and Fats by Supercritical Fluid Chromatography, *J. Am. Oil Chem. Soc. 72*, 1001–1008.
29. Manninen, P., Laakso, P., and Kallio, H. (1995) Separation of γ- and α-Linoleic Acid Containing Triacylglycerols by Capillary Supercritical Fluid Chromatography, *Lipids 30*, 665–671.
30. Kallio, H., and Currie, G. (1993) Analysis of Low Erucic Acid Turnip Rapeseed Oil (*Brassica campestris*) by Negative Ion Chemical Ionization Tandem Mass Spectrometry. A Method Giving Information on the Fatty Acid Composition in Positions *sn*-2 and *sn*-1/3 of Triacylglycerols, *Lipids 28*, 207–215.
31. Currie, G., and Kallio, H. (1993) Triacylglycerols of Human Milk, Rapid Analysis by Ammonia Negative Ion Tandem Mass Spectrometry, *Lipids 28*, 217–222.
32. Demirbüker, M., Blomberg, L.G., Olsson, N.U., Bergqvist, M., and Herslof, B.G. (1992) Characterization of Triacylglycerols in the Seeds of *Aquilegia vulgaris* by Chromatographic and Mass Spectrometric Methods, *Lipids 27*, 436–441.
33. Taylor, D.C., Weber, N., Barton, D.L., Underhill, E.W., Hogge, L.R., Weselake R.J., and Pomeroy, M.K. (1991) Triacylglycerol Bioassembly in Microspore-Derived Embryos of *Brassica napus* L. cv Reston, *Plant. Physiol. 97*, 65–79.
34. Hogge, L.R., Taylor, D.C., Reed, D.W., and Underhill, E.W. (1991) Characterization of Castor Bean Neutral Lipids by Mass Spectrometry/Mass Spectrometry, *J. Am. Oil Chem. Soc. 68*, 863–868.
35. Singleton, J.A., and Pattee, H.E. (1987) Characterization of Peanut Oil Triacylglycerols by HPLC, GLC and EIMS, *J. Am. Oil Chem. Soc. 64*, 534–538.
36. Murata, T., and Takahashi, S. (1977) Qualitative and Quantitative Chemical Ionization Mass Spectrometry of Triglycerides, *Anal. Chem. 49*, 728–731.
37. Lauer, W.M., Aasen, A.J., Graff, G., and Holman, R.T. (1973) Mass Spectrometry of Triglycerides: I. Structural Effects, *Lipids 5*, 861–868.
38. Aasen, A.J., Lauer, W.M., and Holman, R.T. (1973) Mass Spectrometry of Triglycerides: II. Specifically Deuterated Triglycerides and Elucidation of Fragmentation Mechanisms, *Lipids 5*, 869–877.
39. Hites, R.A. (1970) Quantitative Analysis of Triglyceride Mixtures by Mass Spectrometry, *Anal. Chem. 42*, 1736–1740.

40. Barber, M., Merren, T.O., and Kelly, W. (1964) The Mass Spectrometry of Large Molecules I. The Triglycerides of Straight Chain Fatty Acids, *Tetrahedron Letters*, 1063–1067.
41. Lamberto, M., and Saitta, M. (1995) Principal Component Analysis in Fast Atom Bombardment–Mass Spectrometry of Triacylglycerols in Edible Oils, *J. Am. Oil Chem. Soc. 72*, 867–871.
42. Matsubara, T., and Hayashi, A. (1991) FAB/Mass Spectrometry of Lipids, *Prog. Lipid Res. 30*, 301–322.
43. Barber, M., Tetler, L.W., Bell, D., Ashcroft, A.E., Brown, R.S., and Moore, C. (1987) Applications of a Continuous Flow Probe in FAB Mass Spectrometry, *Org. Mass Spectrom. 22*, 647–650.
44. Duffin, K.L., Henion, J.D., and Shieh, J.J. (1991) Electrospray and Tandem Mass Spectrometric Characterization of Acylglycerol Mixtures That Are Dissolved in Nonpolar Solvents, *Anal. Chem. 63*, 1781–1788.
45. Valeur, A., Michelsen, P., and Odham, G. (1993) On-Line Straight-Phase Liquid Chromatography/Plasmaspray Tandem Mass Spectrometry of Glycerolipids, *Lipids 28*, 255–259.
46. Kim, H.-J., and Salem Jr., N. (1987) Application of Thermospray High-Performance Liquid Chromatography/Mass Spectrometry for the Determination of Phospholipids and Related Compounds, *Anal. Chem. 59*, 722–726.
47. Kallio, H., Laakso, P., Huopalahti, R., Linko, R.R., and Oksman, P. (1989) Analysis of Butter Fat Triacylglycerols by Supercritical Fluid Chromatography/Electron Impact Mass Spectrometry, *Anal. Chem. 61*, 698–700.
48. Showell, J.S., Fales, H.M., and Sokoloski, E.A. (1989) Quantitative Variability in the ^{252}Cf Plasma Desorption Mass Spectra of Triglycerides and Waxes, *Org. Mass Spectrom. 24*, 632–636 .
49. Covey, T.R., Lee, E.D., Bruins, A.P., and Henion, J.D. (1986) Liquid Chromatography/Mass Spectrometry, *Anal. Chem. 58*, 1451A–1461A.
50. Lehmann, W.D., and Kessler, M. (1983) Characterization and Quantitation of Human Plasma Lipids from Crude Lipid Extracts by Field Desorption Mass Spectrometry, *Biomed. Mass Spectrom. 10*, 220–226.
51. Byrdwell, W.C., and Emken, E.A. (1995) Analysis of Triglycerides Using Atmospheric Pressure Chemical Ionization Mass Spectrometry, *Lipids 30*, 173–175.
52. Neff, W.E., and Byrdwell, W.C. (1995) Soybean Oil Triacylglycerol Analysis by Reverse-Phase High-Performance Liquid Chromatography Coupled with Atmospheric Pressure Chemical Ionization Mass Spectrometry, *J. Am. Oil Chem. Soc. 72*, 1185–1191.
53. Neff, W.E., and Byrdwell, W.C. (1995) Triacylglycerol Analysis by High Performance Liquid Chromatography–Atmospheric Pressure Chemical Ionization Mass Spectrometry, *Crepis alpina* and *Vernonia galamensis* Seed Oils, *J. Liq. Chromatogr. 18*, 4165–4181.
54. Byrdwell, W.C., and Neff, W.E. (1996) Analysis of Genetically Modified Canola Varieties by Atmospheric Pressure Chemical Ionization Mass Spectrometric and Flame Ionization Detection, *J. Liq. Chrom. & Rel. Technol. 19*, 2203–2225.
55. Byrdwell, W.C., Emken, E.A., Neff, W.E., and Adlof, R.O. (1996) Quantitative Analysis of Triglycerides Using Atmospheric Pressure Chemical Ionization-Mass Spectrometry, *Lipids 31*, 919–935,

56. Kusaka, T., Ishihara, S., Sakaida, M., Mifune, A., Nakano, Y., Tsuda, K., Ikeda, M., and Nakano, H. (1996) Composition Analysis of Normal Plant Triacylglycerols and Hydroperoxidized *rac*-1-Stearoyl-2-Oleoyl-3-Linoleoyl-*sn*-Glycerols by Liquid Chromatography–Atmospheric Pressure Chemical Ionization Mass Spectrometry *J. Chromatogr. 730*, 1–7.
57. Adlof, R.O., Miller, W.R., and Emken, E.A. (1978) Synthesis of *cis* and *trans* Methyl 8- and 13-Octadecenoate-d_2 and d_4 Isomers, *J. Label. Comp. Radiopharm. 15*, 625.
58. Wheeler, D.H., Reimenschneider, R.W., and Sando, C.E. (1940) Preparation, Properties, and Thiocyanogen Absorption of Triolein and Trilinolein, *J. Biol. Chem. 132*, 687–699.
59. Neff, W.E., Mounts, T.L., Rinsch, W.M., Konishi, H., and El-Agaimy, M.A. (1994) Oxidative Stability of Purified Canola Oil Triacylglycerols with Altered Fatty Acid Compositions as Affected by Triacylglycerol Composition and Structure, *J. Am. Oil Chem. Soc. 71*, 1101–1109.
60. Glass, R.L. (1969) Preparation of Milk Fat Methyl Esters by Alcoholysis in an Essentially Nonalcoholic Solution, *J. Dairy Sci. 52*, 1289–1290.
61. Neff, W.E., Zeitoun, M.A.M., and Weisleder, D. (1992) Resolution of Lipolysis Mixtures from Soybean Oil by a Solid-Phase Extraction Procedure, *J. Chromatogr. 589*, 353–357.
62. Daubert, B.F. (1949) The Composition of Fats, *J. Am. Oil Chem. Soc. 26*, 556–558.
63. Bailey, A.E., and Swern, D. (1964) *Bailey's Industrial Oil and Fat Products*, 3rd edn., pp. 965–968, John Wiley and Sons, New York.

Chapter 4
Liquid Chromatography with On-Line Electrospray Mass Spectrometry of Oxidized Diphosphatidylglycerol

M. Bergqvist and A. Kuksis

Banting and Best Department of Medical Research, University of Toronto, Toronto, Canada M5G 1L6

Introduction

Cardiolipin (CL) is the principal polyglycerophospholipid in heart muscle, ranging from 9.0% in humans to 12.6% in oxen of the total lipid phosphorus mass (1). Predominantly found within the inner leaflet of the inner mitochondrial membrane (2), CL is a multifunctional phospholipid, aiding in the association and function of the mitochondrial enzymes (3) and essential for the operation of the electron transport chain (4). As a result, it would be anticipated that CL becomes exposed to oxygenation and that the unsaturated fatty chains become peroxidized (5). In order to identify CL peroxidation products in tissues, reference standards are necessary, along with a knowledge of their chromatographic and mass spectrometric properties. In the present study we have prepared a series of homologous hydroperoxides, core aldehydes, core carboxylates, and the ozonides of bovine heart CL and have characterized them by normal-phase HPLC with on-line electrospray mass spectrometry.

Materials and Methods

Materials

All solvents were HPLC grade (Fisher, Toronto, Canada), and reagents were of the highest grade commercially available. The common phospholipids were available in the laboratory from previous studies (6,7). Cardiolipin (bovine heart) and phospholipase D (*Streptomyces chromofuscus*) were obtained from Sigma Chemical Co., St. Louis, MO. All oxidation reactions were done with the commercial bovine heart cardiolipin without further purification. Fresh cardiolipin was extracted from the hearts of rabbits receiving diets supplemented with menhaden oil or safflower (courtesy of Dr. Howard Parsons, University of Calgary). Total lipid extracts of the lungs of rats were obtained from animals killed in the laboratory during the course of other experiments. Triphenylphosphine and 2,4-dinitrophenylhydrazine were from Aldrich Chemical Co. (Milwaukee, WI). Ferrous sulfate was from Mallinckrodt (Paris, KY). Oxygen used to generate ozone was 99.9% pure.

Preparation of Tissue Extracts

The heart (0.3–0.5 g) was homogenized by a 20 s burst of a Polytron homogenizer (Kinematika, Lucerne, Switzerland) in 5 mL of chloroform/methanol (2:1, v/v) and the extracts collected. Another 5 mL of chloroform/methanol (2:1, v/v) was then added and the residue homogenized. The extracts were combined and centrifuged at full speed for 10 min in a benchtop centrifuge (Model TJ-6, Beckman Instruments, Mississauga, Canada) to pellet the debris. The supernatant was decanted, 5 mL of 0.73% NaCl was added, and the mixture was vortexed for 10 s, and then separated at full speed for 10 min in the benchtop centrifuge, causing phase separation (8). The lower organic layer was removed by a Pasteur pipet and washed with Folch upper phase, and the phases were allowed to separate. The washed organic phase was removed with a Pasteur pipet and dried by passing through a small column of powdered anhydrous sodium sulfate. The dried organic solvent was blown off under dry nitrogen, and the residue was redissolved in a small volume of chloroform/methanol (2:1, v/v) and saved for immediate LC/ES/MS.

Preparation of Hydroperoxides

The hydroperoxidation was performed by suspending 1–3 mg of the CL in 1 mL of 70% *tert*-butyl hydroperoxide solution in water containing 100 µL of 2% taurocholic acid in the presence of 10 µM Fe^{2+} ions (6). The reaction mixture was shaken on a mechanical agitator at 35°C in the dark for 1 to 4 h. At the end of the oxidation 100 µL of 2% EDTA in water and 10 µL of 2% butylated hydroxytoluene in MeOH were added. An aliquot of the reaction mixture was immediately extracted with chloroform/methanol (2:1; v/v); the remainder was treated with 1 mg of dinitrophenylhydrazine in 2 mL of 1 N HCl, and the dinitrophenylhydrazone (DNPH) formation was allowed to proceed for 2 hr at room temperature and overnight at 4°C before total lipid extraction (9).

Preparation of Ozonides

The ozonides of cardiolipin were prepared as previously described for other polyunsaturated glycerophospholipids (7). The lipid ester (1–5 mg) was dissolved in 4 mL of hexane and cooled in a dry ice-acetone bath for 10 min. O_2 gas containing 3–4% ozone was then bubbled through the solution at a rate of 150 mL/min for 5 min. The appearance of a faint blue color indicated the saturation of the hexane solution with ozone, at which time the reaction was stopped. The hexane was evaporated under nitrogen and the residue redissolved in chloroform. The ozonides were isolated by TLC using neutral- and polar-lipid solvent systems as described subsequently.

Triphenylphosphine Reduction of Ozonides

The CL ozonides were reduced to the corresponding aldehydes as previously described for the ozonides of other glycerophospholipids (7). To the CL ozonides

(1–2 mg) in chloroform (4 mL) were added 10 mg of triphenylphosphine and the reaction mixture kept at room temperature for 1 h. At the end of this time the aldehyde and any unreacted ozonides were recovered from the chloroform solution and resolved by TLC on silica gel H (250-µm thick layer, 20 × 20 cm glass plate) using a polar lipid solvent system made up of chloroform/methanol/acetic acid/water (75:45:12:6, by vol) for the development (10). The aldehyde bands were located by spraying the plate with Schiff reagent, which gave a purple color. The aldehydes were recovered from the TLC plate by scraping the gel and extracting it with chloroform/methanol/water (65:25:4, by vol) (11).

Preparation of DNPH Derivatives

Part of the ozonized CL-containing aldehyde function (0.5–1 mg in 1 mL chloroform) was converted into the DNPH derivatives by reaction with dinitrophenylhydrazine in the dark (0.5 mg in 1 mL in 1 N HCl) for 2 h at room temperature and 1 h at 4°C (9). The DNPH derivatives were extracted form the reaction mixture with neutral chloroform/methanol 2:1. The DNPH derivatives of CL were purified by TLC in the phospholipid solvent system as described subsequently.

Hydrolysis of CL with Phospholipase D

The natural CL, CL ozonides, hydroperoxides, and their core aldehydes and core acids were hydrolyzed with phospholipase D (*S. chromofuscus*) (12). To 4 mg cardiolipin in 5 mL of diethyl ether containing traces of ethanol (see below) were added 100 µL of the enzyme solution (1 mg/mL) in 25 mL buffer solution (0.02 M NaAc, 0.008 M $CaCl_2$), and the reaction was allowed to proceed for various periods of time. The reaction was stopped by blowing off the remaining diethyl ether and extracting the aqueous mixture with chloroform. The extraction was completed with two 5-mL portions of chloroform, the organic layer collected, and the solvent evaporated. The residue was dissolved in chloroform and subjected to LC/ES/MS.

Gas Liquid Chromatography (GLC)

High-temperature GLC of the fatty acids and long-chain aldehydes was performed on a Hewlett-Packard (Palo Alto, CA) Model 5889 gas chromatograph equipped with a flexible quartz column (15 m × 0.32 mm i.d.), coated with SP 2380 polar liquid phase, and a flame ionization detector. The column was temperature-programmed nonlinearly from 40 to 250°C; hydrogen was used as the carrier gas at 3 psi head pressure as previously described (13). The fatty acid methyl esters were identified by GLC with reference to standard mixtures of fatty acids. The fatty acid methyl esters of various CL preparations were obtained by transmethylation with 6% H_2SO_4 in methanol and heating at 80°C for 2 h. After the reaction, the methyl esters were extracted twice with hexane. The solvents were blown down under nitrogen and the samples redissolved in hexane.

Thin-Layer Chromatography (TLC)

The CL and CL ozonides were isolated and purified by TLC using two polar solvent systems. The phospholipids were resolved by TLC (20 × 20 cm plates, 250- µm-thick layer of silica gel H) using chloroform/methanol/30% ammonia (65:35:7, by vol) (15). The resolved lipids were located by spraying the TLC plate with 2,7-dichlorofluorescein and viewing it under UV light. CL core aldehydes were located by spraying with Schiff's reagent. Hydroperoxide bands were located by spraying with KI/starch spray. On TLC the phosphatidic acid (R_f 0.07) and phosphatidylglycerol plus cardiolipin (R_f 0.42) were well resolved when the plates were developed with chloroform/methanol/28% ammonia (65:25:5, by vol) (11).

High-Performance Liquid Chromatography (HPLC)

Normal-phase HPLC separations of phospholipids were performed on Spherisorb 3-micron columns (100 × 4.6 mm id; Alltech Associates, Deerfield, Illinois) installed into a Hewlett-Packard Model 1050 Liquid Chromatograph connected to an evaporative light-scattering detector (ELSD) (Model ELSD II, Varex, Rockville, MD). The column was eluted with a linear gradient of 100% Solvent A (chloroform/methanol/30% ammonium hydroxide, 80:19.5:0.5; by vol) to 100% Solvent B (chloroform/methanol/water/30% ammonium hydroxide, 60:34:5.5:0.5, by vol.) in 14 min, then at 100% B for 10 min (16). The flow rate was set at 1 mL/min.

Liquid Chromatography/Mass Spectrometry (LC/MS)

Normal-phase LC/MS was performed by admitting the entire HPLC column effluent to the Hewlett-Packard Model 5988B quadrupole mass spectrometer equipped with a nebulizer-assisted electrospray interface (7). Negative ESI spectra were taken in the mass range 250–1500 Da. The capillary exit voltage was set at 100 V, with the electron multiplier at 1795 V. Selected ion spectra were retrieved from the total ion spectra by computer. The masses given in the tables and in figures are the masses of the [M-H]⁻ ions actually recorded. The molecular species of the various glycerophospholipids were identified on the basis of the molecular mass provided by MS, the knowledge of the fatty acid composition of the phospholipid classes, and the relative HPLC retention time (more saturated species migrating ahead of the unsaturated species) of the phospholipids.

Results and Discussion

LC/ES/MS of Cardiolipin

Figure 1 shows the total negative-ion current profile of the commercial preparation of bovine heart CL as obtained by normal-phase LC/ES/MS. The total negative-ion profile (A) shows a single symmetrical HPLC peak. The single-ion chromatograms

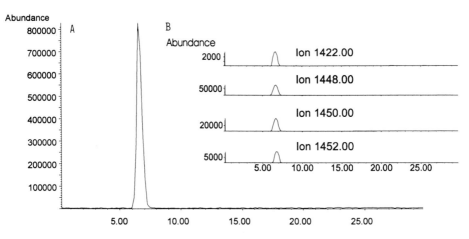

FIG. 1. Total negative-ion profile of a commercial preparation of bovine heart cardiolipin as obtained by normal phase LC/ES/MS. A, total negative-ion current profile; B, single ion chromatograms of major [M-1]⁻ ions. Peak identification: m/z 1448, tetralinoleoyl; m/z 1450, trilinoleoyloleoyl not shown; m/z 1452, dilinoleoyl, dioleoyl; m/z 1426, dilinoleoyloleoyl-palmitoyl (not shown), m/z 1422, trilinoleoylpalmitoleoyl species of CL. LC/ES/MS instrumentation and operating conditions are given under the "Materials and Methods" section. Ion source exit voltage 150 V.

for [M-1]⁻ ions (B) indicate the presence of four major components. In addition to the major tetralinoleoyl (m/z 1448), trilinoleoyloleoyl (m/z 1450), trilinoleoylpalmitoyl (m/z 1424) and dilinoleoyloleoylpalmitoyl (m/z 1426) species, other minor species LLLPo (m/z 1422), LLLA (m/z 1472), and LLOO (m/z 1452) were also noted. Furthermore, in addition to the singly charged ions, each species also gave a doubly charged ion as [M-1]⁻/2. Table 1 gives the quantitative composition of the major molecular species of the commercial sample of bovine heart cardiolipin. The CL preparation also contained small amounts of lyso-cardiolipin, which was obvious from the presence of ions representing the loss of one fatty acid from

TABLE 1
Relative Retention Times and Abundance of Molecular Species of Cardiolipin in a Commercial Preparation from Bovine Heart

Mass (m/z)	Retention time (min)	Tentative identity	Abundance (%)
1448	6.916	LLLL	63.7
1450	6.930	LLLO	34.2
1424	7.022	LLLP	1.5
1426	6.330	LLOP	0.8

Abbreviations: L, linoleic acid; O, oleic acid; P, palmitic acid. See text for additional species.

the CL molecule. Thus, the LLLL had yielded such a lyso derivative (m/z 1185) along with its doubly charged molecular ion (m/z 593). Other CL species also had yielded minor amounts of the lyso compound.

Earlier, Teng and Smith (17) had resolved 11 subfractions of bovine heart cardiolipin on a reversed-phase HPLC column. Eleven molecular species of cardiolipin had been obtained also by Schlame and Otten (18) by reversed-phase HPLC of the 1,3-bisphosphatidyl-2-benzoyl-*sn*-glycerol dimethyl ester). However, no specific CL species were identified. Under the ES conditions (exit voltage 100 V) no fragment ions were observed for any of the CL species, but at higher ionization energy (exit voltage of 300 V), ions corresponding to [M-DG]$^-$ and [PA]$^-$ could be observed at m/z 833 and 695, respectively, and a minor ion for linoleate (m/z 279). The [M-H]$^-$ of m/z 1448 corresponding to the tetralinoleoyl species of CL is identical to that observed by Jensen et al. (19) using negative-ion FAB/MS. The FAB/MS spectrum also contained significant ions at m/z 833 (M-dilinoleoylglycerol), 695 (dilinoleoylglycerophosphate), 415 (phosphatidic acid-RCHCO) and 279 (RCH$_2$COO).

Characterization of Cardiolipin Hydroperoxides

Figure 2 shows the total negative-ion profile of bovine heart CL after treatment with *tert*-butylhydroperoxide, along with the full mass spectrum averaged over the entire HPLC peak as obtained by normal-phase LC/ES/MS. There is no significant resolution between native CL and its hydroperoxides on the normal-phase HPLC

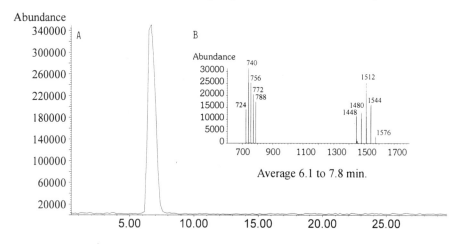

FIG. 2. Negative-ion current profile of commercial bovine heart cardiolipin after treatment with *tert*-butylhydroperoxide. A, total negative-ion current profile; B, total mass spectrum averaged over the entire HPLC peak. Peak identifications: m/z 1448, [M-1]$^-$ (LLLL); m/z 1480, [M-1]$^-$ (LLLL-OOH); m/z 1482, [M-1]$^-$ (LLLO-OOH); m/z 1512, [M-1]$^-$ (LLLL-di-OOH); m/z 1514, [M-1]$^-$ (LLLO-di-OOH); m/z 1544, [M-1]$^-$ (LLLL-tri-OOH); m/z 1576, [M-1]$^-$ (LLLL-tetra-OOH). See Table 2 for abbreviations. Instrumentation and operating conditions are as given in Fig. 1.

column. The major ions correspond to the tetralinoleoyl species of CL (m/z 1448), its monohydroperoxy derivative (m/z 1480), the monohydroperoxy derivative of trilinoleyloleoyl CL (m/z 1482), the dihydroperoxy derivative of tetralinoleoyl CL (m/z 1512), the dihydroperoxy derivative of trilinoleoyoleoyl CL (m/z 1514), the trihydroperoxy derivative of tetralinoleoyl CL (m/z 1544), and the tetrahydroperoxy derivative of tetralinoleoyl CL (m/z 1576). The yields of the various hydroperoxides reflected the proportions of the molecular species in the bovine cardiolipin. The spectrum also shows the presence of these derivatives as the doubly charged molecular ions (m/z 724 to 788).

Figure 3 shows the [M-1]⁻ ion plots for the major hydroperoxide species of bovine CL as obtained by the *tert*-butylhydroperoxide treatment. The major species were: LLLL-OOH (m/z 1480); LLLO-OOH (m/z 1482); LLLL-di-OOH (m/z 1512); LLLO-di-OOH (m/z 1514); LLLL-tri-OOH (1544); and LLLL-tetra-OOH (m/z 1576). Table 2 gives the relative yields of the different derivatives in relation to the residual native molecular species of bovine CL. No specific significance was attached to the relative proportions of the mono-, di- and tri-hydroperoxides, which were anticipated to change with the hydroperoxidation conditions. Table 2 also includes the absolute retention times of the hydroperoxides on the normal-phase HPLC column, which were established with the help of the LC/ES/MS.

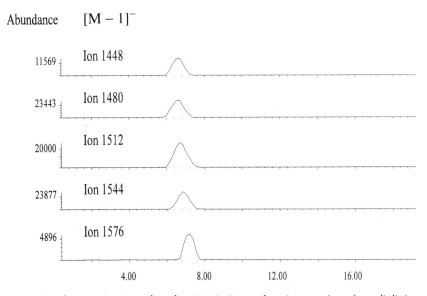

FIG. 3. Single negative-ion plots for [M-1]⁻ ions of major species of cardiolipin hydroperoxides as obtained by normal phase LC/ES/MS. Peak identification: m/z 1448, LLLL; m/z 1480, LLLL-OOH; m/z 1512, LLLL-di-OOH; m/z 1514, LLLO-di-OOH; m/z 1544, LLLL-tri-OOH; and m/z 1576, LLLL-tetra-OOH. Abbreviations are as given in Table 2. Instrumentation and operating conditions are as given in Fig. 1.

TABLE 2
Relative Retention Times and Abundance of the Hydroperoxides of Cardiolipin in a Commercial Sample Subjected to Peroxidation with tert-Butyl Hydroperoxide

Mass (m/z)	Retention time (min)	Tentative identity	Abundance (%)
1480	6.508	LLLL-OOH	19.7
1482	6.601	LLLO-OOH	15.1
1512	6.637	LLLL-di-OOH	29.8
1514	6.729	LLLO-di-OOH	13.7
1544	7.015	LLLL-tri-OOH	19.2
1576	7.263	LLLL-tetra-OOH	2.5

Abbreviations as explained in Table 1; OOH, hydroperoxy group of unspecified location.

In the presence of Fe^{2+} these peroxides yielded small amounts of the corresponding core aldehydes and core acids. There was evidence, however, that the core aldehydes were poorly recovered because they tended to polymerize, as also observed for the triphenylphosphine reduction products of the ozonides (see below). Trace amounts of the core aldehyde ions corresponding to the parent hydroperoxides, however, could be demonstrated by means of the single-ion plots, e.g., m/z 1374 for the 9-oxononanoyl derivative of LLLL. Higher yields of the core aldehydes would be anticipated following prolonged incubation with iron or copper ions, as shown previously for the hydroperoxides of the linoleoyl esters of cholesterol and glycerophosphocholine (20). It is possible that the core aldehydes would have been isolated more readily as the dinitrophenylhydrazones, as also shown by Kamido et al. (20).

Previously, Teng and Smith (17) and Schlame and Otten (18) had recognized peroxidation products in samples of CL stored for extended periods of time in chloroform. These oxidation products of CL were eluted well ahead of the native CL from reversed-phase HPLC columns. The mass spectrometric properties of the hydroperoxides of CL have not been previously determined, but the hydroperoxides of other glycerolipids have been examined. Kuksis et al. (6) used LC/TS/MS and LC/ES/MS to characterize the hydroperoxides and core aldehydes of triacylglycerols, while Myher et al. (21) used flow injection/ES/MS and LC/ES/MS to demonstrate that the hydroperoxides of the choline and ethanolamine phosphatides readily yielded [M-1]⁻ ions in the negative-ion mode. Individual molecular species were identified on basis of the retention times of standards and the molecular weights of these solutes provided by the mass spectrometer. Zhang et al. (22) used thermospray LC/MS to assay PC hydroperoxides and to clarify a confusion about the nature of the hydroperoxides identified on the basis of fluorescence. In all instances the hydroperoxides appeared to be sufficiently stable for HPLC and mass spectrometry with thermospray and electrospray ionization.

Characterization of the Ozonides of Cardiolipin

Figure 4 shows the total negative-ion current profile of the bovine heart cardiolipin following 2 h (A) and 6 h (B) of spontaneous post-ozonization degradation as obtained

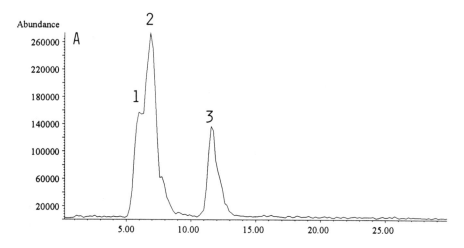

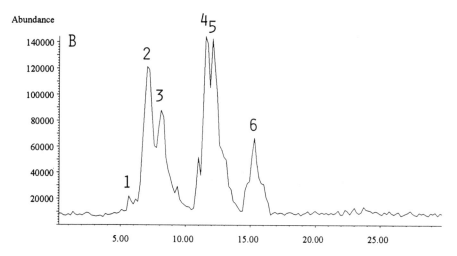

FIG. 4. Total negative-ion profile of bovine heart cardiolipin following 2 h (A) and 6 h (B) post-ozonization degradation as obtained by normal phase LC/ES/MS. Peak identification as given in text and in Figs. 5 and 6. Peak A1, ozonides; Peak A2, mixed ozonides/aldehydes; Peak A3, mixed ozonides/aldehydes/monocarboxylates. Peak B1, ozonides; Peaks B2 and B3, mixed ozonides/aldehydes; Peaks B4 and B5, mixed ozonides/aldehydes/monocarboxylates; Peak B6, mixed oznides/aldehydes/dicarboxylates. Complete peak identification in Table 3. Instrumentation and operating conditions are as given in Fig. 1.

by normal phase LC/ES/MS. Detailed examination of the mass spectra of the separate LC/MS peaks revealed an orderly sequence of elution for the oxidized species of each series of the octa-, hepta-, hexa-, penta-, tetra-, tri-, and di-ozonide series. Peak 4A1 yielded a major ion at m/z 1832 (octaozonide of LLLL) and a minor ion at m/z 1786 (heptaozonide of LLLO). Peak 4A2 gave several major ions, including m/z 1832; m/z 1660 (unknown), m/z 1628 (hexaozonide-9ald,); m/z 1690 (unknown); m/z 1486 (unknown); m/z 1424 (tetraozonide-di-9ald); and m/z 1031 (tri-9ald, 9CA). Peaks 4B4 and 4B5 yielded a major ion at m/z 1645 (hexaozonide-9CA), m/z 1441 (tetraozonide 9ald 9CA), m/z 1665 (pentaozonide, LLOP), m/z 1598 (pentaozonide 9CA), m/z 1732 (heptaozonide, 12ozCA), and others. Peak 4B6 contained m/z 1456 (tetraozonide, di-9CA), m/z 1544 (pentaozonide, 9CA, 12ozCA), m/z 1478 (triozonide, 9CA), m/z 1252 (diozonide, 9ald, di-9CA), m/z 1340 (triozonide, 9ald, 9CA, 12ozCA), and others. Figure 5 shows the full mass spectra averaged over Peaks A1 and A2 in Fig. 4A, whereas Fig. 6 shows the full mass spectra averaged over Peaks B4, B5 and B6 in Fig. 4B. Table 3 gives the masses, relative retention times, and proposed identities of the ozonization products of the different CL preparations as obtained by normal phase LC/ES/MS. Table 3 also includes the uncorrected estimates of the relative proportions of various ozonization and spontaneous decomposition products. Note that ions of the same mass may appear at different retention times. The major components are hexaozonides with 9CA, and hexaozonides with 9ald, followed by pentaozonides with 9CA and 12ozCA and tetraozonides with di-9CA.

Although the ozone can react with linoleic acid to form a diozonide and two isomeric monoozonides, only the diozonide was readily detected under the present conditions of ozonization. Furthermore, the core aldehydes and core acids were largely of the nine carbon type, although some species contained also the 12-oxo-9,10-trioxolandodecanoic acid. The apparent absence of partially ozonized CL was consistent with the rapid and complete ozonization obtained previously for other unsaturated glycerophospholipids (15). Furthermore, no fragment ions were

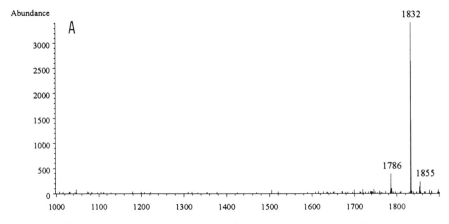

FIG. 5. Full mass spectra averaged over Peak A1 (Fig. 4) (A) and Peak A2 (Fig. 4) (B) [M-1]$^-$ ions identified as explained in the text and as listed in Table 4. Instrumentation and operating conditions are as given in Fig. 1.

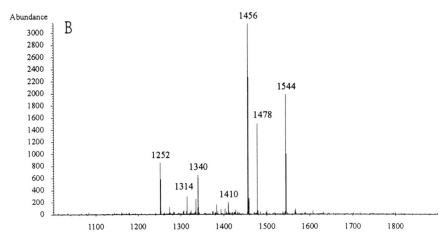

FIG. 6. Full mass spectra averaged over Peak B4 and B5 (Fig. 4) (A) and Peak B6 (Fig. 4) (B) [M-1]⁻ ions are identified as explained in the text and as listed in Table 4. Instrumentation and operating conditions are as given in Fig. 1.

observed for any of the ozonides at the 100-V exit voltage ionization conditions. Increasing the exit voltage to 300 V would have been expected to produce fragment ions of the type proposed for the tetralinoleoyl species of CL (see above).

The parallel formation of significant amounts of the oxo- and carboxy-derivatives during ozonization has been recently reported also by Harrison and Murphy (23) during ozonization of phosphatidylcholine. They (23) have suggested that a possible mechanism leading to the formation of these two ions proceeds by initial homolytic cleavage of the peroxide bridge of the trioxolane, followed by rearrangement to yield either the aldehyde functionality or the carboxylic acid functionality at carbon 9. Apparently all the ozonide species decomposed to the corresponding core aldehydes and acids in proportion to their masses in the mixture. Harrison and Murphy (23) also reported the CID fragmentation of the 16:0/18:2-diozonide of GPC, which resulted in ions from a homolytic cleavage of the peroxide bond and formation of ω-carboxy and ω-aldehyde species at the carbon atom in the original double bond closest to the phosphocholine charge site. It is not known whether or not such remote charge fragmentation also applies to the acidic glycerophospholipids.

Figure 7 shows the total negative-ion current profile of the triphenylphosphine reduction products of the 6-h post ozonization sample of CL. Total mass spectra averaged over the various HPLC peaks revealed a great complexity in molecular species, much of which resulted from the incompleteness of the reduction of the ozonides to the aldehydes. Four or more major peaks or clusters of molecular species can be recognized in the HPLC column eluate, the first eluted peak being clearly resolved into two peaks. The aldehyde/carboxylate-containing remnants of CL,

TABLE 3
Relative Retention Time and Abundance of Ozonides Obtained during Ozonization of a Commercial Sample of Bovine Heart Cardiolipin

Mass (m/z)	Retention time (min)	Tentative identity	Abundance (%)
1832	5.753	LLLL (octa)	3.5
1786	5.200	LLLO (hepta)	0.2
1712	6.502	LLLP (hexa)	0.3
1666	11.472	LLOP (penta)	1.5
1628	7.226	LLL (hexa, 9ald)	19.7
1582	7.199	LLO (penta, 9ald)	3.3
1508	8.207	LLP (tetra, 9ald)	0.2
1424	8.212	LL (tetra, di-9ald)	2.7
1220	10.178	L (di, tri-9ald)	0.05
1732	11.093	LLL (hepta, 12ozCA)	5.2
1686	11.082	LLO (hexa, 12ozCA)	1.2
1644	11.709	LLL (hexa, 9CA)	24.0
1598	11.7	LLO (penta, 9CA)	0.9
1528	11.710	LL (penta, 9ald, 12ozCA)	4.2
1524	12.027	LLP (tetra, 9CA)	0.3
1482	10.342	LO (tetra, 9ald, 12ozCA)	0.1
1440	12.559	LL (tetra, 9ald, 9CA)	10.8
1394	13.653	LO (tri, 9ald, 9CA)	0.1
1324	12.509 (14.895)	L (tri, di-9ald,12ozCA)	0.2
1320	12.763	LP (di, 9ald,9CA)	1.3
1236	13.268	L (di, di-9ald, 9CA)	0.6
1544	14.845	LL (penta, 9CA, 12ozCA)	5.7
1456	8.406	LL (tetra, 9CA, 9CA)	5.9
1410	15.310	LO (tri, di-9CA)	0.6
1340	15.431	L (tri, 9ald, 9CA, 12ozCA)	4.0
1336	16.254	LP (di, di-9CA)	0.2
1252	16.240	L (di, 9ald, di-9CA)	3.2

Abbreviations are as given in Table 1; mono-, di-, tri-, tetra-, penta-, hexa-, hepta-, and octa- refer to the number of ozonide groups in the molecule without specifying their position; 9ald, 9-oxononanoic acid; 9CA, 9-carboxynonanoic (azelaic) acid; 12ozCA, 12-oxo-9,10-trioxolandodecanoic acid.

resulting from the reduction, are eluted in order of increasing polarity over a wide range of retention times. Roughly, noncarboxylic reduction products were eluted between 6 and 10 min, monocarboxylic reduction products were eluted between 11 and 14 min, and dicarboxylic reduction products were eluted between 14 and 17 min.

Figure 8 shows selected ion chromatograms of the major ion masses ranging from tetra-[9-oxo]nonanoyl (m/z 1016) to the 9ald, tri-9CA (m/z 1064) and from 12ozCA, tri-9ald (m/z 1074) to O-monoozonide, di-9ald, 9CA (m/z 1190) species of CL. Although the ion intensities are low, the ions appear at the correct retention times. It was anticipated that the reduction of the ozonides would yield comparable mixtures of core aldehydes and carboxylates of cardiolipin, but this expectation was not realized, because of the extremely low yields of the core aldehydes arising from the reduction

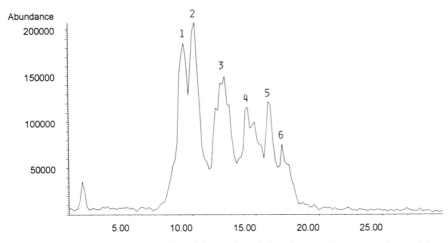

FIG. 7. Negative-ion current profile of the triphenylphosphine reduction products of the ozonides of commercial cardiolipin of bovine heart as obtained by normal-phase LC/ES/MS. Peak 1, residual ozonides; Peaks 2 and 3, mixed ozonides/aldehydes; Peak 3, mixed ozonides/aldehydes/monocarboxylates; Peak 4, mixed ozonides/aldehydes/dicarboxylates; Peaks 5 and 6, polyaldehydes and polycarboxylates. More detailed component identification is given in Fig. 8 and in Table 4. Instrumentation and operating conditions are as given in Fig. 1.

metal ion scission of the hydroperoxides. On the basis of similar single-ion plots for the various incompletely reduced ozonides it was possible to arrive at plausible identities for a large number of the components of the reaction mixture. Table 4 lists these compounds together with their retention times. Table 4 also includes the amounts estimated for these and a number of other molecular species found in the mixture.

Phospholipase D Digestion of Oxidized Cardiolipin

Phospholipase D (*S. chromofuscus*) released about 10–12% phosphatidylglycerol and 30% phosphatidic acid in about 8 min of digestion of the tetralinoleoyl species of bovine heart CL, as shown by the ions at m/z 769 and m/z 695, respectively, detected on LC/ES/MS. The bulk of the phosphatidic acid, however, was found in the form of its ethyl ester (m/z 723). The ethanol apparently originated from the small amounts of alcohol in the diethyl ether that was used to solubilize the substrate during the enzyme digestion. Phospholipase D is known to yield phosphatidylethanol as a result of transphosphatidylation. Figure 9 shows the single-ion chromatograms for the phospholipase D degradation products of CL as obtained by normal-phase LC/ES/MS. Similar [M-1]$^-$ ions for the phosphatidylglycerol and phosphatidic acid have been previously reported by Jensen et al. (19) using FAB/MS

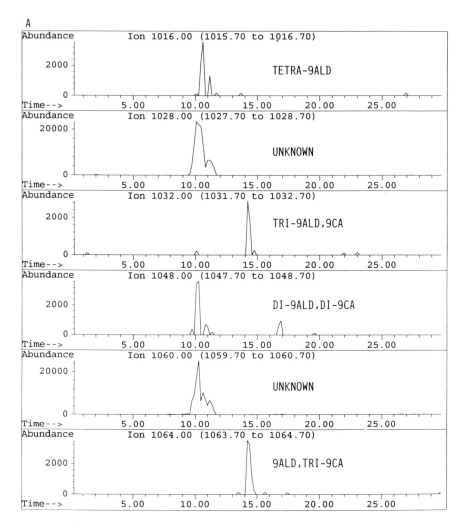

FIG. 8A. Selected ion plots for the major oxo-cardiolipins as obtained following triphenylphosphine reduction of ozonides. A, [M-1]⁻ ions of m/z 1016 to 1064; Peak identification is given in the figure. Further peak characterizations and quantitations are given in Table 4. Instrumentation and operating conditions are as given in Fig. 1.

and by Han and Gross (24) using flow/ES/MS. The oxidized CL was also hydrolyzed by the phospholipase D preparation to yield complex mixtures of phosphatidylglycerols and phosphatidic acids which gave ions anticipated for the hydroperoxide, core aldehyde, and ozonide-containing derivatives (results not shown).

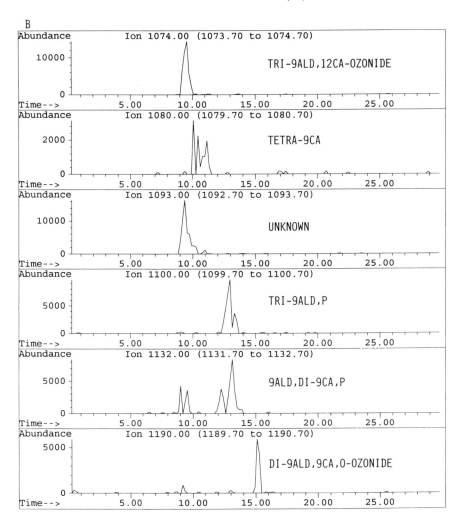

FIG. 8B. Selected ion plots for the major oxo-cardiolipins as obtained following triphenylphosphine reduction of ozonides. B, [M-1]$^-$ ions of m/z 1074 to 1190. Peak identification is given in the figure. Further peak characterizations and quantitations are given in Table 4. Instrumentation and operating conditions are as given in Fig. 1.

Search for Cardiolipin Oxidation Products in Tissues

The new methodology was applied in a search for the ozonide, hydroperoxide, and core aldehyde-containing CL in rabbit heart and rat lung. Minor ions, corresponding to the synthetic mono-, di-, and trihydroperoxides, were found in an isolated perfused rabbit

TABLE 4
Relative Retention Time and Abundance of Triphenylphosphine Reduction Products of the Ozonides of a Commercial Preparation of Bovine Heart Cardiolipin

Mass (m/z)	Retention Time (min)	Tentative Identity	Abundance (%)
1304	12.198	LP (di, di-9ald)	0.9
1258	12.295	OP (mono, di-9ald)	1.1
1220	11.735	L (di, tri-9ald)	0.7
1174	12.447	O (mono, tri-9ald)	1.4
1100	12.974	P (tri-9ald)	23.6
1016	11.238	(tetra-9ald)	4.7
1644	18.783	LLL (hexa, 9CA)	0.1
1598	24.871	LLO (penta, 9CA)	0.1
1524	12.517 (25.764)	LLP (tetra, 9CA)	0.2
1478	19.676	LOP (tri, 9CA)	0.1
1440	16.815	LL (tetra, 9ald, 9CA)	0.1
1394	16.995	LO (tri, 9ald, 9CA)	0.1
1320	16.432	LP (di, 9ald, 9CA)	0.2
1274	19.840	OP (mono, 9ald, 9CA)	0.3
1236	12.729 (18.783)	L (di, di-9ald, 9CA)	0.4
1190	15.163	O (mono, di-9ald, 9CA)	8.4
1116	15.791	P (tri-9ald)	2.8
1032	14.351	(tri-9ald, 9CA)	4.0
1456	8.125 (21.627)	LL (Tetra, di-9CA)	0.2
1410	4.424	LO (Tri, di-9CA)	0.1
1336	11.840	LP (di, di-9CA)	0.2
1290	7.954 (12.439)	OP (mono, di-9CA)	0.7
1252	8.663 (19.840)	L (di, 9ald, di-9CA)	0.4
1206	8.587 (17.553)	O (mono, 9ald, di-9CA)	1.2
1132	13.181	P (9ald, di-9CA)	17.4
1048	10.269 (16.935)	(di-9ald, di-9CA)	5.5
1268	8.939 (12.540)	L (di, tri-9CA)	1.4
1222	14.930	O (mono, tri-9CA)	8.4
1148	8.916 (12.517)	P (tri-9CA)	3.3
1064	14.387	(9ald, tri-9CA)	6.7
1080	10.183	(tetra-9CA)	4.8

Abbreviations are as given in Table, and Table 3.

heart tissue and in a rat lung, but the corresponding ozonides were not immediately recognized in either tissue (Parsons, Kuksis, Bergqvist, and Ravandi, 1997; in preparation).

Conclusions

Conventional methods of preparation of hydroperoxides, ozonides, core aldehydes, and core carboxylates of cardiolipin yield extremely complex mixtures, which are difficult to fractionate and yield only small amounts of specific derivatives. The oxidizing agents carry out an uncontrolled simultaneous attack upon all four linoleoyl moieties of the bovine heart cardiolipin molecule. Pure oxo-cardiolipin species for future studies should probably be prepared by chemical synthesis.

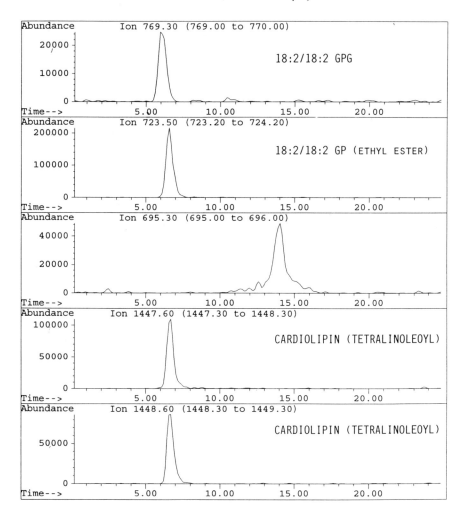

FIG. 9. [M-1]⁻ ion chromatograms for the phospholipase D digestion products of cardiolipin as obtained by normal phase LC/ES/MS. Peak identification: m/z 695, phosphatidic acid; m/z 769, phosphatidylglycerol; m/z 723, tetralinoleoyl cardiolipin ($m/2$); m/z 1447, [M-1]⁻ of nominal mass of tetralinoleyl CL (monoisotopic ion) ; m/z 1448, [M-1]⁻ of tetralinoleoyl CL. All species are dilinoleoyl derivatives as obtained from tetralinoleoyl species of bovine heart cardiolipin. Instrumentation and operating conditions are as given in Fig. 1.

Nevertheless, the present studies provide a wealth of information on the chromatographic and mass spectrometric properties of oxidized cardiolipin. Fresh samples of cardiolipin appeared to be free of any oxidation products, possibly because of the high affinity binding to proteins, which protects them from peroxidation. Liposomes containing cardiolipin were readily peroxidized *in vitro*.

Acknowledgment

These studies were supported by funds from the Heart and Stroke Foundation of Ontario, Toronto, Ontario, the Medical Research Council of Canada, Ottawa, Ontario, and the Swedish Institute, Stockholm, Sweden.

References

1. Simon, G., and Rouser, G. (1969) Species Variation in Phospholipid Class Distribution of Organs: II. Heart and Skeletal Muscle, *Lipids 4*, 607–614.
2. Daum, G. (1985) Lipids of Mitochondria, *Biochim. Biophys. Acta 822*, 1–42.
3. Robinson, N.C. (1993) Functional Binding of Cardiolipin to Cytochrome c Oxidase, *J. Bioenerg. Biomembranes 25*, 153–163.
4. Ohtsuka, T., Nishijima, M., and Akamatsu, Y. (1993) A Somatic Cell Mutant Defective in Phosphatidylglycerolphosphate Synthase, with Impaired Phosphatidylglycerol and Cardiolipin Biosynthesis, *J. Biol. Chem. 268*, 22908–22913.
5. Nielsen, H. (1978) Reaction Between Peroxidized Phospholipid and Protein: I. Covalent Binding of Peroxidized Cardiolipin to Albumin, *Lipids 13*, 253–258.
6. Kuksis, A., Myher, J.J., Marai, L., and Geher, K. (1993) Analyses of Hydroperoxides and Core Aldehydes of Triacylglycerols in *Proceedings of the 17th Nordic Lipid Symposium* (Malki, Y., ed.), A Lipidforum Publication, Imatra, Finland, pp. 188–195.
7. Ravandi, A., Kuksis, A., Myher, J.J., and Marai, L. (1995) Determination of Lipid Ester Ozonides and Core Aldehydes by High–Performance Liquid Chromatography with On-Line Mass Spectrometry, *J. Biochem. Biophys. Methods 30*, 271–285.
8. Shaikh, N.H. (1994) Assessment of Various Techniques for the Quantitative Extraction of Lysophospholipids from Myocardial Tissues, *Anal. Biochem. 216*, 313–321.
9. Esterbauer, H., and Cheeseman, K.H. (1990) Determination of Aldehydic Peroxidation Products: Malonaldehyde and 4-Hydroxynonenal, *Methods Enzymol. 186*, 407–421.
10. Skipski, V.P., Petersen, R.F., and Barclay, M. (1964) Quantitative Analysis of Phospholipids by Thin-Layer Chromatography, *Biochem. J. 90*, 374–378.
11. Kuksis, A., and Marai, L. (1967) Determination of the Complete Structure of Natural Lecithins, *Lipids 2*, 217–224.
12. Imamura, S., and Horiuti, Y. (1979) Purification of *Streptomyces chromofuscus* Phospholipase D by Hydrophobic Affinity Chromatography on Palmitoyl Cellulose, *J. Biochem. 85*, 79–95.
13. Myher, J.J., and Kuksis, A. (1984) Determination of Plasma Total Lipid Profiles by Capillary Gas-Liquid Chromatography, *J. Biochem. Biophys. Methods 10*, 13–23.
14. Kamido, H., Kuksis, A., Marai, L., and Myher, J.J. (1993) Identification of Core Aldehydes Among In-Vitro Peroxidation Products of Cholesteryl Esters, *Lipids 28*, 331–336.
15. Privett, O.S., and Blank, M.L. (1963) A Method for the Structural Analysis of Triglycerides and Lecithins, *J. Am. Oil Chem. Soc. 40*, 70–75.
16. Becart, J., Chevalier, C., and Biesse, J.P. (1990) Quantitative Analysis of Phospholipids by HPLC with a Light Scattering Evaporating Detector—Applications to Raw Materials for Cosmetic Use, *J. High Resol. Chromatogr. 13*, 126–129.
17. Teng, J.I., and Smith, L.L. (1985) High Performance Liquid Chromatography of Cardiolipin, *J. Chromatogr. 339*, 35–44.

18. Schlame, M., and Otten, D. (1991) Analysis of Cardiolipin Molecular Species by High-Performance Liquid Chromatography of its Derivative 1,3-Bisphosphatidyl-2-Benzoyl-*sn*-Glycerol Dimethyl Ester, *Analyt. Biochem. 195*, 290–295.
19. Jensen, N.J., Tomer, K.B., and Gross, M.L. (1987) FAB MS/MS for Phosphatidylinositol, -Glycerol, -Ethanolamine and Other Complex Phospholipids, *Lipids 22*, 480–489.
20. Kamido, H., Kuksis, A., Marai, L., and Myher, J.J. (1995) Identification of Glycerolipid-Bound Aldehydes in Copper-Oxidized Low Density Lipoproteins, *J. Lipid Res. 36*, 1876–1886.
21. Myher, J.J., Kuksis, A., Ravandi, A.H.K., and Cocks, N. (1994) Identification of Polyoxyglycerophospholipids by Normal Phase LC/MS with Electrospray, *INFORM 5*, 478–479.
22. Zhang, J.R., Cazers, A.R., Lutzke, B.S., and Hall, E. D. (1995) HPLC-Chemiluminescence and Thermospray LC/MS Study of Hydroperoxides Generated from Phosphatidylcholine, *Free Rad. Biol. Med. 18*, 1–10.
23. Harrison, K.A., and Murphy, R.C. (1996) Direct Mass Spectrometric Analysis of Ozonides: Application to Unsaturated Glycerophosphocholine Lipids, *Anal. Chem. 68*, 3224–3230.
24. Han, X., and Gross, R.W. (1994) Electrospray Ionization Mass Spectrometric Analysis of Human Erythrocyte Plasma Membrane Phospholipids, *Proc. Natl. Acad. Sci. USA 91*, 10635–10639.

Chapter 5

Stereospecific Analysis of Docosahexaenoic Acid-Rich Triacylglycerols by Chiral-Phase HPLC With Online Electrospray Mass Spectrometry

J.J. Myher[a], A. Kuksis[a], and P.W. Park[b]

[a]Banting and Best Department of Medical Research, University of Toronto, Toronto, Canada,
[b]Crown Nutritionals, Inc., 6780 Caballo Street, Las Vegas, Nevada 89119.

Introduction

Designer oils containing more than one n-3 and n-6 polyunsaturated fatty acid per acylglycerol molecule have recently been introduced for dietary supplementation (1). Because of the large difference in the intestinal absorption of the long-chain fatty acids between free acids, 2-monoacylglycerols (2,3), and triacylglycerols (4), there is a need to determine the positional distribution of long-chain fatty acids in these dietary oils. Furthermore, since the tissue lipases (5) and acyltransferases (6) exhibit both fatty acid specificity and acylglycerol stereospecificity, it is necessary to determine the exact positional placement of the long-chain acids in the circulating and stored acylglycerol molecules in the body. In the past, stereospecific analyses of fats and oils containing one long-chain polyunsaturated fatty acid per triacylglycerol molecule have been found to be compromised by their resistance to enzymatic hydrolysis (7,8). Recent work has shown (9) that the presence of two long-chain fatty acids in a triacylglycerol molecule leads unexpectedly to additional chromatographic resolution of the acylglycerol intermediates. This complicates the purification of intermediates and may result in selective losses of molecular species during conventional stereospecific analyses.

In the present study we have undertaken a detailed examination of the chromatographic and mass-spectrometric behavior during stereospecific analyses of natural triacylglycerols containing two or three docosahexaenoic acids per molecule. The results show that the molecular species with two or three long-chain fatty acids can be reliably identified and estimated by parallel analyses of appropriate standards and meticulous attention to chromatographic resolution, and chiral-phase HPLC with online mass spectrometry of the enantiomeric diacylglycerol intermediates.

Experimental Procedures

Materials

Two polyunsaturated oils (Oil A and Oil B), of microalgal origin and rich in docosahexaenoic acid, were gifts from Martek Biosciences Corporation (Columbia, MD).

Synthetic diarachidonoyl and didocosahexaenoylglycerols were obtained from Nu-Check Co., Elysian, MN. The DG were isomerized to the corresponding 1,3-DG in presence of dilute $HClO_4$. Reference sn-1,2- and sn-2,3-DG dinitrophenylurethane (DNPU) derivatives prepared from standard TG were available in the laboratory from previous studies (10,11). 3,5-Dinitrophenyl isocyanate was purchased from Fluka Chemical Co. (Switzerland). Ethyl magnesium bromide and boric acid were obtained from Aldrich Chemical Co. (Milwaukee, WI). Pancreatic lipase was obtained from Sigma Chemical Co. (St. Louis, MO).

Purification of TG

Five drops of each oil were dissolved in 1 mL of chloroform, and 250 µL of the solution was applied to a 20 × 20 cm TLC plate coated with a 250 µm thin layer of Silica Gel H (Merck, Darmstadt, Germany). The plate was developed with hexane/ethyl acetate (88:11, v/v) (12). The TLC bands corresponding to standard steryl ester, TG, and phosphatidylcholine were scraped off and eluted with chloroform/methanol (2:1, v/v) and the solvents blown off under nitrogen. Care was taken in collecting the TG fraction to avoid loss of short- and long-chain TG that might have trailed or preceded the bulk of the species. The TG fraction was subjected to immediate stereospecific analysis, while the other fractions were saved for later analyses. Aliquots were taken for polar capillary gas chromatography (GLC) of fatty acids and reversed-phase high-performance liquid chromatography (HPLC) of the intact lipid esters, described later.

Partial Deacylation of TG

Ten mg of the purified TG were dissolved in dry diethyl ether (1 mL), freshly diluted 0.5 M ethyl magnesium bromide in dry ethyl ether (250 µL) was added, and the mixture was shaken for 1 min (vortexed) before glacial acetic acid (6 µL) in hexane (5 mL) and water (2 mL) were added to stop the reaction (5). The organic layer was washed twice with water (2 mL) and dried over anhydrous sodium sulfate and bicarbonate contained in a Pasteur pipette. After evaporating the solvent in a stream of nitrogen at room temperature, the mixture of deacylation products was separated by borate TLC (13) using chloroform/acetone (88:12, v/v). The sn-1,2(2,3)- and 1,3-DG were separately recovered. Alternatively, the TG were partially deacylated by digestion with pancreatic lipase using another established procedure (14). The digestion was performed in the presence of gum arabic for 30 s and the digestion products extracted with diethyl ether. The sn-1,2(2,3)- and X-1,3-DG and 2-MG were resolved by boric acid TLC and separately recovered as described above. For preparation of representative 2-MG the digestion was prolonged to 2–3 min. After aliquots of the DG fractions were taken for fatty acid analyses, the remainder of the sn-1,2(2,3)- and 1,3-DG was converted into the DNPU derivatives for chiral-phase HPLC resolution, as detailed later.

Preparation of 3,5-DNPU

The sn-1,2(2,3)- and 1,3-DG (1 mg) were dissolved in dry toluene (400 µL) containing 40 µL of dry pyridine and about 2 mg of 3,5-dinitrophenyl isocyanate. This was allowed to stand for 1 h at room temperature (5). The resulting 3,5-DNPU derivatives were purified by TLC on a silica gel plate (Kodak, Rochester, NY). Prior to use, the plate was activated at 110–120°C for 1 h. The reaction mixture, dissolved in 400 µL of chloroform, was spotted on the plate and developed up to a height of 15 cm using petroleum ether/1,2-dichloroethane/ethanol (40:10:3, by vol) as the developing solvent. Bands were located under ultraviolet light and the 3,5-DNPU fractions (R_f = 0.5–0.6) were scraped off the plate. The pure 3,5-DNPU were recovered from the adsorbent by extraction with diethyl ether.

Chiral-Phase HPLC

The enantiomeric sn-1,2- and sn-2,3-DG in the sn-1,2(2,3)- and in the 1,3-DG fractions were resolved using the chiral (R)-(+)-1-(1-naphthyl)ethylamine phase bonded to 300 Å wide-pore silica (YMC Pack A-KO3; YMC Inc., Kyoto, Japan) on a 25 cm × 4.6 mm i.d. column installed in a Hewlett-Packard (Palo Alto, CA) Model 1090 liquid chromatograph (10,11). The analysis was performed isocratically at 23°C using hexane/1,2-dichloroethane/ethanol (40:10:1, by vol) as the mobile phase at a constant flow rate of 0.5 mL/min. Peaks were monitored at 254 nm using a variable-wavelength detector. The corresponding sn-1,2- and sn-2,3-DAG from the two separations were collected and combined for fatty acid analysis.

GLC of Fatty Acids

The fatty acid composition of the purified TAG and their deacylation products was determined following transmethylation with 6% H_2SO_4-MeOH (12). The methyl esters were resolved by polar capillary GLC and the component fatty acids were quantitated by flame ionization detection (12).

Reversed-Phase High-Performance Liquid Chromatography

Intact TG were resolved by reversed-phase HPLC using a Supelcosil LC-18 column (250 mm × 4.6 mm ID, Supelco, Bellefonte, PA) in combination with two solvent systems (System A and System B). In System A, the column was eluted with 20–80% isopropanol in methanol over a period of 30 min, then isocratic at 80% isopropanol to 40 min (0.85 mL/min). HPLC peaks were detected by online ES/MS. In System B, the column was eluted with a linear gradient of 20–80% isopropanol in acetonitrile for 30 min, then isocratic at 80% isopropanol for 10 min (0.85 mL/min). The peaks were detected by online ES/MS as described below. In both instances, the column was installed in a Hewlett-Packard Model 1090 liquid chromatograph.

Liquid Chromatography/Electrospray/Mass Spectrometry (LC/ES/MS)

When intact TG were resolved by reversed-phase HPLC with online ES/MS, a postcolumn flow of 0.2 mL/min of 1% ammonia in isopropanol was used to enhance the nebulizer-assisted electrospray ionization system, which was interfaced with a Hewlett-Packard 5985 MS Engine quadrupole mass spectrometer (16). At an exit voltage (Cap_{ex}) of 170 V there was little fragmentation; the pseudomolecular $[M+NH_4]^+$ (M+18) and $[M+Na]^+$ (M+23) ions had the only major intensities. At Cap_{ex} 215 V (ES/CID/MS), there was considerable fragmentation of the $[M+NH_4]^+$ to the DG type ions. At Cap_{ex} 250 V, the $[M+NH_4]^+$ were nearly completely lost and replaced by the DG type ions, while the $[M+Na]^+$ ions remained unaffected.

Chiral-phase LC/ES/MS of the DAG DNPU was performed using the YMC Pack A-KO3 column installed in a Hewlett-Packard Model 1090 liquid chromatograph. The separations of the *sn*-1,2-, *sn*-2,3-, and 1,3-DG fractions were obtained with hexane/dichloromethane/ethanol (40:10:1) at 0.5 mL/min with postcolumn addition of 0.3 mL isopropanol containing 1% aqueous ammonia. Using the ES equipment described in the previous paragraph with Cap_{ex} 170 V, the ions were recorded in the negative mode in the mass range (*m/z*) 600 to 1200.

Calculations

The mole % fatty acid composition of the *sn*-1-, *sn*-2-, and *sn*-3- positions of the TG molecules were calculated by subtraction of the mole % fatty acid compositions as follows (10):

$$sn\text{-}2 = (2\text{-MG}), \text{ from pancreatic lipase deacylation} \quad (1)$$
$$sn\text{-}1 = 2 \times (sn\text{-}1,2\text{-DG}) - (sn\text{-}2\text{-MG}) \quad (2)$$
$$sn\text{-}3 = 2 \times (sn\text{-}2,3\text{-DG}) - (sn\text{-}2\text{-MG}) \quad (3)$$

Alternatively, the positional fatty acid distribution was calculated as follows:

$$sn\text{-}1 = 3 \times (\text{total TG}) - 2 \times (sn\text{-}2,3\text{-DG}) \quad (4)$$
$$sn\text{-}3 = 3 \times (\text{total TG}) - 2 \times (sn\text{-}1,2\text{-DG}) \quad (5)$$

The final composition of the *sn*-1- and *sn*-3-positions was obtained by averaging the results of the two calculations.

Results and Discussion

Preliminary Characterization

TLC of the oil samples in a neutral solvent system revealed the presence of a large TG band preceded by a faint band corresponding to steryl esters. Small amounts of polar material were left at the origin of the plate. The oils contained trace amounts of free sterols or free DG, which were also removed by TLC (chromatoplates not

TABLE 1
Fatty Acid Composition of Purified Triacylglycerols
from Two Docohexaenoic Acid-Rich Oils

Fatty acids	Oil A	Oil B
10:0	0.6	0.75
12:0	4.5	6.47
14:0	19.2	19.76
16:0	13.0	17.08
16:1n-7	2.4	1.86
18:0	0.5	0.24
18:1n-9	16.8	9.01
18:2n-6	1.5	0.41
18:3n-3	0.3	0.14
22:5n-3	0.4	0.4
22:6n-3	40.5	43.5
Other		

shown). The silica gel containing the pure TG band was scraped off and extracted with chloroform. An aliquot of the purified TG was transmethylated; Table 1 gives the fatty acid composition of the purified TG fractions. Both oils contain very similar proportions of 22:6n-3 and 14:0 acids, but differ in the proportions of 16:0 and 18:1n-9, which are the other two major fatty acids present in the samples. Oil A contains more of the 18:1n-9 and less of the 16:0 than Oil B. The oils are free of 20:4n-6 and 20:5n-3 acids. The high content of 22:6n-3 (40–44%) indicates that a high proportion of the TG contain two or three long-chain fatty acids per molecule. There appeared to be no chain-length segregation of the TG species on the TLC plate, but this was not a problem once the upper limit of migration of the polyunsaturates was established by means of standards.

Deacylation of TG and Isolation of DGs

On the basis of previous work, it was clear that the formation of random sn-1,2(2,3)-DG was of critical importance for accurate stereospecific analysis, as was the isolation of representative samples during the purification of the reaction products. Both lipase digestion and Grignard degradation had been suspected to generate nonrandom mixtures of sn-1,2(2,3)-DG. In fact, allyl magnesium bromide had been proposed for positional analysis of fish oil TG because of its higher reactivity when compared to the ethyl or methyl magnesium bromide usually employed (16). Since we felt that the major problem was the overlapping of isomeric acylglycerols of medium and long-chain length on borate-TLC, we proceeded with the use of ethyl magnesium bromide in the Grignard reaction. Figure 1 compares the borate-TLC separation of the reaction products of corn oil and an oil rich in docosahexaenoic acid. Corn oil yields the typical pattern of separation obtained for many vegetable oils (Fig. 1A). All of the lipid classes are clearly resolved and can be recovered without loss or contamination. The band structure obtained for the polyunsaturated

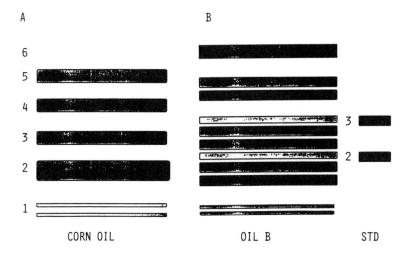

FIG. 1. Separation of TG deacylation products by borate TLC. The mixture of deacylation products in chloroform was applied as a band 2.5 cm from the lower edge of a 20 × 20 cm plate (250 μM thin-layer) of Silica Gel G containing 5% borate. The chromatoplate was developed using chloroform/acetone (88:12, v/v) as the solvent (13). (A) deacylation products of corn oil TG: 1, monoacylglycerols; 2, *rac*-1,2-diacylglycerols; 3, 1,3-diacylglycerols; 4, *tert*-alcohols; 5, triacylglycerols; 6, solvent front. (B) deacylation products of Oil B triacylglycerols: 2, *rac*-1,2-didocosahexaenoylglycerol; 3, 1,3-didocosahexaenoylglycerol. TLC bands were detected by spraying with 2,7-dichlorofluorescein in methanol.

oil is much more complex (Fig. 1B). Each lipid fraction is split up into a number of bands. The use of standard DHA-containing DG, however, provides a good indication of the upper limit of migration of the polyunsaturates. The standard *sn*-1,2(2,3)-di-DHA species clearly migrates ahead of all the other *sn*-1,2(2,3)-DG species and overlaps with the medium-chain 1,3-DG band. By judicious choice of the TLC bands, it was possible to scrape off nearly representative *sn*-1,2(2,3)- and 1,3-DGs from the DHA-rich oils. Any remaining overlapping of isomers could be corrected for following chiral-phase LC/ES/MS (see below). We also considered loss of 22:6 due to peroxidation, incomplete conversion of the DG to DG-DNPU, and even possible discrimination by the Grignard reagent. None of these problems were valid, as can be seen from the excellent recoveries obtained with standard polyunsaturated TG taken through the entire analytical routine (see below).

Chiral-phase LC/ES/MS of DG-DNPU

Figure 2 shows the chiral-phase LC/ES/MS profiles of the DNPU derivatives of the isomeric diarachidonoylglycerols and didocosahexaenoylglycerols. Conditions for LC/ES/MS of DAG DNPU are described in the Experimental Procedures section of

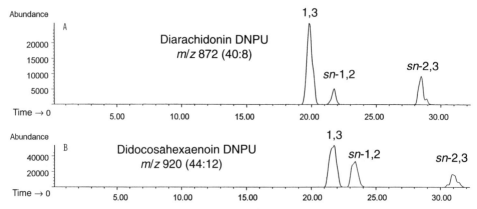

FIG. 2. Chiral-phase LC/ES/MS of DNPU of total DG derived by isomerization of diarachidonoyl (A) and didocosahexaenoyl (B) glycerol. Total negative ion profiles: sn-1,2-, sn-1,2-DG; sn-2,3-DG; 1,3-, 1,3-DG.

this chapter. In both instances the 1,3-isomers are eluted ahead of the sn-1,2-enantiomers, which are followed considerably later by the sn-2,3-enantiomers during isocratic elution from this chiral column. All three classes of the isomeric diarachidonoylglycerols (40:8) are eluted ahead of the corresponding isomers of the didocosahexaenoylglycerols (44:12), as would be expected on the basis of the difference in the effective carbon numbers (24 vs. 20). The recoveries of the long-chain DG derivatives are complete, as indicated by the peak areas. Clearly, there was no loss of the long-chain acid derivatives during deacylation of the TG, derivatization or chromatography. There was also no evidence of any readily detectable peroxidation during chromatography. Using the chiral-phase LC/ES/MS method it was possible to establish that the loss of the polyunsaturated sn-1,2-DG resulted from their migration into the medium-chain sn-1,3-diacylglycerol fraction during borate TLC. This conclusion was confirmed by chiral-phase LC/MS of the 1,3-DG fraction, which was recognized to be contaminated with the long-chain polyunsaturates of both sn-1,2 and sn-2,3-configuration, as they are not separated on borate TLC. When the cuts were made just above the standard sn-1,2-DG, there was very little overlap of sn-1,2- and 1,3-isomers, because the long-chain polyunsaturated sn-1,2(2,3)-DG migrated faster when compared to medium-chain sn-1,2(2,3)-DG.

We have previously determined (9) the fatty acid composition of the original and reconstituted TG for Oil A and have compared it to the corrected and calculated composition of the sn-1,2(2,3)-DG derived from total TG and the 2-MG recovered from pancreatic lipase digestion. There was reasonably good agreement between the corrected experimental sn-1,2-(2,3)-DG and the sn-1,2(2,3)-DG calculated from the original TG and the 2-MG. This fatty acid composition could be reconstituted to the composition of the original TG. However, with the use of the polyunsaturated DG as markers, it became possible to scrape off representative sn-1,2(2,3)- and 1,3-DG isomers, and there was no need to correct for loss of the polyunsaturated species of *sn*-1,2(2,3)-DG in the 1,3-DG fraction from oils enriched in polyunsaturated long-chain acids after chiral-phase LC/ES/MS analysis of the *sn*-1,2- and 1,3-DG fractions.

Because of the chain-length segregation and background contamination, it was also difficult to obtain representative samples from the small 2-MG fraction produced during Grignard reaction. However, representative 2-MG could be recovered by delayed extraction of the products of pancreatic lipolysis, which yields 2-MG as the final reaction products. The isomerization of the secondary to the primary esters appears to be slow in relation to the production of the 2-isomers, even from long-chain TGs. In this case, the TLC band was so strong that a representative 2-MG fraction was readily obtained without contamination and undesirable segregation based on chain length. Pancreatic lipase is known to discriminate against the long-chain polyunsaturated fatty acids during brief digestion periods, leading to nonrepresentative formation of the DG intermediates.

Figure 3 shows the chiral-phase LC/ES/MS separation of the *sn*-1,2- and *sn*-2,3-DG derived from Oil B. The resolution of the *sn*-1,2- and the *sn*-2,3-DG is excellent as established by separately run reference standards of *sn*-1,2- and *sn*-2,3-DG prepared from corn oil and synthetic didocosahexaenoylglycerols. The chiral-phase LC/ES/MS profile of the DNPU derivatives of the 1,3-DG is shown in Figure 4. In view of the retention times established above for the standard (Fig. 2), the peak emerging at 21.00 min might be due to the 1,2-didocosahexaenoylglycerol species. However, there are significant amounts of the *m/z* 920 ion also in the total mass spectrum averaged over the *sn*-1,3-DG DNPU peak, as shown in Fig. 5.

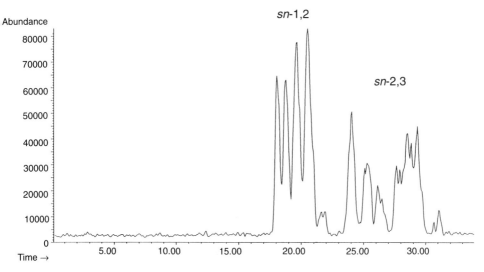

FIG. 3. Chiral-phase LC/ES/MS of DNPU of *sn*-1,2(2,3)-DG derived from Oil B TG. Total negative ion profiles: *sn*-1,2-, *sn*-1,2-DG; *sn*-2,3-, *sn*-2,3-DG. Chiral-phase HPLC and online ES/MS conditions are as given in the text.

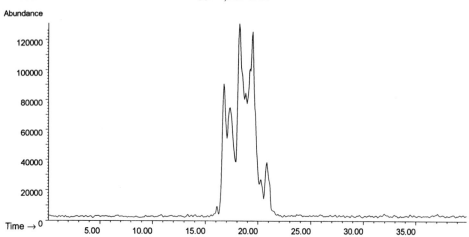

FIG. 4. Chiral-phase LC/ES/MS of DNPU of 1,3-DG derived from Oil B TG. Total negative ion profile. Chiral-phase HPLC and online ES/MS conditions are as given in the text.

Figure 6 gives the total mass spectrum averaged over all the sn-1,2-DG peaks. In this instance the small amounts of the m/z 920 ion represent the true content of the sn-1,2-di-docosahexaenoylglycerol, as does the much higher intensity for the m/z 920 ion in the sn-2,3-DG (see below). The other species represent sn-1,2-DG DNPU with carbon number/double bond number (CN/NDB) ranging from 28:0 to 38:6.

Figure 7 gives a comparable mass spectrum averaged over the entire sn-2,3-DG peak. This spectrum includes ions corresponding to CN/DBN of 26:0 to 44:12. The sn-

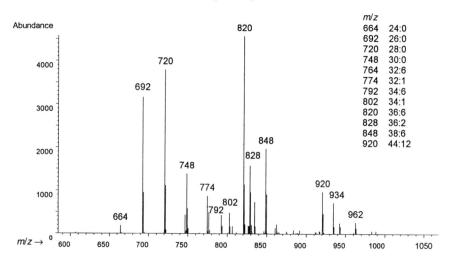

FIG. 5. Total negative ion spectrum averaged over the 1,3-DG DNPU peak in Fig. 4. The [M-1]⁻ ions are identified in the figure by carbon/double bond number. Chiral-phase HPLC and online ES/MS conditions are as given in the text.

Stereospecific analyses of triacylglycerols 109

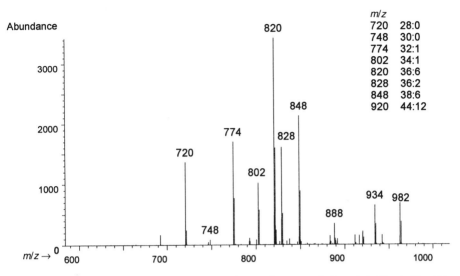

FIG. 6. Total negative ion spectrum averaged over the sn-1,2-DG peak in Fig. 3. The [M-1]⁻ ions are identified in the FIG. by carbon/double bond number. Chiral-phase HPLC and online ES/MS conditions are as given in the text.

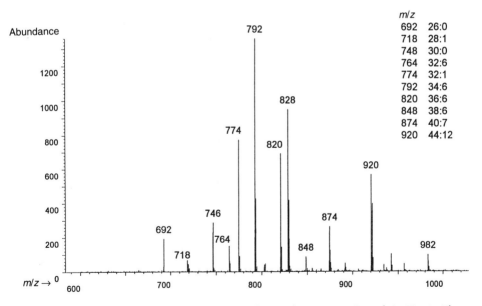

FIG. 7. Total negative ion spectrum averaged over the sn-2,3-DG peak in Fig. 3. The [M-1]⁻ ions are identified in the FIG. by carbon/double bond number. Chiral-phase HPLC and online ES/MS conditions are as given in the text.

2,3-DG DNPU include the ion for the didocosahexaenoylglycerol species (m/z 920), but the site of elution cannot be ascertained from the averaged mass spectra alone. A more complete identification of the molecular species and the determination of their elution sequence is possible by an examination of the single-ion chromatograms.

Figure 8 shows the single-ion chromatograms for the major sn-1,2- and sn-2,3-DG species of Oil B as retrieved from the chiral-phase LC/ESI/MS profiles. By refer-

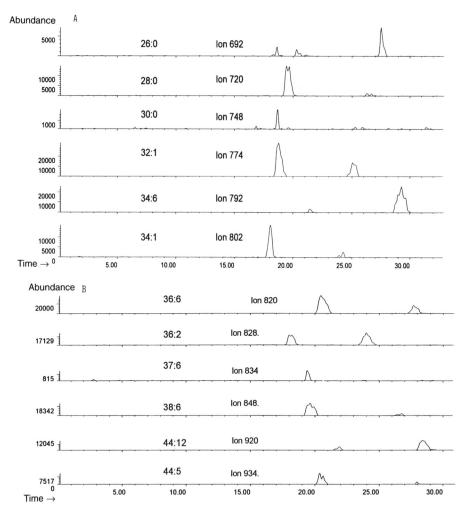

FIG. 8. [M-1]$^-$ ion plots for the major species of sn-1,2- and sn-2,3-DG DNPU as extracted from total negative ion spectra recorded by chiral-phase LC/ES/MS. (A) [M-1]$^-$ ions m/z 692 to m/z 802 identified by carbon/double bond number; (B) [M-1]$^-$ ions m/z 820 to m/z 934 identified by carbon/double bond number. Chiral-phase HPLC and online ES/MS conditions are as given in the text.

ence to the elution times of the individual peaks in the total ion current profile, it is possible to associate the various ions with the individual peaks emerging from the chiral column. Thus, the ion at m/z 920 is seen to emerge from the chiral column with two retention times. The earliest-emerging peak (21.5–22.2 min) is apparently due to the sn-1,2-22:6/22:6 species, whereas the last-emerging peak (28.0–29.0 min) is due to sn-2,3-22:6/22:6 species. It may be noted that the correspondence of the retention times between the peaks of the standards and unknowns is not exact, due to a slight chromatographic drift that takes place over time. For the corresponding 1,3-DG fraction the peak for the m/z 920 ion occurs at 20.2 min (not shown). Similarly, it is possible to sort out the other ions according to their origin in the sn-1,2- or sn-2,3-DG DNPU of Oil A. As already noted for the 1,3-DG DNPU, the elution order of the species is affected by both chain length and degree of unsaturation. The retention time increases with decreasing chain length as well as with increasing unsaturation. The identified molecular species of the sn-1,2- and sn-2,3-DG were quantified from the peak areas of the single-ion chromatograms, assuming that all molecular species give comparable responses on ES ionization (9). It is seen that the two enantiomeric DG populations possess markedly different molecular species, which confirm the nonrandom distribution of the fatty acids in the parent TG molecules, as concluded from the stereospecific distribution of the fatty acids (see below).

The molecular species identification of the enantiomeric DG resolved by the chiral-phase LC/ES/MS might be confirmed by LC/ES/CID/MS increasing the capillary exit voltage: Although the cleavage of the fatty acid moieties could be demonstrated, it was not sufficient to identify the DG parent molecules with certainty in the absence of significant lyso-DG DNPU ions.

Table 2 reproduces the stereospecific distribution of fatty acids in the TGs of Oil B (9). This calculation was based on the fatty acid composition of the original TG and the sn-1,2- and sn-2,3-DG enantiomers resolved with reference to the polyunsaturated DG standards. The oil is characterized by a preferential association of the saturated fatty acids with the sn-1 and sn-3 positions, somewhat greater proportions being found in the sn-1 position. Oleic acid was greatly preferred for the sn-2 position, which also favored the DHA, but the latter acid was also found in a high proportion in the sn-3-position and somewhat smaller amounts in the sn-2 position. This distribution of fatty acids is identical to the values reported for a DHA-rich oil in our earlier paper (9).

Determination of TG Structure

Another aliquot of the purified TG was subjected to HPLC with online ES/MS in order to obtain some indication about the molecular association of the fatty acids in the TG molecules. A reversed-phase column in combination with two solvent systems compatible with online mass spectrometry were employed as described earlier in the Experimental Procedures section. Figure 9A shows the total LC/ESI/MS positive ion profile (Cap$_{ex}$ 170 V) of the purified TG from Oil A obtained by a linear gradient of

TABLE 2
Stereospecific Distribution Of Fatty Acids In A High DHA-
Containing TG (mole%) (Oil B)

Fatty acids	sn-1-	sn-2-	sn-3	Total
10:0	1.12	0.00	1.12	0.75
12:0	8.76	0.44	10.22	6.47
14:0	30.53	4.75	23.99	19.76
15:0	−0.29	0.11	0.15	0.01
16:0	29.81	2.72	18.70	17.08
16:1n-7	1.83	2.03	1.42	1.86
18:0	0.84	0.40	−0.57	0.24
18:1n-9	3.90	22.86	−0.26	9.01
18:2n-6	−0.35	0.72	0.86	0.41
20:0	−0.18	0.00	0.25	0.14
18:3n-3	0.22	0.09	0.11	0.02
20:1ω-9	0.09	0.00	0.2	0.06
20:4n-6	0.09	0.09	0.07	0.08
24:1	0.2	0.00	0.12	0.08
22:6	23.58	64.62	42.34	43.51
Other	0.78	1.06	0.33	

20–80% isopropanol in methanol (System A). This solvent system yielded a total of 23 major HPLC peaks. Figure 9B shows the effect of online ES/CID/MS on the HPLC profile. The ES/CID/MS generated small amounts of DG type ions, which allowed the recognition of additional TG species. Table 3 gives the retention times for the identified species along with the estimated peak areas.

System B, made up of a linear gradient of 20–80% isopropanol in acetonitrile gave much improved resolution and allowed the recording of some 33 HPLC peaks. Figure 10 gives the positive ion current profile of Oil B (A) along with the full mass spectrum (B) averaged over the entire HPLC run as obtained at Cap_{ex} 215 V. There was some change in the total ion current profile resulting from the increase in the voltage from 170 to 215 V. The increase over the basal voltage led to a relative decrease of the ammonia adducts, appearance of the DG type of ions, and retention of the sodium adducts. This change in the ion composition brought about a change in the relative distribution of the ion intensities among the HPLC peaks. The DG type of ions helped to identify the TG components of each HPLC peak. The suggested ion identities are given in the legend to Fig. 10B and in Table 4.

Figure 11 shows the full mass spectrum averaged over peak 20 of the total ion current profile of Oil B. The ions represent the 14:0/16:0/22:6 TG species as indicated by $[M+18]^+$ (m/z 868), $[M+23]^+$ (m/z 873), and the $[M-RCOO]^+$ (DG ions). The DG-type ions are due to 30:0 (m/z 523); 36:6 (m/z 595); and 38:6 (m/z 623). Peak 26 in the total ion current profile of Oil B (Fig. 12) contains two TG components. The 14:0/14:0/18:1 TG species is represented by $[M+18]^+$ (794) and the DG type ions 28:0 (m/z 495); 32:1 (m/z 549) and 34:1 (m/z 577). The other TG species of the same mass (16:0/16:0/16:1) is represented by the DG type of ions m/z 521 (30:0) and m/z 549

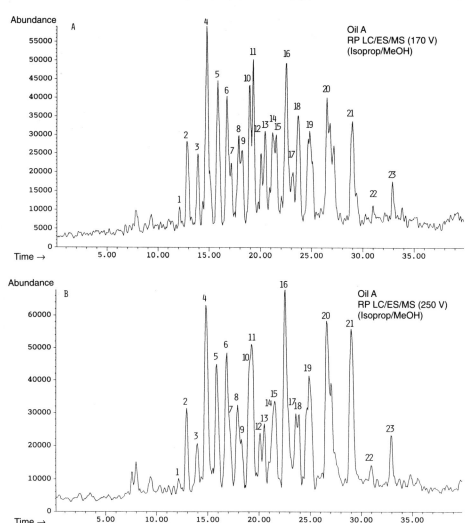

FIG. 9. Positive ion current profile of Oil A TG as obtained by reversed-phase LC/ES/MS. (A) total positive ion current profile; peak identification as given in Table 3; (B) selected [M+NH$_4$]$^+$ ion plots; peaks are identified by carbon/double bond number. Reversed-phase HPLC conditions are as described in the text, using Solvent System A.

(32:1). The identified molecular species of TG are listed in Table 4 along with their HPLC retention times and peak areas.

The type of fragmentation achieved by increasing the exit voltage in a single-stage quadrupole mass spectrometer with electrospray is similar to that obtained previously by Duffin et al. (19), who used ES/MS/MS to examine the mass spectra of synthetic MG, DG, and TG. Byrdwell and Emken (20) and Byrdwell et al. (21)

TABLE 3
Reversed-Phase LC/ES/MS Resolution and Identification of Molecular Species of TG in Oil A Using 20–80% Isopropanol in Methanol in 30 min

Peak no.[a]	Molecular species	CN/DBN	Partition number	Retention time (min)	Oil A (area%)	m/z [M+18]+
1	8:0/22:6/22:6	52:12	28	12.14	0.66	856
2	10:0/12:0/22:6	44:6	32	12.87	3.41	756
3	10:0/22:6/22:6	54:12	30	13.93	2.65	884
4	10:0/14:0/22:6	46:6	34	14.82	8.96	784
5	12:0/22:6/22:6	56:12	32	15.88	6.23	912
6	12:0/14:0/22:6	48:6	36	16.76	5.29	812
7	22:6/22:6/22:6	66:18	30	17.18	2.03	1040
8	14:0/22:6/22:6	58:12	34	17.90	3.16	940
9	14:0/22:6/22:6	58:12	34	18.22	2.97	940
10	12:0/18:1/22:6	52:7	38	18.97	4.86	866
11	14:0/14:0/22:6	50:6	42	19.32	4.95	840
12	16:0/22:6/22:6	60:12	36	20.08	2.12	968
13	12:0/14:0/16:0	42:0	42	20.48	3.50	740
14	16:0/16:1/22:6	54:7	40	21.21		894
14	14:0/18:1/22:6	54:7	40	22.13	4.84	894
15	14:0/16:0/22:6	52:6	40	21.56	2.86	868
16	16:1/18:1/22:6	56:8	40	22.05	0.77	920
16	14:0/14:0/18:1	46:1	44	22.56	7.75	794
17	18:1/18:3/18:1	54:5	44	23.18	2.18	898
18	16:0/16:0/22:6	54:6	42	23.72	5.36	896
19	14:0/16:0/18:1	48:1	46	24.85	6.69	822
20	18:1/18:1/18:1	54:3	48	26.53	4.23	902
20	16:0/18:1/18:1	52:2	48	26.82	3.30	876
20	16:0/16:0/18:1	50:1	48	27.19	2.99	850
21	18:0/18:1/18:1	54:2	50	29.01	4.84	904
22	14:0/18:1/24:1	56:2	52	31.00	0.63	932
23	16:0/18:1/24:1	58:2	54	32.92	1.32	960
Other					6.26	

[a]Figure 9A.

used APCI/MS to obtain both molecular and fragment ions of TG, and Laakso and Kallio (22) recently optimized the mass spectrometric analysis of natural TG using NICI with ammonia. The advantages and disadvantages of the ES/CID/MS approach using a single quadrupole instrument have been discussed along with applications to TG (18) and glycerophospholipids (23).

Tables 3 and 4 compare the absolute retention times of the different TG species on the reversed-phase column using the two different solvent systems. As noted above, the different species showed significant differences in the retention times, resulting in several reversals of elution. Table 4 also shows the composition of TG species determined by LC/ESI/MS and that obtained by calculation by the 1-random 2-random 3-random method from the stereospecific fatty acid distribution. Except for a few coincidences, the agreement between the two estimates is not very

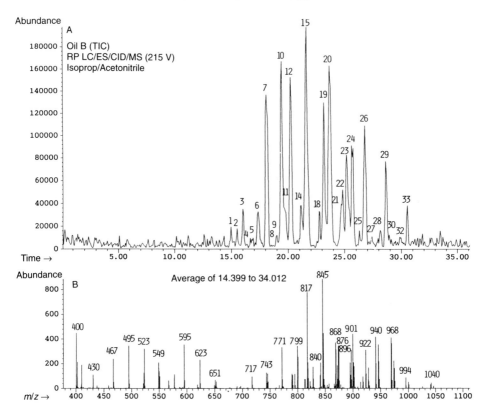

FIG. 10. Positive ion current profile of Oil B TG as obtained by reversed-phase LC/ES/MS at exit voltage of 215 V. (A) total ion current profile; peak identification as given in Table 4; (B) full mass spectrum averaged over total ion profile; DG type ions: m/z 495 (14:0/14:0), 523 (14:0/16:0), 549 (16:0/16:1), 595 (14:0/22:6), 623 (16:0/22:6), 651 (18:0/22:6); [TG+18]$^+$ ions: m/z 868 (14:0/16:0/22:6), 876 (16:0/18:1/18:1), 896 (16:0/16:0/22:6), 922 (16:0/18:1/22:6), 940 (14:0/22:6/22:6), 968 (16:0/22:6/22:6), 994 (18:1/22:6/22:6) and 1040 (22:6/22:6/22:6); [TG+23]$^+$ ions: m/z 743 (12:0/14:0/16:1), 771 (12:0/14:0/18:1), 799 (14:0/14:0/18:1), 817 (12:0/14:0/22:6), 845 (14:0/14:0/22:6), 901 (16:0/16:0/22:6). Reversed-phase column and online LC/ES/MS conditions are as given in the text, using Solvent System B.

good, although both methods appear to agree on the major components. The LC/ES/MS estimates did not provide any indication of the enantiomer composition. Furthermore, no attempt was made to arrive at a differentiation between the fatty acid composition of the primary and secondary ester groups in the glycerol molecule, as recently reported by Kusaka et al. (24). Silver-ion HPLC with online APCI/MS has recently been reported for the separation of isomers of isobaric TG (25). The discrepancies among the total LC/ES/MS estimates of the major molecu-

TABLE 4

Reversed-Phase LC/ES/MS Resolution and Identification of Molecular Species of TG in Oil B Using 20–80% Isopropanol in Acetonitrile in 30 min

Peak no.[a]	Molecular species	CN/DBN	PN	Retention time (min)	Oil B (area%)	Oil B (1R/2R/3R)	m/z [M+18]+
1	22:6/22:6/22:6	66:18	30	14.98	1.00	6.45	1040
2	10:0/12:0/22:6	44:6	32	15.52	0.77	0.14	756
3	12:0/22:6/22:6	56:12	32	16.04	1.67	4.00	912
6	10:0/14:0/22:6	46:6	34	17.38	2.28	1.04	784
7	14:0/22:6/22:6	58:12	34	18.13	8.51	12.48	940
9	12:0/16:1/22:6	50:7	36	19.01	0.54	0.53	838
10	12:0/14:0/22:6	48:6	36	19.45	10.82	4.12	812
11	18:1/22:6/22:6	62:13	36	19.75	1.75	3.39	994
12	16:0/22:6/22:6	60:12	36	20.25	8.19	11.28	968
14	12:0/18:1/22:6	52:7	38	21.17	2.05	2.67	866
15	14:0/14:0/22:6	50:6	38	21.63	12.91	8.89	840
18	12:0/14:0/16:1	42:1	40	22.79	1.42	0.49	738
19	14:0/18:1/22:6	54:7	40	23.19	6.54	5.87	894
20	14:0/16:0/22:6	52:6	40	23.72	11.94	9.68	868
22	12:0/14:0/18:1	44:1	42	24.81	3.54	1.64	766
23	16:0/18:1/22:6	56:7	42	25.22	5.31	4.51	922
24	16:0/16:0/22:6	54:6	42	25.71	5.59	4.18	896
25	12:0/18:1/18:1	48:2	44	26.34	0.56	0.44	820
26	14:0/14:0/18:1	46:1	44	26.80	6.41	3.13	794
27	14:0/14:0/16:0	44:0	44	27.43	0.75	0.99	768
28	14:0/18:1/18:1	50:2	46	28.18	1.03	0.48	848
29	14:0/16:0/18:1	48:1	46	28.68	4.14	3.16	822
31	18:1/18:1/18:1	54:3	48	29.54	0.24		902
32	16:0/18:1/18:1	52:2	48	29.94	0.63	0.20	876
33	16:0/16:0/18:1	50:1	48	30.54	1.41	1.32	850
Other							

[a]Figure 10A.

lar species were considerably smaller than the differences between the determined and calculated values and must have represented either true differences in the composition or, more likely, the absence of calibration of the mass spectrometric response of the widely differing lipid species.

Conclusions

This study demonstrates that online ES/MS provides an important improvement for the stereospecific analysis of complex natural triacylglycerols. In addition to peak identification, and quantitation, it helps to police the cross-contamination of *sn*-1,2(2,3)- and 1,3-diacylglycerols during borate TLC. Due to lack of appropriate fragmentation, it was not possible, however, to utilize chiral-phase LC/ES/CID/MS for the identification of the molecular species of the diacylglycerols beyond the carbon/double bond number.

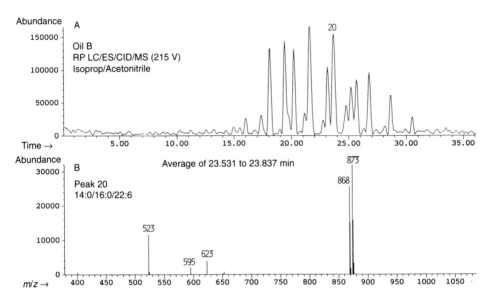

FIG. 11. Reversed-phase LC/ES/CID/MS of Oil B TG. (A) total positive ion current profile of Oil B TG; (B) ES/CID/MS of Peak 20 (14:0/16:0/22:6): m/z 868, [M+18]$^+$; m/z 873, [M+23]$^+$; m/z 523, [M-RCOO]$^+$ (30:0 DG); m/z 595, [M-RCOO]$^+$ (36:6 DG); m/z 623, [M-RCOO]$^+$ (38:6 DG). Instrumentation and operating conditions are as given in the text. Exit voltage, 215 V.

The present study also shows that reversed-phase HPLC with online ES/CID/MS provides an effective method for the identification of the molecular species of the resolved triacylglycerols. The possibility of a differential cleavage of fatty acids from the primary and secondary positions of the triacylglycerol molecule was not fully explored, because of the complexity of the composition of the peaks resolved by reversed-phase HPLC. It is possible that a prefractionation of the triacylglycerol mixture by reversed-phase and silver-ion HPLC would produce simple enough triacylglycerols for such a positional analysis by mass spectrometry.

Acknowledgment

The studies by the authors and their collaborators were supported by the Heart and Stroke Foundation of Ontario, Toronto, Ontario, and the Medical Research Council of Canada, Ottawa, Ontario, Canada. We thank the Martek Biosciences Corporation for supplying samples of triacylglycerols rich in arachidonic and docosahexaenoic acids, including Oil A used in the present experiments.

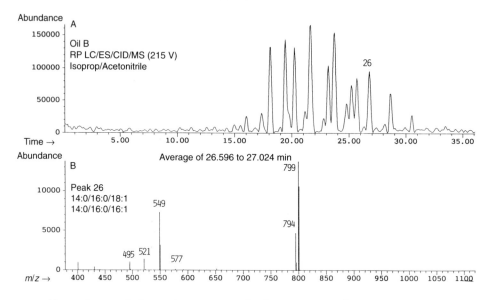

FIG. 12. Reversed phase LC/ES/CID/MS of Oil B TG. (A) Total positive ion current profile of Oil B TG; (B) ES/CID/MS of peak 26 (14:0/14:0/18:1; 14:0/16:0/16:1): m/z 794, [M+18]$^+$; m/z 799, [M+23]$^+$; m/z 495, [M-RCOD]$^+$ (28:0 DG); m/z 521, [M-RCOO]$^+$ (30:1 DG); m/z 549, [M-RCOO]$^+$ (32:1 DG); m/z 577, [M-RCOO]$^+$ (34:1 DG). Instrumentation and operating conditions are as given in the text.

References

1. Boswell, K., Koskelo, E.-K., Carl, L., Glaza, S., Hensen, D.J., Williams, K.D., and Kyle, D.J. (1996) The Preclinical Evaluation of Single Cell Oils Which Are Highly Enriched in Arachidonic Acid and Docosahexaenoic Acid, *Food Chem. Toxicol. 34*, 585–593.
2. Yang, L.Y., and Kuksis, A. (1987) Size and Composition of Lymph Chylomicrons Following Feeding Corn Oil or Its Fatty Acid Methyl Esters, *Biochem. Cell Biol. 65*, 514–524.
3. Yang, Y.L., Kuksis, A., and Myher, J.J. (1990) Intestinal Absorption of Menhaden and Rapeseed Oil and Their Fatty Acid Ethyl Esters in the Rat, *Biochem. Cell Biol. 68*, 480–491.
4. Christensen, M.S., Mullertz, A., and Hoy, C.-E. (1995) Absorption of Triglycerides with Defined or Random Structure by Rats with Biliary and Pancreatic Diversion, *Lipids 30*, 521–526.
5. Paultauf, F. (1983) Ether Lipids as Substrates for Lipolytic Enzymes, in *Ether Lipids, Biochemical and Biomedical Aspects* (Mangold, H.K., and Paltauf, F., eds.), Academic Press, New York, pp. 211–229.
6. Lehner, R., and Kuksis, A. (1996) Biosynthesis of Triacylglycerols, *Progr. Lipid Res. 35*, 169–201.

7. Bottino, R., Vandenburg, G.A., and Reiser, R. (1967) Resistance of Certain Long-Chain Polyunsaturated Fatty Acids of Marine Oils to Pancreatic Lipase Hydrolysis, *Lipids 2*, 489–493.
8. Yang, L.Y., Kuksis, A., and Myher, J.J. (1990) Lipolysis of Menhaden Oil Triacylglycerols and the Corresponding Fatty Acid Alkyl Esters by Pancreatic Lipase *In Vitro*: A Reexamination, *J. Lipid Res. 31*, 192–204.
9. Myher, J.J., Kuksis, A., Geher, K., Park, P.W., and Diersen-Schade, D.A. (1996) Stereospecific Analysis of Triacylglycerols Rich in Long-Chain Polyunsaturated Fatty Acids, *Lipids 31*, 207–215.
10. Yang, L.-Y., Kuksis, A., and Myher, J.J. (1995) Biosynthesis of Chylomicron Triacylglycerols by Rats Fed Glyceryl or Alkyl Esters of Menhaden Oil Fatty Acids, *J. Lipid Res. 36*, 1046–1057.
11. Lehner, R., Kuksis, A., and Itabashi, Y. (1993) Stereospecificity of Monoacylglycerol and Diacylglycerol Acyltransferases from Rat Intestine as Determined by Chiral-Phase High-Performance Liquid Chromatography, *Lipids 28*, 29–34.
12. Myher, J.J., and Kuksis, A. (1982) Resolution of Natural Diacylglycerols by Gas–Liquid Chromatography on Polar Capillary Columns, *Can. J. Biochem. 60*, 638–650.
13. Thomas, A.E., Scharoun, J.E., and Ralston, H. (1965) Quantitative Estimation of Isomeric Monoglycerides by Thin–Layer Chromatography, *J. Am. Oil Chem. Soc. 42*, 789–792.
14. Myher, J.J., Kuksis, A., and Yang, L.-Y. (1990) Stereospecific Analysis of Menhaden Oil Triacylglycerols and Resolution of Complex Polyunsaturated Diacylglycerols by Gas–Liquid Chromatography on Polar Capillary Columns, *Biochem. Cell Biol. 68*, 336–344.
15. Marai, L., Kuksis, A., and Myher, J.J. (1994) Reversed-Phase Liquid Chromatography–Mass Spectrometry of the Uncommon Triacylglycerol Structures Generated by Randomization of Butteroil, *J. Chromatogr. A, 672*, 87–99.
16. Myher, J.J., Kuksis, A., Ravandi, A., and Cocks, N. (1994) Identification of Polyoxyglycerophospholipids by Normal Phase LC/MS with Electrospray, *INFORM 5*, 478–479.
17. Becker, C.C., Rosenquist, A., and Holmer, G. (1993) Regiospecific Analysis of Triacylglycerols Using Allyl Magnesium Bromide, *Lipids 28*, 147–149.
18. Kuksis, A., and Myher, J.J. (1995) Application of Tandem Mass Spectrometry for the Analysis of Long-Chain Carboxylic Acids, *J. Chromatogr. B, 671*, 35–70.
19. Duffin, K.L., Henion, J.D., and Shieh, J.J. (1991) Electrospray and Tandem Mass Spectrometric Characterization of Acylglycerol Mixtures That Are Dissolved in Nonpolar Solvents, *Anal. Chem. 63*, 1781–1788.
20. Byrdwell, W.C., and Emken, E.A. (1995) Analysis of Triglycerides Using Atmospheric Pressure Chemical Ionization–Mass Spectrometry, *Lipids 30*, 173–175.
21. Byrdwell, W.C., Emken, E.A., Neff, W.E., and Adlof, R.O. (1996) Quantitative Analysis of Triglycerides Using Atmospheric Pressure Chemical Ionization–Mass Spectrometry, *Lipids 31*, 919–935.
22. Laakso, P., and Kallio, H. (1996) Optimization of the Mass Spectrometric Analysis of Triacylglycerols Using Negative Ion Chemical Ionization with Ammonia, *Lipids 31*, 33–42.

23. Kim. H.Y., Wang, T.-C., and Ma, Y.-C. (1994) Liquid Chromatography/Mass Spectrometry of Phospholipids Using Electrospray Ionization, *Anal. Chem. 66*, 3977–3982.
24. Kusaka, T., Ishihara, S., Sakaida, M., Mifune, A., Nakanao, Y., Tsuda, K., Ikeda, M., and Nakanao, H. (1996) Composition Analysis of Normal Plant Triacylglycerols and Hydroperoxidized *rac*-1-Stearoyl-2-Oleoyl-3-Linoleoyl-*sn*-Glycerols by Liquid Chromatography Atmospheric Pressure Chemical Ionization Mass Spectrometry, *J. Chromatogr. A730*, 1–7.
25. Laakso, P., and Voutilainen, P. (1996) Analysis of Triacylglycerols by Silver Ion High–Performance Atmospheric Pressure Chemical Ionization Mass Spectrometry, *Lipids 31*, 1311–1322.

Chapter 6

NMR Characterization of Fatty Compounds Obtained via Selenium Dioxide–Based Oxidations

G. Knothe

National Center for Agricultural Utilization Research, Agricultural Research Service, U.S. Department of Agriculture, 1815 N. University St., Peoria, IL 61604

Introduction

Several years ago, the microbial synthesis of a novel hydroxy fatty acid, 7,10-dihydroxy-8(*E*)-octadecenoic acid, from oleic acid or olive oil by the *Pseudomonas* strains PR3 and 42A2 was reported independently by two research groups (1,2). This microbial hydroxylation is unique, because it involves a shift of the double bond and a double allylic hydroxylation to give 2-ene-1,4-diols.

Selenium dioxide (SeO_2) is the classical allylic hydroxylation reagent in organic chemistry (3). Allylic hydroxylation with SeO_2 always yields *trans* double bond configuration, regardless of the double bond configuration of the starting material (3). The mechanism has been elucidated (4). In 1977, *tert*-butyl hydroperoxide (TBHP) was introduced as a reoxidant for toxic colloidal selenium, which was formed during the reaction and had been difficult to remove during workup (5). SeO_2 and SeO_2/TBHP usually have been used to introduce allylic hydroxy groups at one specific site in a molecule. In fatty acid chemistry, SeO_2 has been used only sparingly. Methyl oleate and oleic acid have been reacted with SeO_2 to give mixtures of various hydroxy compounds (6–8). While the combination of SeO_2 with TBHP affords allylically hydroxylated products, the combination of SeO_2 with H_2O_2 gives vicinally dihydroxylated products (9).

The aforementioned microbial formation of an allylically dihydroxylated fatty acid provided an incentive to seek chemical means for producing fatty 2-ene-1,4-diols. Thus, reaction of SeO_2/TBHP with various fatty acids, methyl esters, alcohols, and symmetrical alkenes afforded allylic mono- and dihydroxy products (10–13). When the double bond is close ($\Delta 5$, $\Delta 6$) to C1 (the carbon carrying the original acid, ester, or OH group) in methyl esters and acids, only one monohydroxy product is obtained, because lactonization occurs when the hydroxy group is between C1 and the double bond (10–12). Some products have been hydrogenated to give the corresponding saturated compounds. For other reports on the synthesis or occurrence of allylic hydroxy compounds, see Ref. 11. With SeO_2/H_2O_2, a novel intermediate in the formation of vicinal dihydroxy compounds, consisting of a cyclic selenite moiety bound to a fatty compound, is obtained (14). This hydroxylation thus probably proceeds via an epoxide and the selenite to the diol (14).

The NMR characterization of the novel fatty compounds obtained by SeO_2-based hydroxylations is particularly interesting. Especially, ^{13}C-NMR of fatty compounds has received considerable attention (15). Positional isomers of allylic monohydroxy compounds and diastereomers of allylic dihydroxy can be distinguished by both ^{1}H- and ^{13}C-NMR (10–12). The NMR spectra of mono- and dihydroxy compounds derived from symmetrical alkenes provide base values for the evaluation (16). The chemical shifts and their differences of olefinic carbons and protons can be evaluated mathematically in terms of rational or logarithmic functions (17–18). The selenite-containing fatty compounds were characterized by ^{1}H-, ^{13}C-, and ^{77}Se-NMR (14). The NMR features of the various compounds are discussed in detail in the present chapter.

Discussion

The allylic mono- and dihydroxy compounds were obtained by reacting monoenoic fatty acids, esters, alcohols (10–12), and symmetrical alkenes (13) with SeO_2 and TBHP in CH_2Cl_2. Subsequent purification was carried out by high-performance liquid chromatography (HPLC). For experimental details and product yields see Refs. 10–13. Selenites were obtained by reacting fatty compounds with SeO_2 and 50% H_2O_2 in CH_2Cl_2 at ambient temperature (14). Purification again was carried out by HPLC. Experimental details and yields are given in Ref. 14. All NMR spectra discussed here were obtained with $CDCl_3$ as solvent unless noted otherwise (cosolvent CD_3OD).

Definitions

The structures of the compounds discussed here are depicted in Figs. 1 and 2. For allylic monohydroxy compounds, the following definition is used (11): The monohydroxy compound in which the hydroxy group is located between the double bond and the functional group at C1 is termed the position I compound (i.e., 2(*E*)-ene-1-ol). The monohydroxy compound in which the hydroxy group is located on the side of the terminal methyl group is termed the position II compound (i.e., 1(*E*)-ene-3-ol).

In the ^{13}C-NMR spectra of monounsaturated, unsubstituted fatty compounds, the two signals of the unsaturated carbons degenerate into one signal at a certain distance from the double bond (usually around C11). This position, where only one signal occurs for the two unsaturated carbons, will be called the position of shift equivalence (POSE) (17).

The following discussion is organized by the types of compounds.

Allylic monohydroxy compounds. ^{1}H-NMR (10–13). The olefinic protons of the allylic monohydroxy compounds show one doublet of doublets and one doublet of triplets in the region of 5.4–5.7 ppm; the former set of signals is attributed to the olefinic carbon adjacent to the hydroxy-bearing carbon.

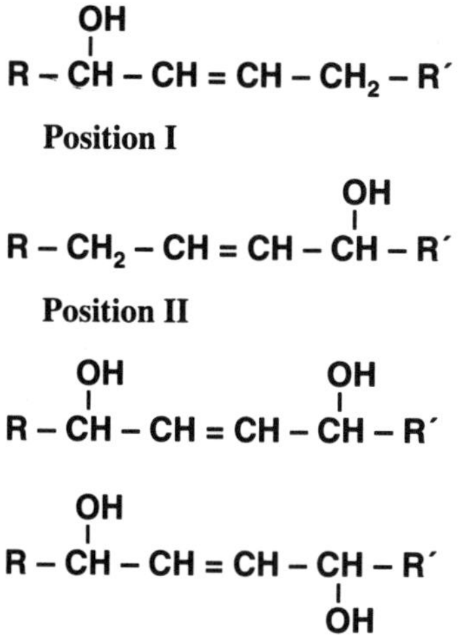

OH
|
R – CH – CH = CH – CH$_2$ – R'

Position I

OH
|
R – CH$_2$ – CH = CH – CH – R'

Position II

OH   OH
| |
R – CH – CH = CH – CH – R'

OH
|
R – CH – CH = CH – CH – R'
                      |
                      OH

R = (CH$_2$)$_x$ CO$_2$H, (CH$_2$)$_x$ CO$_2$Me, (CH$_2$)$_x$ CH$_2$OH
R' = (CH$_2$)$_y$ CH$_3$

Fig. 1. Structure of compounds obtained by allylic hydroxylation of fatty compounds with SeO$_2$/TBHP.

The *trans* configuration of the allylic monohydroxy compounds was confirmed by the coupling constants of the olefinic protons, which were in the region of $J = 15$ to 15.5 Hz. Shift data are discussed in a subsequent section as a mathematical evaluation in terms of the small shift differences between the two sets of signals.

Allylic monohydroxy compounds. ^{13}C-NMR. ^{13}C-NMR data (11–12) for the olefinic carbons of the allylic monohydroxy compounds are given in Table 1. The data show that the resonance differences for the position II monohydroxy compounds

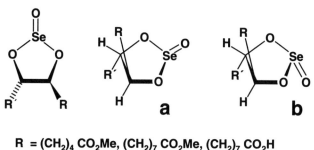

R = (CH$_2$)$_4$ CO$_2$Me, (CH$_2$)$_7$ CO$_2$Me, (CH$_2$)$_7$ CO$_2$H
R' = (CH$_2$)$_{10}$ CH$_3$, (CH$_2$)$_7$CH$_3$

Fig. 2. Compounds obtained by hydroxylation of fatty compounds with SeO$_2$/H$_2$O$_2$.

TABLE 1
^{13}C-NMR Signals of the Olefinic Carbons of trans Monoenoic Hydroxy Compounds Obtained from Fatty Acids, Esters, and Alcohols

Compound[a]	CHOH=CH[b]	CHOH=CH[b]
16,Δ9,CO$_2$H,8-OH	132.73 (9)	132.36 (10)
16,Δ9,CO$_2$H,11-OH	132.96 (10)	132.07 (9)
18,Δ6,CO$_2$H,8-OH	133.42 (7)	131.22 (6)
18,Δ9,CO$_2$H,8-OH	132.82 (9)	132.36 (10)
18,Δ9,CO$_2$H,11-OH	133.05 (10)	132.05 (9)
18,Δ11,CO$_2$H,10-OH	132.88 (11)	132.27 (12)
18,Δ11,CO$_2$H,13-OH	132.92 (12)	132.18 (11)
19,Δ7,CO$_2$H,6-OH	132.61 (7)	132.53 (8)
19,Δ7,CO$_2$H,9-OH	133.22 (8)	131.76 (7)
19,Δ10,CO$_2$H,9-OH	132.78 (10)	132.38 (11)
19,Δ10,CO$_2$H,12-OH	132.86 (11)	132.23 (10)
20,Δ5,CO$_2$H,7-OH	134.14 (6)	130.46 (5)
20,Δ8,CO$_2$H,7-OH	132.66 (8)	132.51 (9)
20,Δ8,CO$_2$H,10-OH	133.07 (9)	132.00 (8)
20,Δ11,CO$_2$H,10-OH	132.75 (11)	132.30 (12)
20,Δ11,CO$_2$H,13-OH	132.92 (12)	132.18 (11)
22,Δ13,CO$_2$H,12-OH	132.74 (13)	132.23 (14)
22,Δ13,CO$_2$H,15-OH	132.89 (14)	132.24 (13)
16,Δ9,CO$_2$CH$_3$,8-OH	132.95 (9)	132.21 (10)
16,Δ9,CO$_2$CH$_3$,11-OH	133.09 (10)	131.87 (9)
18,Δ6,CO$_2$CH$_3$,8-OH	133.58 (7)	131.21 (6)
18,Δ9,CO$_2$CH$_3$,8-OH	132.72 (9)	132.05 (10)
18,Δ9,CO$_2$CH$_3$,11-OH	133.07 (10)	131.78 (9)
18,Δ11,CO$_2$CH$_3$,10-OH	132.98 (11)	132.09 (12)
18,Δ11,CO$_2$CH$_3$,13-OH	133.01 (12)	132.08 (11)
19,Δ7,CO$_2$CH$_3$,6-OH	132.74 (7)	132.40 (8)
19,Δ7,CO$_2$CH$_3$,9-OH	133.13 (8)	131.53 (7)
19,Δ10,CO$_2$CH$_3$,9-OH	132.94 (10)	132.29 (11)
19,Δ10,CO$_2$CH$_3$,12-OH	133.05 (11)	132.13 (10)
20,Δ5,CO$_2$CH$_3$,7-OH	134.20 (6)	130.41 (5)
20,Δ8,CO$_2$CH$_3$,7-OH	132.84 (7)	132.32 (8)
20,Δ8,CO$_2$CH$_3$,10-OH	133.18 (8)	131.93 (7)
20,Δ11,CO$_2$CH$_3$,10-OH	132.90 (11)	132.05 (12)
20,Δ11,CO$_2$CH$_3$,13-OH	133.01 (12)	131.96 (11)
22,Δ13,CO$_2$CH$_3$,12-OH	132.95 (13)	132.15 (14)
22,Δ13,CO$_2$CH$_3$,15-OH	133.03 (14)	132.11 (13)
18,Δ6,OH,5-OH	132.80 (6)	132.42 (7)
18,Δ6,OH,8-OH	133.30 (7)	131.65 (6)
18,Δ9,OH,8-OH	132.95 (9)	132.21 (10)
18,Δ9,OH,11-OH	133.07 (10)	132.03 (9)
18,Δ11,OH,10-OH	132.97 (11)	132.18 (12)
18,Δ11,OH,13-OH	133.01 (12)	133.12 (11)
20,Δ11,OH,10-OH	132.99 (11)	132.18 (12)
20,Δ11,OH,13-OH	133.05 (12)	132.03 (11)
22,Δ13,OH,12-OH	132.99 (13)	132.23 (14)

[a]The compounds are coded as in the following example: 8-hydroxy-9(E)-octadecenoic acid is 18,Δ9,CO$_2$H,8-OH. First the number of carbon atoms in the chain is given, followed by the location of the double bond (always trans), the functional group at C1, and finally the location of the introduced hydroxy group.
[b]The numbers in parentheses indicate the chain number of the carbon atom assigned to the signal.

are greater than for the corresponding position I compounds. The shift differences of the position II compounds can also be evaluated mathematically, as discussed subsequently. That this effect is not visible in the position I compounds is presumably due to the position I hydroxy group, which blocks the influence of the functional group at C1. 2D heteronuclear correlation assigned the downfield olefinic carbon to the doublet of triplets (11), which is the reverse of a previous evaluation without the aid of 2D correlation (18).

Allylic dihydroxy compounds. ^{1}H-NMR. ^{1}H-NMR data (11–12) for the olefinic protons and the protons attached to the hydroxy-bearing carbons of the allylic dihydroxy compounds are presented in Table 2. The *trans* configuration of the double bonds was confirmed by a simulation, which showed that the separation of two small peaks in the outlying parts of the peak region should be separated by 32 Hz in case of *trans* and 22 Hz or less for *cis*. The present compounds show differences of of 32–33 Hz.

For distinguishing the *erythro/threo* diastereomers of these compounds, it is convenient to select the signals of the two C2 protons (which cause a triplet at about 2.30 ppm in the fatty acids and esters) as standard for comparing differences in the olefinic proton shifts. The olefinic proton shift of the *erythro* diastereomer (absolute configurations *R,S* and *S,R*) is downfield compared to the *threo* (absolute

TABLE 2
Selected ^{1}H-NMR Signals of Allylic Dihydroxy Compounds Obtained from Fatty Acids, Esters, and Alcohols[a]

Compound	CH=CH	H_2-2 (t)
18,Δ9,CO$_2$H,e-8,11-diOH	5.68	2.33
18,Δ9,CO$_2$H, t-8,11-diOH	5.66	2.33
18,Δ11,COH ,e-10,13-diOH[b]	5.58	2.22
18,Δ11,CO$_2$H, t-10,13-diOH	5.59	2.29
19,Δ7,CO$_2$H, e-6,9-diOH	5.51	2.18
19,Δ7,CO$_2$H, t-6,9-diOH	5.59	2.29
19,Δ10,CO$_2$H, e-9,12-diOH	5.57	2.21
19,Δ10,CO$_2$H, t-9,12-diOH	5.53	2.23
20,Δ8,CO$_2$H, e-7,10-diOH[b]	5.48	2.14
20,Δ8,CO$_2$H, t-7,10-diOH	5.52	2.23
18,Δ9,CO$_2$CH$_3$,e-8,11-diOH	5.68	2.29
18,Δ9,CO$_2$CH$_3$, t-8,11-diOH	5.64	2.28
18,Δ11,CO$_2$CH$_3$,e-10,13-diOH	5.68	2.28
18,Δ11,CO$_2$CH$_3$, t-10,13-diOH	5.64	2.28
18,Δ6,OH,e-5,8-diOH[b]	5.61	3.57
18,Δ6,OH,t-5,8-diOH	5.6	3.58
18,Δ9,OH,e-8,11-diOH[b]	5.57	3.51
18,Δ9,OH,t-8,11-diOH	5.64	3.61
18,Δ11,OH,e-10,13-diOH[b]	5.62	3.55
18,Δ11,OH,t-10,13-diOH	5.59	3.58

[a]Compounds coded as in Table 1. Additionally, here *erythro* and *threo* are abbreviated e and t, respectively.
[b]Spectrum obtained with CD$_3$OD as cosolvent.

configurations R,R and S,S) congener. Application of this rule to the ^{1}H-NMR spectra of the microbially produced 7,10-dihydroxy-8(E)-octadecenoic acid (2) shows that this compound possesses *threo* configuration, likely R,R absolute configuration (19). The *erythro/threo* assignments are confirmed by the melting points (*threo* diastereomers have lower melting points than the *erythro* isomers) and by surface tension studies (20).

The main signals of the olefinic protons appear as a doublet of doublets consisting of an outer, stronger signal pair and an inner, weaker signal pair. These signals pose a second possibility for distinguishing the diastereomers (11). The separation of the outer pair is 5–6 Hz, and that of the inner pair is 1.7–2.1 Hz. Comparison of the two diastereomers shows that the separation of these peaks is greater for the *threo* isomers than for the *erythro* counterparts.

A third possibility for distinguishing the diastereomers is the multiplet caused by the protons on the hydroxy-bearing carbons. Better resolution on a 400-MHz spectrometer for the 2(E)-1,4-diols obtained from symmetrical alkenes (13) showed that the overall width of this multiplet for *erythro* diastereomers (*meso* in the case of those compounds derived from symmetrical alkenes) was 17.61–17.87 Hz, whereas for *threo* it was 18.65–19.28 Hz.

Previously, the *erythro/threo* diastereomers of long-chain compounds had been distinguished by ^{1}H-NMR when vicinal hydroxy groups were present (21–23).

Allylic dihydroxy compounds. ^{13}*C-NMR.* Table 3 contains the ^{13}C-NMR data (11–12) for the olefinic carbons and the hydroxy-bearing carbons.

The ^{13}C-NMR signals of the olefinic and hydroxy-bearing carbons can be used to distinguish the *erythro/threo* diastereomers. These signals appear downfield regularly in the *threo* diastereomers when compared to the *erythro*. The ^{13}C signals of the olefinic carbons originally were not assigned to one of the two carbons (11). A comparison utilizing the spectra of *threo* diastereomers derived from symmetrical alkenes (13) and *threo*-6,9-dihydroxy-7(E)-nonadecenoic acid (where the shift differences are greatest due to the proximity of the functional group at C1), showed that the downfield olefinic carbon is on the side of the terminal methyl group (corresponding to position II in allylic monohydroxy compounds) (16). The difference between the signals decreased with increasing chain length.

Saturated mono- and dihydroxy compounds. ^{13}C-NMR data are shown for saturated monohydroxy products (11,12) in Table 4 and for saturated dihydroxy compounds in Table 5.

The saturated monohydroxy compounds do not display any features in their ^{13}C-NMR spectra noteworthy in the present discussion. The saturated dihydroxy compounds (1,4-diols), however, display effects analogous to the unsaturated allylic dihydroxy compounds: downfield shift of the olefinic and hydroxy-bearing carbons, in the threo diastereomers.

TABLE 3
^{13}C-NMR Signals of Allylic Dihydroxy Compounds Obtained from Fatty Acids, Esters, and Alcohols

Compound	Olefinic		OH-bearing	
16,Δ9,CO$_2$H,e-8,11-diOH	133.45,	133.27	72.05,	71.96
16,Δ9,CO$_2$H,t-8,11-diOH	133.85,	133.72	72.39,	72.28
18,Δ9,CO$_2$H,e-8,11-diOH	133.72,	133.52	72.34,	72.26
18,Δ9,CO$_2$H,t-8,11-diOH	133.67,	133.49	72.31,	72.23
18,Δ11,CO$_2$H,e-10,15-diOH[b]	133.07,	133.02	71.70,	71.67
18,Δ11,CO$_2$H,t-10,15-diOH	133.85,	133.79	72.54,	72.49
19,Δ7,CO$_2$H, e-6,9-diOH	133.28,	132.80	71.68,	71.35
19,Δ7,CO$_2$H, t-6,9-diOH	133.87,	133.48	72.31,	71.91
19,Δ10,CO$_2$H, e-9,12-diOH	133.25,	133.14	71.84,	71.81
19,Δ10,CO$_2$H, t-9,12-diOH	133.84,	133.76	72.34,	72.29
20,Δ8,CO$_2$H,e-7,10-diOH	133.14,	132.88	71.66,	71.49
20,Δ8,CO$_2$H,t-7,10-diOH	133.87,	133.62	72.30,	72.11
20,Δ11,CO$_2$H,e-10,13-diOH[b]	133.37,	133.29	72.00,	71.97
20,Δ11,CO$_2$H,t-10,13-diOH[b]	133.71,	133.66	72.22,	72.18
22,Δ13,CO$_2$H,e-12,15-diOH[b]	133.18[c]		71.86[c]	
22,Δ13,CO$_2$H,t-12,15-diOH[b]	133.73[c]		72.25[c]	
16,Δ9,CO$_2$CH$_3$,e-8,11-diOH	133.58,	133.38	72.14,	72.09
16,Δ9,CO$_2$CH$_3$,t-8,11-diOH	133.86,	133.67	72.42,	72.38
18,Δ9,CO$_2$CH$_3$,e-8,11-diOH	133.70,	133.46	72.26[c]	
18,Δ9,CO$_2$CH$_3$,t-8,11-diOH	133.87,	133.67	72.41,	72.37
18,Δ11,CO$_2$CH$_3$,e-10,13-diOH	133.60,	133.52	72.24[c]	
18,Δ11,CO$_2$CH$_3$,t-10,13-diOH	133.81,	133.75	72.43[c]	
20,Δ11,CO$_2$CH$_3$,e-10,13-diOH	133.61,	133.52	72.24[c]	
20,Δ11,CO$_2$CH$_3$,t-10,13-diOH	133.84,	133.77	72.43[c]	
22,Δ13,CO$_2$CH$_3$,e-12,15-diOH	133.55[c]		72.24[c]	
22,Δ13,CO$_2$CH$_3$,t-12,15-diOH	133.82[c]		72.44[c]	
18,Δ6,OH,e-5,8-diOH	133.54,	133.18	71.96,	71.78
18,Δ6,OH,t-5,8-diOH	133.89,	133.61	72.36,	72.05
18,Δ9,OH,e-8,11-diOH	133.89,	133.61	71.78,	71.75
18,Δ9,OH,t-8,11-diOH	133.85,	133.74	72.43[c]	
18,Δ11,OH,e-10,13-diOH[b]	133.34,	133.28	71.91[c]	
18,Δ11,OH,t-10,13-diOH	133.95,	133.92	72.45[c]	
20,Δ11,OH,e-10,13-diOH	133.18,	133.13	71.77[c]	
20,Δ11,OH,t-12,15-diOH	133.82,	133.80	72.30[c]	
22,Δ13,OH,e-12,15-diOH	133.17[c]		71.80[c]	
22,Δ13,OH,t-12,15,diOH	133.87[c]		72.32[c]	

[a]For compound coding, see Tables 1 and 2.
[b]Spectrum obtained with CD$_3$OD as cosolvent.
[c]One signal.

Enones. Allylic keto compounds (enones) are side products in allylic hydroxylations with SeO$_2$ and were detected in the hydroxylation of fatty compounds (11). ^{13}C-NMR data of some selected enones are given in Table 6.

The *trans* configuration of the enones was confirmed by ^{1}H-NMR through coupling constants slightly below 16 ppm (11).

TABLE 4
13C-NMR Signals of Saturated Monohydroxy Fatty Acids and Saturated Diols[a]

Compound	OH-bearing	Neighboring	
16,CO$_2$H,8-OH	72.02 (8)	37.48 (9),	37.32 (7)
18,CO$_2$H,8-OH	72.00 (8)	37.52 (9),	37.35 (7)
18,CO$_2$H,10-OH	71.79 (10)	37.22 (11),	37.17 (9)
18,CO$_2$H,11-OH	72.10 (11)	37.44 (12),	37.40 (10)
18,CO$_2$H,13-OH	72.13 (13)	37.41 (12,14)	
20,CO$_2$H,10-OH	72.08 (10)	37.48 (11),	37.41 (9)
20,CO$_2$H,13-OH	72.12 (13)	37.46 (12,14)	
22,CO$_2$H,12-OH	72.12 (12)	37.46 (11,13)	
22,CO$_2$H,15-OH[b]	71.79 (15)	37.12 (14,16)	
18;1,5-diOH	71.90 (5)	37.56 (6),	37.02 (4)
18;1,8-diOH[b]	71.81 (8)	37.34 (9),	37.23 (7)
18;1,10-diOH	72.01 (10)	37.49 (11),	37.45 (9)
18;1,11-diOH	72.01 (11	37.50 (10,12)	
18;1,13-diOH	72.02 (13)	37.47 (14),	37.45 (12)
20;1,10-diOH	71.87 (10)	37.33 (11),	37.29 (9)
22;1,13-diOH	72.02 (13)	37.48 (12,14)	

[a]For compound coding, see Table 1. Compounds with OH at C1 are given as diols. For numbers in parentheses, see Footnote b, Table 1.

In the 13C-NMR spectra of enones, position I compounds had greater shift differences than position II compounds. This is the reverse of the corresponding hydroxy compounds. The carbons α to the carbonyl group are shifted slightly downfield for the position II compounds. This coincides with literature values for 9-oxo-10(*E*)-octadecenoic acid and 10-oxo-8(*E*)-octadecenoic acid (24).

Hydroxy compounds derived from symmetrical alkenes. ^{1}*H-NMR.* The features in the ^{1}H-NMR of the dihydroxy compounds derived from symmetrical alkenes (13) are similar to those of the compounds derived from the fatty acids, esters, and alcohols discussed above. Table 7 presents ^{1}H-NMR data for the 2(*E*)-1,4-diols from symmetrical alkenes. The downfield shift in the *meso* (*erythro*) form may again be noted (signals of the terminal methyl group selected as reference signals).

Hydroxy compounds derived from symmetrical alkenes. ^{13}*C-NMR.* ^{13}C-NMR data for the olefinic carbons and the hydroxy-bearing carbons (13) are given in Table 8 for the monohydroxy compounds and Table 9 for the dihydroxy compounds.

The differences in the shifts of the olefinic carbons of the monohydroxy products are 0.42–0.92 ppm. These differences are nearly identical to those found in C1-functionalized compounds in which the sole hydroxy group is on the side of the C1 functional group (position I). This is a further indication of the blocking effect of the position I hydroxy group on functional groups at C1.

TABLE 5
^{13}C-NMR Signals of Saturated Dihydroxy Acids and Esters and Saturated Triols[a]

Compound	OH-bearing C	C adjacent to C–OH
16,CO$_2$H,e-8,11-diOH	71.62 (11), 71.46 (8)	37.12 (12), 36.96 (7), 32.82/31.81 (9,10)
16,CO$_2$H,t-8,11-diOH	72.21 (11), 72.04 (8)	37.53 (12), 37.39 (7), 33.91 (9,10)[b]
18,CO$_2$H,e-8,11-diOH	71.41 (11), 71.29 (8)	36.98 (12), 36.83 (7), 32.68 (9,10)[b]b
18,CO$_2$H,t-8,11-diOH	71.90 (11), 71.77 (8)	37.38 (12), 37.33 (7), 33.65 (9,10)[b]
18,CO$_2$H,e-10,13-diOH	71.70 (13), 71.60 (10)	37.18 (9,14), 32.90 (11,12)[b]
18,CO$_2$H,t-10,13-diOH	71.88 (13), 71.84 (10)	37.32 (9,14), 33.93 (11,12)[b]
20,CO$_2$H,e-10,13-diOH	71.63 (13), 71.54 (10)	37.18 (14), 37.11 (9), 32.83 (11,12)[b]
20,CO$_2$H,t-10,13-diOH	72.24 (13), 72.11 (10)	37.65 (14), 37.57 (9), 33.96/33.81 (11,12)
22,CO$_2$H,e-12,15-diOH	71.48 (12,15)[b]	37.03 (11,16), 32.70 (13,14)[b]
22,CO$_2$H,t-12,15-diOH	72.11 (12,15)[b]	37.56 (16), 37.51 (11), 33.93/33.88(13,14)
16,CO$_2$CH$_3$,e-8,11-diOH	71.96 (11), 71.85 (8)	37.51 (12), 37.40 (7), 33.32/33.26 (9,10)
16,CO$_2$CH$_3$,t-8,11-diOH	72.32 (11), 72.20 (8)	37.75 (12), 37.65 (7), 33.99/33.94 (9,10)
18,CO$_2$CH$_3$,e-8,11-diOH	71.87 (11), 71.74 (8)	37.43 (12), 37.28 (7), 33.13/33.07 (9,10)
18,CO$_2$CH$_3$,t-8,11-diOH	72.28 (11), 72.16 (8)	37.78 (12), 37.64 (7), 34.09/34.04 (9,10)
18,CO$_2$CH$_3$,e-10,13-diOH	71.95 (13), 71.91 (10)	37.53 (9,14), 33.33 (11,12)[b]
18,CO$_2$CH$_3$,t-10,13-diOH	72.28 (13), 72.24 (10)	37.77 (9,14)[b], 34.03 (11,12)[b]
20,CO$_2$CH$_3$,e-10,13-diOH	71.86 (13), 71.81 (10)	37.46 (14), 37.42 (9), 33.15 (11,12)[b]
20,CO$_2$2CH$_3$,t-10,13-diOH	72.29 (13), 72.24 (10)	37.78 (14), 37.73 (9), 34.05 (11,12)[b]
22,CO$_2$CH$_3$,e-12,15-diOH	71.95 (12,15)[b]	37.57 (11,16), 33.33 (13,14)[b]
22,CO$_2$CH$_3$,t-12,15-diOH	72.28 (12,15)[b]	37.80 (11,16), 34.03 (13,14)[b]
18,OH,e-5,8-diOH	71.52 (8), 71.22 (5)	37.12 (9), 36.51 (4), 32.84/32.78 (6,7)
18,OH,t-5,8-diOH	72.22 (8), 71.89 (5)	37.70 (9), 37.05 (4), 34.16/34.02 (6,7)
18,OH,e-8,11-diOH	71.45 (11), 71.38 (8)	37.03 (12), 36.93 (7), 32.70 (9,10)[b]
18,OH,t-8,11-diOH	72.18 (11), 72.09 (8)	37.66 (12), 37.55 (7), 33.99/33.95 (9,10)
18,OH,e-10,13-diOH	71.61 (11), 71.57 (8)	37.16 (9,14), 32.85 (11,12)[b]
18,OH,t-10,13-diOH	72.31 (11), 72.28 (8)	37.74 (9,14), 34.00 (11,12)[b]
20,OH,e-10,13-diOH	71.36 (10,13)[b]	36.92 (14), 36.88 (9), 32.58 (11,12)[b]
20,OH,t-10,13-diOH	71.99 (13), 71.96 (10)	37.46 (14), 37.43 (9), 33.78 (11,12)[b]
22,OH,e-12,15-diOH	71.74 (12,15)[b]	37.35 (16,11)[b], 33.02 (13,14)[b]
22,OH,t-12,15-diOH	72.12 (12,15)[b]	37.58 (16,11)[b]), 33.91 (13,14)[b]

[a]For compound coding, see Tables 1 and 2. For numbers in parentheses, see Footnote b in Table 1. Solvents: CDCl$_3$-CD$_3$OD cosolvents for acids and triols, CDCl$_3$ for esters.
[b]One signal.

TABLE 6
^{13}C-NMR Signals of Enones Obtained as Side Products in Allylic Hydroxylations[a]

Compound	Olefinic carbons	C adjacent to C=O
18,Δ9,CO$_2$H,8-oxo	147.52 (10), 130.22 (9)	39.89 (7)
18,Δ9,CO$_2$H,11-oxo	147.16 (9), 130.32 (10)	40.14 (12)
18,Δ9,OH,8-oxo	147.40 (10), 130.23 (9)	39.99 (7)
18,Δ9,OH,11-oxo	147.23 (9), 130.27 (10)	40.09 (12)

[a]For compound coding, see Tables 1 and 2. For numbers in parentheses, see Footnote b in Table 1.

TABLE 7
Selected ^{1}H-NMR Signals of Unsaturated 1,4-Diols Obtained from Symmetrical Alkenes

Compound	CH=CH	CHOH	CH$_3$
m-4(E)-Octene-3,6-diol[a]	5.68	4.04	0.93
t-4(E)-Octene-3,6-diol	5.6	3.97	0.91
m-5(E)-Decene-4,7-diol	5.66	4.09	0.9
t-5(E)-Decene-4,7-diol	5.55	4.01	0.88
m-6(E)-Dodecene-5,8-diol	5.67	4.09	0.88
t-6(E)-Dodecene-5,8-diol	5.6	4.05	0.87
m-7(E)-Tetradecene-6,9-diol	5.68	4.1	0.87
t-7(E)-Tetradecene-6,9-diol	5.57	4.02	0.87
m-8(E)-Hexadecene-7,10-diol	5.68	4.1	0.86
t-8(E)-Hexadecene-7,10-diol	5.64	4.08	0.87
m-9(E)-Octadecene-8,11-diol	5.68	4.11	0.86
t-9(E)-Octadecene-8,11-diol	5.63	4.07	0.86

[a]m = meso, t = threo.

The unsaturated 1,4-diols show similar effects in ^{13}C-NMR as those derived from fatty acids, esters, and alcohols. The olefinic and hydroxy-bearing carbons are shifted downfield in the *threo* isomers (absolute configuration when derived from symmetrical alkenes R,R and S,S) compared to *meso* (absolute configuration R,S). Similar observations hold for the corresponding saturated 1,4-diols (see Tables 10 and 11). The methylene group α to the hydroxy-bearing carbon is shifted slightly upfield in the *threo* diastereomer for the unsaturated diols and downfield in the saturated diols (13).

Selenites. The NMR characterization of fatty compounds containing a selenite moiety (14) in the chain is listed in Table 12.

The ^{1}H-NMR of the selenites showed two broadened doublets or triplets of triplets, at 4.3–4.4 ppm and 3.8–3.9 ppm. The existence of two isomeric selenites was evidenced by two very closely spaced singlets of the CO$_2$Me group (shift difference 0.004 ppm, integration ratio approximately 3:1) and four ^{13}C-NMR signals of the two functionalized carbons, two closely spaced at approximately 88.0–88.5 and two closely spaced at 85.0–85.5 ppm. 2D heteronuclear correlation assigned the downfield proton to the upfield carbon signals and vice versa. The ^{13}C-NMR signals of cyclic selenite esters range from 70.6 ppm for unsubstituted 1,3,2-dioxaselenolan-2-oxide to 90.8 ppm for 4,4,5,5-tetramethyl-1,3,2-dioxaselenolan-2-oxide (25).

The ^{77}Se-NMR spectra of the selenites derived from oleic acid and methyl petroselinate showed signals at 1338.5 and 1338.9 ppm (37 Hz separation; integration ratio 3:1, the upfield signal being stronger), corresponding to the existence of two isomers. These signals are in the region of selenites, which is 1284–1430 ppm (26).

The isomers of the selenite esters are depicted in Fig. 2. The selenite is likely planar (27). Analogous to sulfites (28), in ^{1}H-NMR the proton *cis* to the Se=O bond

TABLE 8
^{13}C-NMR Signals of Unsaturated 1,4-Diols Obtained from Symmetrical Alkenes[a]

Compound	CH=CH	CHOH	CH$_2$	CH$_3$
m-4(E)-Octene-3,6-diol[b]	133.4	73.49	30.03	9.64
t-4(E)-Octene-3,6-diol	133.88	73.88	29.9	9.7
m-5(E)-Decene-4,7-diol	133.44	71.87	39.32 (3,8)	13.93
			18.58 (2,9)	
t-5(E)-Decene-4,7-diol	134.09	72.29	39.14 (3,8)	13.95
			18.64 (2,9)	
m-6(E)-Dodecene-5,8-diol	133.52	72.21	36.92 (4,9)	14.01
			27.54 (3,10)	
			22.58 (2,11)	
t-6(E)-Dodecene-5,8-diol	134	72.53	36.78 (4,9)	14.02
			27.59 (3,10)	
			22.55 (2,11)	
m-7(E)-Tetradecene-6,9-diol	133.54	72.26	37.21 (5,10)	14.01
			31.73 (3,12)	
			25.05 (4,11)	
			22.59 (2,13)	
t-7(E)-Tetradecene-6,9-diol	133.98	72.54	37.06 (5,10)	13.98
			31.68 (3,12)	
			25.08 (4,11)	
			22.57 (2,13)	
m-8(E)-Hexadecene-7,10-diol	133.57	72.24	37.27 (6,11)	14.06
			31.76 (3,14)	
			29.19 (4,13)	
			25.32 (5,12)	
			22.57 (2,15)	
t-8(E)-Hexadecene-7,10-diol	133.84	72.5	37.22 (6,11)	14.05
			31.77 (3,14)	
			29.16 (4,13)	
			25.36 (5,12)	
			22.57 (2,15)	
m-9(E)-Octadecene-8,11-diol	133.54	72.26	37.25 (7,12)	14.06
			31.78 (3,16)	
			29.49	
			29.23	
			25.37 (6,13)	
			22.62 (2,17)	
t-9(E)-Octadecene-8,11-diol	133.82	72.52	37.20 (7,12)	14.06
			31.78 (3,16)	
			29.46	
			29.24	
			25.42 (6,13)	
			22.62 (2,17)	

[a]For numbers in parentheses, see Footnote b in Table 1.
[b]m = meso, t = threo.

TABLE 9
^{13}C-NMR signals of Monohydroxy Products Obtained from Symmetrical Alkenes[a]

Compound	CH=CH	CHOH	CH$_2$	CH$_3$
4(E)-Octen-3-ol	132.62 (4)	74.59	34.22 (2)	13.57 (8)
	132.20 (5)		29.98 (6)	9.69 (1)
			22.28 (7)	
5(E)-Decen-4-ol	132.99 (5)	72.92	39.42 (3)	13.95
	132.07 (6)		31.83	13.87
			31.82	
			22.15 (9)	
			18.65 (2)	
6(E)-Dodecen-5-ol	132.91 (6)	73.23	36.95 (4)	14.01[b]
	132.22 (7)		32.12 (8)	
			31.33 (10)	
			28.83 (9)	
			27.63 (3)	
			22.59	
			22.47	
7(E)-Tetradecen-6-ol	132.91 (7)	73.19	37.18 (5)	14.01
	132.12 (8)		32.14 (9)	13.95
			31.71	
			31.65	
			29.10	
			28.77	
			25.11 (4)	
			22.56 (2,13)[b]	
8(E)-Hexadecen-7-ol	132.92 (8)	73.29	37.29 (6)	14.07[b]
	132.28 (9)		32.17 (10)	
			31.82 (3,14)[b]	
			29.21	
			29.17	
			29.14	
			29.09	
			25.43 (5)	
			22.64	
			22.59	
9(E)-Octadecen-8-ol	132.88 (9)	73.33	37.26 (7)	14.09[b]
	132.34 (10)		32.18 (11)	
			31.87	
			31.80	
			29.50	
			29.44	
			29.27[b]	
			29.16[b]	
			25.47 (6)	
			22.65 (2,17)[b]	

[a]For numbers in parentheses, see Footnote b in Table 1.
[b]Signal of two carbons.

TABLE 10
Selected ^{1}H-NMR Signals of Saturated 1,4-Diols.

Compound	CHOH	CH_3
m-Decane-4,7-diol[a]	3.64	0.92
t-Decane-4,7-diol	3.63	0.89
m-Dodecane-5,8-diol	3.6	0.89
t-Dodecane-5,8-diol	3.63	0.88
m-Tetradecane-6,9-diol	3.6	0.88
t-Tetradecane-6,9-diol	3.6	0.88
m-Hexadecane-7,10-diol	3.63	0.87
t-Hexadecane-7,10-diol	3.61	0.87
m-Octadecane-8,11-diol	3.62	0.86
t-Octadecane-8,11-diol	3.6	0.87

[a] m = meso, t = threo.

likely absorbs at lower field than the *trans* proton. Lanthanide-induced shifts (using Eu(fod)$_3$) confirmed this assignment. The major isomer is assigned the *trans* configuration from the lanthanide shift experiment (14).

Evaluation of ^{13}C-NMR spectra

The signal positions and shift differences of unsaturated carbons in ^{13}C-NMR spectra of fatty compounds are described mathematically by rational (17, 29–30) or logarithmic functions (29–30; O.W. Howarth, personal communication). Rational functions possess the general formula:

$$R_x = z_x \pm (a_x/u^b) \tag{1}$$

in which u represents the position of the unsaturation (for example, a double or triple bond at C3–C4 receives, analogous to nomenclature, the value $u = 3$), R is the shift position of the signal, z is a base value for calculating the shifts of the individual carbons, b is the power of u, and a is a factor as determined below. As an example, such functions are plotted in Fig. 3 as they were obtained for allylic hydroxy fatty acids and esters.

Shift differences of the unsaturated carbons are defined by

$$s = R_2 - R_1 \tag{2}$$

in which s denotes the separation of the unsaturated carbon signals in ppm, R_1 is the shift of the carbon closer to the functional group at C1 (analogous to position I in allylic hydroxy compounds) and R_2 is the shift of the unsaturated carbon closer to the terminal methyl group (analogous to position II). Functions with the general formula

$$s = (a/u^b) \pm f \tag{3}$$

approximate s.

TABLE 11
^{13}C-NMR Signals of Saturated 1,4-Diols[a]

Compound	CHOH	CH$_2$	CH$_3$
m-Decane-4,7-diol	71.66	39.70 (3,8)	14.09
		33.29 (5,6)	
		18.90 (2,9)	
t-Decane-4,7-diol	72.05	39.93 (3,8)	14.1
		34.01 (5,6)	
		18.90 (2,9)	
m-Dodecane-5,8-diol	71.93	37.21 (4,9)	14.06
		33.25 (6,7)	
		27.90 (3,10)	
		22.72 (2,11)	
t-Dodecane-5,8-diol	72.31	37.47 (4,9)	14.06
		34.00 (6,7)	
		27.93 (3,10)	
		22.72 (2,11)	
m-Tetradecane-6,9-diol	71.94	37.47 (5,10)	14.03
		33.23 (7,8)	
		31.86 (3,12)	
		25.40 (4,11)	
		22.63 (2,13)	
t-Tetradecane-6,9-diol	72.33	37.75 (5,10)	14.04
		34.01 (7,8)	
		31.87 (3,12)	
		25.40 (4,11)	
		22.63 (2,13)	
m-Hexadecane-7,10-diol	71.96	37.54 (6,11)	14.08
		33.28 (8,9)	
		31.81 (3,14)	
		29.33 (4,13)	
		25.68 (5,12)	
		22.61 (2,15)	
t-Hexadecane-7,10-diol	72.32	37.78 (6,11)	14.08
		33.98 (8,9)	
		31.81 (3,14)	
		29.33 (4,13)	
		25.68 (5,12)	
		22.61 (2,15)	
m-Octadecane-8,11-diol	71.88	37.47 (7,12)	14.08
		33.15 (9,10)	
		31.81 (3,16)	
		29.63	
		9.27	
		25.74 (6,13)	
		22.64 (2,17)	
t-Octadecane-8,11-diol	72.33	37.78 (7,12)	14.09
		33.98 (9,10)	
		31.81 (3,16)	
		29.63	
		29.28	
		25.72 (6,13)	
		22.64 (2,17)	

[a]For numbers in parentheses, see Footnote *b* in Table 1.

TABLE 12
NMR Characterization of Selenites Obtained from the Reaction of Fatty Compounds with SeO_2 and H_2O_2[a]

Compound	Selenite carbons (ppm)	Selenite protons (ppm)	Selenite (ppm)
C18,CO_2H;9,10-SeO_3	88.48, 88.40, 85.48, 85.41	4.42, 3.91	1338.9, 1338.5
C18,CO_2CH_3;6,7-SeO_3	88.42, 88.35, 85.40, 85.35	4.33, 3.83	1338.9, 1338.5
C18,CO_2CH_3;9,10-SeO_3	88.40, 88.02, 85.39, 85.09	4.39, 3.89	

[a]For compound coding, see Tables 1 and 2. Selenite moiety is referred to as SeO_3.

Equations of types (1) and (3) are valid only for unsaturations between C1 and POSE ($s = 0$).

For allylic hydroxy acids, the following equations were obtained (see also Fig. 3):

$$R_1 = 132.40 - (240/u^3) \tag{4}$$

$$R_2 = 132.80 + (160/u^3) \tag{5}$$

$$s = (400/u^3) + 0.40 \tag{6}$$

and for methyl esters, the equations are

$$R_1 = 132.30 - (240/u^3) \tag{7}$$

$$R_2 = 132.90 + (160/u^3) \tag{8}$$

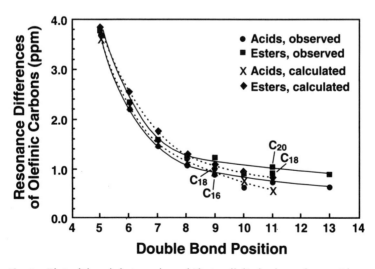

Fig. 3. Plot of the olefinic carbon shifts in allylic hydroxy fatty acids and esters.

$$s = (400/u^3) + 0.60 \tag{9}$$

Similar equations are obtained for unsubstituted octadecenoic acids (16) and also hold for unsaturated triacylglycerols (30). However, in the case of octadecynoic acids, s, R_1, and R_2 are proportional to u^{-2} instead of u^{-3} (17).

In ^{1}H-NMR similar equations are obtained for s of the olefinic protons (29). For allylic hydroxy fatty acids the equation is

$$s = (5/u^3) + 0.185 \tag{10}$$

and for methyl esters the equation is

$$s = (3/u^3) + 0.185 \tag{11}$$

The rational functions described here coincide with results within the framework of the electric field model for explaining ^{13}C-NMR shifts of fatty compounds (31–32). Recently, a model challenging the electric field model has been proposed that argues for through-bond σ-inductive effects (33–34). With this model, logarithmic functions, in which s is proportional to a factor raised to the uth power, are obtained (33, O.W. Howarth, personal communication). Generally, the quality of approximation of the shifts is comparable in both models.

References

1. Parra, J.L., Pastor, J., Comelles, F., Manresa, A., and Bosch, P. (1990) Studies of Biosurfactants Obtained from Olive Oil, *Tenside, Surfactants, Deterg. 27,* 302–306.
2. Hou, C.T., Bagby, M.O., Plattner, R.D., and Koritala, S. (1991) A Novel Compound, 7,10-Dihydroxy-8(E)-octadecenoic Acid from Oleic Acid by Bioconversion, *J. Am. Oil Chem. Soc. 68,* 99–101.
3. Page, P.C.B., and McCarthy, T.J. (1991) Oxidation Adjacent to C=C Bonds, in Comprehensive Organic Synthesis, in *Comprehensive Organic Synthesis,* Trost, B.M., Fleming, I., and Ley, S.V. (Vol. Ed.), Pergamon, Oxford, vol. 7, pp. 83–117.
4. Sharpless, K.B., and Lauer, R.F. (1972) Selenium Dioxide Oxidation of Olefins. Evidence for the Intermediacy of Allylseleninic Acids, *J. Am. Chem. Soc. 94,* 7154–7155.
5. Umbreit, M.A., and Sharpless, K.B. (1977) Allylic Oxidation of Olefins by Catalytic and Stoichiometric Selenium Dioxide with *tert.*-Butyl Hydroperoxide, *J. Am. Chem. Soc. 99,* 5526–5528.
6. Watanabe, Y., Ito, Y., and Matsuura, T. (1957) Oxidation of Unsaturated Fatty Acids. I. Reaction of Methyl Oleate with Selenium Dioxide, *J. Sci. Hiroshima Univ. Ser. A 20,* 203–208.
7. Tubul-Peretz, A., Ucciani, E., and Naudet, M. (1964) Preparation of Hydroxy Allylic Derivatives of Monounsaturated Fatty Acids. II. Oxidation with Selenium Dioxide, *Rev. Franç. Corps Gras 13,* 155–163.
8. Tubul-Peretz, A., Ucciani, E., and Naudet, M. (1966) Structure of Products Formed During the Reaction of SeO$_2$ on Oleic Acid, *Bull. Soc. Chim. France,* 2331–2336.
9. Haines, A.H. (1991) Addition Reactions with Formation of Carbon-Oxygen Bonds: (iii) Glycol Forming Reactions, in Comprehensive Organic Synthesis, in

Comprehensive Organic Synthesis, Trost, B.M., Fleming, I., and Ley, S.V. (Vol. Ed.), Pergamon, Oxford, vol. 7, pp. 437–448.
10. Knothe, G., Weisleder, D., Bagby, M.O., and Peterson, R.E. (1993) Hydroxy Fatty Acids Through Hydroxylation of Oleic Acid with Selenium Dioxide/*tert.*-Butylhydroperoxide, *J. Am. Oil Chem. Soc. 70,* 401–404.
11. Knothe, G., Bagby, M.O., Weisleder, D., and Peterson, R.E. (1994) Allylic Mono- and Di-hydroxylation of Isolated Double Bonds with Selenium Dioxide - *tert.*-Butyl Hydroperoxide. NMR Characterization of Long-Chain Enols, Allylic and Saturated 1,4-Diols, and Enones, *J. Chem. Soc., Perkin Trans. 2,* 1661–1669.
12. Knothe, G., Bagby, M.O., Weisleder, D., and Peterson, R.E. (1995) Allylic Hydroxy Fatty Compounds with $\Delta 5$-, $\Delta 7$-, $\Delta 8$-, and $\Delta 10$-Unsaturation, *J. Am. Oil Chem. Soc. 72,* 703–706.
13. Knothe, G., Bagby, M.O., and Weisleder, D. (1995) Fatty Alcohols Through Hydroxylation of Symmetrical Alkenes with Selenium Dioxide/*tert.*-Butylhydroperoxide, *J. Am. Oil Chem. Soc. 72,* 1021–1026.
14. Knothe, G., Glass, R.S., Schroeder, T.B., Bagby, M.O., and Weisleder, D. (1997) Reaction of Isolated Double Bonds with Selenium Dioxide/Hydrogen Peroxide: Formation of Novel Selenite Esters, *Synthesis,* 57–60.
15. Gunstone, F.D. (1993) High Resolution ^{13}C NMR Spectroscopy of Lipids, in *Advances in Lipid Methodology—Two,* Christie, W.W., The Oily Press, Dundee, pp. 1–68.
16. Knothe, G., and Bagby, M.O. (1996) Assignment of ^{13}C Nuclear Magnetic Resonance Signals in Fatty Compounds with Allylic Hydroxy Groups, *J. Am. Oil Chem. Soc. 73,* 661–663.
17. Knothe, G., and Bagby, M.O. (1995) ^{13}C NMR Spectroscopy of Unsaturated Long-Chain Compounds: an Evaluation of the Unsaturated Carbon Signals as Rational Functions, *J. Chem. Soc., Perkin Trans. 2,* 615–620.
18. Frankel, E.N., Garwood, R.F., Khambay, B.P.S., Moss, G.P., and Weedon, B.C.L. (1984) Stereochemistry of Olefin and Fatty Acid Oxidation. Part 3. The Allylic Hydroperoxides from the Autoxidation of Methyl Oleate, *J. Chem. Soc., Perkin Trans. 1,* 2233–2240.
19. Knothe, G., Bagby, M.O., Peterson, R.E., and Hou, C.T. (1992) 7,10-Dihydroxy-8(E)-Octadecenoic Acid: Stereochemistry and a Novel Derivative, 7,10-Dihydroxyoctadecanoic Acid, *J. Am. Oil Chem. Soc. 69,* 367–371.
20. Knothe, G., Dunn, R.O., and Bagby, M.O. (1995) Surface Tension Studies on Novel Allylic Mono- and Dihydroxy Fatty Compounds. A Method to Distinguish *erythro/threo* Diastereomers, *J. Am. Oil Chem. Soc. 72,* 43–47.
21. Ewing, D.F., and Hopkins, C.Y. (1967) Optical and Geometric Isomers of Some Fatty Acids with Vicinal Hydroxy Groups, *Can. J. Chem. 45,* 1259–1265.
22. Kannan, R., Subbaram, M.R., and Achaya, K.T., (1974) NMR Studies of Some Oxygenated, Halogenated and Sulphur-Containing Fatty Acids and Their Derivatives, *Fette, Seifen, Anstrichm. 76,* 344–350.
23. Kuranova, I.L., and Balykina, L.V. (1978) Determination of *threo-* and *erythro-*Configuration of Vicinal Diastereomeric Derivatives of Higher Fatty Acids Using Proton NMR Spectra, Khim. Prir. Soedin. 299–305; *Chem. Abstr.* (1979) 90:5793d.
24. Porter, N.A., and Wujek, J.S. (1987) Allylic Hydroperoxide Rearrangement: β-Scission or Concerted Pathway?, *J. Org. Chem. 52,* 5085–5089.

25. Denney, D.B., Denney, D.Z., Hammond, P.J., and Hsu, Y.F. (1981) Preparation and NMR Studies of Tetraalkoxyselenuranes and Tetraalkoxytelluranes, *J. Am. Chem. Soc. 103,* 2340–2347.
26. Duddeck, H. (1995) Selenium-77 Nuclear Magnetic Resonance Spectroscopy, *Prog. Nucl. Magn. Reson. 27,* 1–323.
27. Arbuzov, B.A., Naumov, V.A., Zaripov, N.M., and Pronicheva, L.D. (1970) Electron Diffraction Study of the Structure of Ethylene Sulfite and Ethylene Selenite Molecules, *Dokl. Akad. Nauk. 195,* 1333–1336; Chem. Abstr. (1971) 74:92295.
28. Green, C.H., and Hellier, D.G. (1973) Chemistry of the S=O Bond. Part II. Nuclear Magnetic Resonance and Infrared Studies on Ethylene Sulphites (1,3,2-Dioxathiolan-2-oxides), *J. Chem. Soc., Perkin Trans. 2,* 243–252.
29. Knothe, G., Bagby, M.O., and Weisleder, D. (1996) Evaluation of the Olefinic Proton Signals in the ^{1}H-NMR Spectra of Allylic Hydroxy Groups in Long-Chain Compounds, *Chem. Phys. Lipids 82,* 33–37.
30. Knothe, G., Lie Ken Jie, M.S.F., Lam, C.C., and Bagby, M.O. (1995) Evaluation of the ^{13}C-NMR Signals of the Unsaturated Carbons of Triacylglycerols, *Chem. Phys. Lipids 77,* 187–191.
31. Batchelor, J.G., Prestegard, J.H., Cushley, R.J., and Lipsky, S.R. (1973) Electric Field Effects in the ^{13}C Nuclear Magnetic Resonance Spectra of Unsaturated Fatty Acids. A Potential Tool for Conformational Analysis, *J. Am. Chem. Soc. 95,* 6358–6364.
32. Batchelor, J.G., Cushley, R.J., and Prestegard, J.H. (1974) Carbon-13 Fourier Transform Nuclear Magnetic Resonance. VIII. Role of Steric and Electric Field Effects in Fatty Acid Spectra, *J. Org. Chem. 39,* 1698–1705.
33. Bianchi, G., Howarth, O.W., Samuel, C.J., and Vlahov, G. (1995) Long-range σ-Inductive Interactions Through Saturated C-C Bonds in Polymethylene Chains, *J. Chem. Soc., Perkin Trans. 2,* 1427–1432.
34. Howarth, O.W., Samuel, C.J., and Vlahov, G. (1995) The σ-Inductive Effects of C=C and CYC Bonds: Predictability of NMR Shifts at sp^2 Carbon in Non-Conjugated Polyenoic Acids, Esters, And Glycerides, *J. Chem. Soc., Perkin Trans. 2,* 2307–2310.

Chapter 7

Supercritical Fluid Chromatography: A Shortcut in Lipid Analysis[1]

J.W. King and J.M. Snyder

Food Quality and Safety Research Unit, National Center for Agricultural Utilization Research, Agricultural Research Service, USDA, 1815 N. University Street, Peoria, IL 61604

Introduction

Supercritical fluid chromatography (SFC) offers many advantages to the lipid analyst, particularly in the area of applied fat/oil technology, where the speed of analysis is critical. Commercial development of capillary SFC equipment in the early 1980s provided the analytical chemist with a unique tool that permitted chromatography of lipid moieties, up to and exceeding 1000 daltons in molecular weight, that could be detected and quantified using the universal flame ionization detector (FID). By programming the density of the mobile phase, the analyst could change the solvent power of the mobile phase, thereby effecting high-resolution separations, particularly between oligomeric species or members of a homologous series of compounds. SFC can permit very rapid analysis of many compounds and mixtures having the above characteristics and typically employs a mobile phase of supercritical carbon dioxide (SC–CO_2) at temperatures that minimize thermal degradation of temperature-sensitive compounds.

In the 1990s the use of SFC as an analytical technique became preferred for reasons not originally envisioned in the 1980s. With concerns about the reduction of solvent use in the analytical laboratory (1), SFC became very competitive with high-performance liquid chromatography (HPLC) for certain types of lipid analysis, since the technique reduces solvent usage relative to HPLC considerably. In addition, as has been noted by King (2) and Borch–Jensen (3), SFC reduces the need for derivatization of lipid moieties prior to their analysis, as well as extensive sample preparation prior to the determinative chromatographic step. This is particularly true for cases in which SFC can be combined with analytical-scale supercritical fluid extraction (SFE) in the on-line mode (4) to yield simplified or fractionated solventless extracts for chromatographic assay.

SFC has been characterized and compared with other analytical techniques for a wide variety of sample matrices (5,6). The utilization of packed and capillary columns for the analysis of complex lipid samples has been reviewed by Laakso (7). Advances

---
[1]Names are necessary to report factually on available data: however, the USDA neither guarantees nor warrants the standard of the product, and the use of a name by USDA implies no approval of the products to the exclusion of others that may also be suitable.

in column packings and supercritical fluid technology make SFC suitable even for the separation of polar lipid analytes (8). The advantages and flexibility of SFC for the characterization of lipid compounds has also been documented by others (9–11). The recent implementation of SFC-compatible evaporative light scattering detectors (ELSD) should increase the various modes of SFC even more for lipid analysis.

In this chapter, the uniqueness of capillary SFC as an analytical technique will be demonstrated as it has been applied in our research program. The chromatography in these studies was accomplished using primarily 50 to 100 µm i.d., 10 to 15-meter capillary columns. Specific chromatographic conditions—temperature, density or pressure programming characteristics, and column type—are given for each analysis problem as it is discussed in the text that follows. Many of the SFC analyses were accomplished on a SB-Octyl capillary column (15 m × 50 µm i.d.; 0.25 µm film thickness) (Dionex, Inc., Sunnyvale, CA), held isothermally at 120°C, using a density programming from 0.28 g/mL to 0.66 g/mL at 0.006 g/mL/min. This has proven to be a very facile program and set of conditions that is amenable to a wide variety of lipid analysis problems we have encountered. Although this is a relatively long program, for reasons that will be discussed shortly, the analysis time can be shortened by altering the density or temperature program once the retention pattern of the solutes has been established and compounds of interest identified for quantitation.

It should be noted that SFC is a natural technique for our laboratory, which is heavily involved in the exploitation of supercritical fluid technology for purposes of extraction, reaction, and fractionation. Hence, the philosophy of implementing SFC that follows is designed to address these analysis problems. Such research has included the analysis of lipid-containing mixtures extracted or fractionated using SC–CO_2; the deformulation of cosmetic products; detection and quantitation of trace lipid moieties; and, particularly in the last two years, the monitoring of reactions involving lipid reactants and products.

The Features of SFC

Before proceeding with some examples that highlight the convenience and versatility of SFC, it is worth discussing some of the salient features that make SFC somewhat unique among the chromatographic techniques applicable to lipids. Mobile-phase programming is perhaps the most important variable in capillary SFC; choice of stationary phase for the separation of lipid-type solutes plays a secondary role in effecting the desired separation. Figure 1 illustrates a typical SFC density program and the resultant separation of minor lipid constituents (α-tocopherol and cholesterol) from the triglyceride profile constituting fish oil. It should be noted that this SFC separation was performed on a fish oil concentrate contained in a nutraceutical capsule. Sample preparation prior to SFC consisted of squeezing the oil out of the capsule and diluting it with *n*-hexane before injection onto the capillary SFC column.

A SB-Methyl column, 50 µm i.d., 15 m in length (Dionex Corporation, Sunnyvale, CA), using the density program described in the previous section was used to effect the separation of the lipid components. Such density programs are

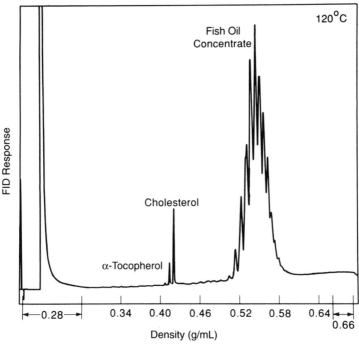

FIG. 1. Capillary SFC of fish oil from a nutraceutical capsule.

usually designed to provide a isoconfertic (constant-density) segment in the beginning of the program to allow the solvent peak to elute from the column. This is then followed by a ramping of the density or pressure as a function of time to a constant density or pressure, where the density or pressure is then held to ensure elution of the last solutes off of the column. The density program described (which is also reproduced on the horizontal axis of the chromatogram in Fig. 1) is fairly long (approximately 90 min), but such time-extensive programs are essential to understanding the molecular complexity of the sample being chromatographed. Once this is established, the programs may be further truncated with respect to time, or temperature programming may be brought into play, to shorten the analysis time or improve the resolution between the chromatographic peaks. One feature of the chromatographic separation illustrated in Fig. 1 is that the higher-molecular-weight components, which were of no interest in this assay (i.e., the fish oil triglycerides), can be density-programmed out of the column after the target analytes, cholesterol and α-tocopherol, have eluted. This eliminates further sample preparation or cleanup on the front end of this assay.

To illustrate the versatility provided by SFC mobile-phase programming, Fig. 2 shows the capillary SFC separation of a lipophilic additive, p,p' = DOPA (i.e., p,p' = dioctyldiphenylamine) from the main constituents in a lubricant designed for use in a space communication satellite. The major component in this formulation is Apiezon

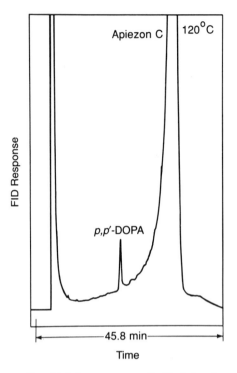

FIG. 2. Capillary SFC of a lubricant formulation for a space communications satellite.

C, which is a vacuum-distilled fraction consisting of a mixture of high-molecular-weight hydrocarbons. This separation was achieved without any prior sample preparation by dissolving the lubricant formulation in hexane, followed by a specially designed inverse asymptotic density program (12,13) to separate the amine additive from the Apiezon C mixture in only 45 min. The density program started at 0.35 g/mL and finished at 0.66 g/mL using an analysis temperature of 120°C. The SB-Methyl column described in the previous example was used in this assay. The described separation could not have been achieved in this time frame without resorting to SFC.

There is no doubt that the ability of combining the FID with SFC offers the analyst a unique universal sensitivity to most lipid compounds, similar to that experienced when using gas chromatography (GC)–FID, but with the additional advantage of having the ability to access higher molecular-weight compounds. In the early development of capillary SFC, it was often felt that quantitation of solutes primarily suffered because of the poor injection reproducibility the technique afforded. However, we have found for most of our samples that we can quantitatively reproduce our capillary SFC analyses without having to resort to more elaborate injection protocols described in the literature (14). We have also observed good linearity for the FID when used in conjunction with capillary SFC. Figure 3 illustrates the resultant calibration curve obtained from injecting millimolar (mM) quantities of the ester, octyl laurate, onto a SB-Octyl column. One can see that the calibration curve is very linear and that peak area can be used to calibrate the detector in terms of mM

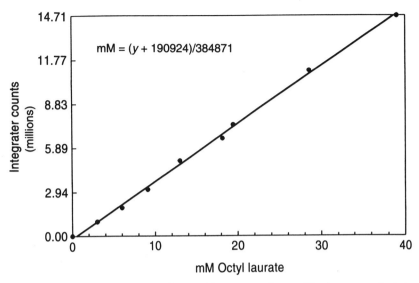

FIG. 3. FID calibration curve for octyl laurate synthesized by lipase catalysis in the presence of SC–CO$_2$.

response to octyl laurate. It is worth noting that the octyl laurate described above is an ester synthesized by lipase catalysis in the presence of flowing SC–CO$_2$, which is used to solvate and transport the acid and alcohol reactants over the catalyst bed (i.e., a supercritical fluid reaction).

Retention Trends in Capillary SFC

Understanding the retention trends of solutes in capillary SFC is important, since such data plays a key role in the identification of lipid-type compounds and optimization of the separation pattern. To a first approximation, solutes undergoing pressure or density programming in capillary SFC elute according to increasing molecular weight; that is, higher molecular-weight components elute later than their lower-molecular-weight homologs (15,16). This principle is adequately illustrated in Fig. 4, where a mixture of linear, lipophilic alcohols are separated according to chain length with increasing density of the mobile phase. The aforementioned SB-Octyl column was used to effect this separation along with the density program previously described. The high resolution of the capillary column also allows isomers of the long-chain alcohols to be ascertained. Molecular weight determination of such homologs is possible by constructing plots of elution density vs. solute molecular weight (16).

The same elution principle applies in the case of coconut oil triglycerides pictured in Fig. 5. Since coconut oil is primarily composed of saturated triglycerides ranging from C_{18} to C_{54}, density-programmed SFC of the oil yields an attractive chromatogram with equally spaced triglyceride peaks. This idealized

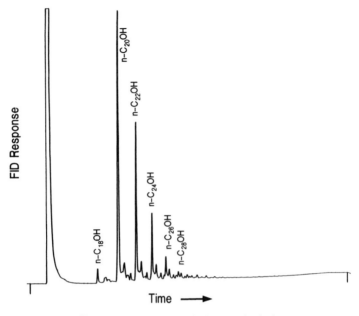

FIG. 4. Capillary SFC separation of a linear alcohol mixture.

separation was accomplished using the SB-Octyl column and the previously described density program from 0.28 to 0.66 g/mL. Such standard profiles for a variety of industrially used oils can be used as a relatively fast method for quality control. It should be noted that since many vegetable oils consist of mixed saturated and unsaturated triglyceride moieties, peak overlap is inevitable, making quantitation of the individual triglycerides difficult. The situation can be improved

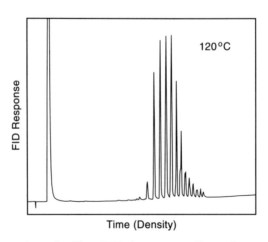

FIG. 5. Capillary SFC of a coconut oil sample.

somewhat by employing inverse temperature programming or a change in the column stationary phase (17).

The elution pattern for mixed lipid species in capillary SFC using nonpolar columns has been characterized by Borch-Jensen (3) and is given in Fig. 6. Once again, the overall retention pattern for a variety of lipid solutes is molecular weight-dependent; however, as can be seen in Fig. 6, there is adequate resolution between the different lipid groups, to permit useful separations and quantitation to be attained. In general, fatty acids elute first from nonpolar capillary SFC stationary phases, because of their lower molecular weights and higher polarity; they are followed by hydrocarbonaceous solutes, such as squalane, squalene, and the fat-soluble vitamin group, and then sterols, such as cholesterol. The higher molecular-weight wax esters occupy an intermediate elution position and can overlap with diglyceride moieties if both are present in a sample matrix. Again, depending on the molecular weights of the individual species and their presence in a sample matrix, one finds that cholesteryl esters, ether-containing lipids, and triglycerides tend to emerge late in the elution profile. It is has been our experience that many of the common cholesteryl esters elute adequately before major triglycerides found in animal- and vegetable-derived oils. Thus, Fig. 6 should be regarded as the worst possible scenario for the capillary SFC of complex lipid mixtures, since most lipid-containing mixtures will contain only some of the individual species within the groups shown in Fig. 6. It should be noted that fatty acid methyl esters (FAMEs), a commonly encountered lipid species, usually elute before the corresponding free fatty acid moieties, based on vapor pressure considerations.

An excellent example of the capillary SFC of a complex lipid mixture is shown in Fig. 7 (18), where the components present in deodorizer distillate have been nicely

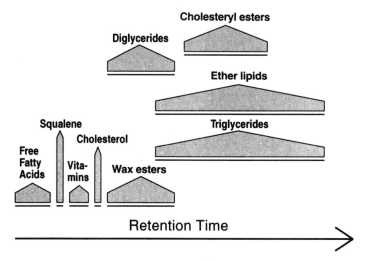

FIG. 6. Retention pattern for lipid classes and solutes for capillary SFC on nonpolar stationary phases (Ref. 3).

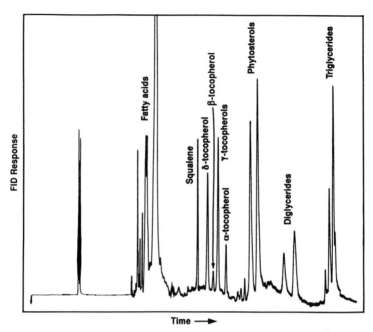

FIG. 7. Separation of lipid components in deodorizer distillate by capillary SFC.

separated using a SB-Octyl-50 column employing both pressure and temperature programming simultaneously. One can see that the retention pattern described by Borch–Jensen holds: fatty acids followed by squalene, the tocopherols (vitamin E precursors), sterols, diglycerides, and finally triglycerides. For the pressure program, the initial pressure was set at 100 atm for 10 min and then was increased at a rate of 5 atm/min to 180 atm. At that point the ramp rate was changed to 2 atm/min until a pressure of 220 atm was reached. Then the pressure increase rate was changed to 5 atm/min until a pressure of 350 atm was attained. The corresponding temperature program as a function of time consisted of holding at 100°C for 10 min, then increasing the temperature to 180°C at a rate of 5°C/min, followed by decreasing the oven temperature from 180 to 100°C at a rate of −5°C/min. This is a excellent example of how the versatility of SFC programming allows a separation pattern to be optimized.

The authors know of no other chromatographic technique that can perform such a separation encompassing so many different lipid species over such a large molecular weight range. Obviously, such methods could be used for the assay of tocopherol antioxidant mixtures as well as vitamins.

SFC in Comparison with Other Techniques

There are few comparisons in the literature where results derived from SFC are compared to those determined by other chromatographic or chemical analysis methods. A particularly thorough study has been conducted by Lee, Bobik, and Malone (19) for the determination of mono- and diglycerides in commercial emulsifier samples, using SFC on both derivatized and underivatized samples and comparing these results with data derived from a GC-derivatization method. The agreement between methods was found to be very satisfactory, and relative standard deviations for the various methods varied between 1.88 and 3.98%. Similarly, we have used capillary SFC for the analysis of the diglyceride content in a randomized lard sample developed for use in a feeding study. These results are presented in Table 1, where they are compared to those derived from GC–FID analysis, three HPLC-based methods, and a TLC assay of the randomized lard sample. It is interesting to note that there is considerable variation between the six listed methods with respect to the sample's diglyceride content; hence SFC is no better or worse then any of the other listed lipid analysis techniques. Results for the major component in the sample, the triglycerides, agree much better from one method to the other. It is worth noting that capillary SFC was competitive with the other listed methods with respect to analysis time; the SFC assay took only 25 min, whereas GC–FID, HPLC–ELSD, and TLC methods required 30 min for completion.

We have also used capillary SFC to determine the mixed tocopherol content in a deodorizer distillate sample and a commercial antioxidant formulation using the SFC method previously described for the deodorizer distillate matrix. Table 2 compares the results that we obtained with those derived from a GC method that required derivatization of the tocopherols to effect chromatography. There are some differences between the results derived by capillary SFC and the GC-derivatization method; in general, SFC yields slightly higher results then those found via the GC method. This could reflect the loss of analyte in the derivatization step of the GC method. Both chromatographic methods yielded similar relative standard deviations, indicating that the precision experienced using both methods is nearly identical. Although there is no inherent reason to expect SFC to be any less precise or accurate

TABLE 1
Analysis of Glyceride Content of a Randomized Lard Sample by Different Chromatographic Methods

Analysis Method	%MG[a]	%DG[a]	%TG[a]	Time
SFC–FID[b]	0.2	9.6	90.1	25 min
GC–FID[c]	0.1	6.9	92.9	30 min
HPLC–FID[d]		14.5	86.5	1 h
HPLC–ELSD		8.0	92.0	30 min
Silica column[e]	1.0	7.7	91.3	2 h
TLC[f]	2.0	11.0	87.0	30 min

[a]MG = monoglycerides; DG = diglycerides; TG = triglycerides.
[b]SFC method used for SFE/SFR.
[c]GC analysis accomplished by a high-temperature column.
[d]HPLC column.
[e]Silica column.

TABLE 2
Comparison of Methods for Determination of Tocopherol in Deodorizer Distillate and Commercial Antioxidant Samples

Type of tocopherol	Deodorizer distillate		Antioxidant	
	SFC[a]	GC[a]	SFC[a]	GC[a]
α-tocopherol	2.37 (6.95)[b]	1.59 (3.62)	7.21 (2.04)	8.98 (1.54)
β-tocopherol	0.58 (7.55)	0.36 (10.89)	0.52 (20.39)	0.24 (21.15)
γ-tocopherol	8.56 (4.20)	7.14 (3.32)	23.47 (2.49)	22.39 (1.46)
δ-tocopherol	6.24 (2.68)	4.89 (5.62)	25.24 (1.79)	24.49 (1.90)

[a]Average of five analyses; values in mg/100 mg.
[b]Numbers in parentheses are relative standard deviations.

then any other analytical technique, it suffices to say that more studies of this type need to be carried out in the future.

Applications of Capillary SFC to Lipid-Containing Samples

Monitoring SFE and SFF (Supercritical Fluid Fractionation) Processes

As noted previously, SFC is natural analysis tool for monitoring the results obtained from SFE and SFF experiments. The rationale behind this statement is that if solutes can be solubilized in supercritical fluids (particularly $SC-CO_2$), then they probably are amenable to chromatography using the same supercritical fluid media. Only in the case of extractions or fractionations carried out at pressures or temperatures beyond the instrumental capabilities of analytical SFC should a degree of caution be exercised. We have primarily used capillary SFC to analyze extracts and fractions after their collection from the process under study (i.e., off-line analysis) as opposed to integrating the SFC for on-line, real-time analysis. Several examples of this type of SFC analyses follow.

We have extracted evening primrose oil from the seeds of *Oenothera biennis* L., since the oil is a source of γ-linolenic acid, a fatty acid entity having reported therapeutic value (20). Whereas it is well known that the fractionation of lipid moieties can be effected by changing the pressure and temperature at which process SFE is conducted, we were also interested in monitoring whether fractionation of the oil's triglycerides was occurring during the course of the SFE conducted at a constant pressure and temperature. Table 3 shows the triglyceride composition of the collected oil according to carbon number from both SFE and conventional Soxhlet extraction as determined by capillary SFC. From this data it is apparent that the SFE conducted at 70 MPa and 50°C yields an equivalent extract in terms of the triglyceride composition as determined by SFC. However, capillary SFC analysis shows that this is not the case for the two extracts obtained at 20 MPa and 40°C. Here a time-based fractionation effect is apparent, since the triglyceride composition has changed between the first and last fractions (particularly for the C_{52} and C_{56} triglycerides). It is interesting to note that GC analysis of fatty acid methyl esters (FAMEs)

TABLE 3
Normalized Triglyceride Composition of Evening Primrose Oil Extracts as Determined by Capillary Supercritical Fluid Chromatography

	% Triglycerides			
Extraction type	C_{50}	C_{52}	C_{54}	C_{56}
Soxhlet (hexane)	1.23	16.95	74.89	6.93
SC–CO_2 at 70 MPa and 50°C (first fraction)	1.27	16.92	75.01	6.80
SC–CO_2 at 20 MPa and 40°C (first fraction)	1.55	18.83	74.80	4.82
SC–CO_2 at 20 MPa and 40°C (last fraction)	0.67	13.04	75.20	11.09

on the above fractions did not show any disparity between the fatty acid compositions of the collected fractions. This indicates that SFC analysis provides somewhat different information on the fractions obtained from the SFE experiment.

Chromatographic profiling of an extract composition by capillary SFC can also provide valuable information as to how the SFE process is affecting the resultant extracts. As shown in Fig. 8, significant changes occur in the capillary SFC profile of a wool grease extract obtained via extraction with SC–CO_2 at 520 bar (52 MPa) and 80°C (21). The individual profiles in Fig. 8 represent in descending order extracts collected under these conditions at 210, 215, 217, and 290 min, respectively. It is apparent that as the extraction of the wool grease proceeds, the cholesterol content of the extract decreases, while the unresolved cluster of peaks, representing the wax ester content of the wool grease, also changes as a function of processing time. These changes manifest themselves in altering the melting point of the individual fractions (the melting point difference between the first and last fractions shown in Fig. 8 is approximately 17°C). These graphic SFC profiles were obtained using the aforementioned SB-Octyl column along with the density program previously described, which goes from 0.28 g/mL to 0.66 g/mL. The observed trends were also confirmed by thin-layer chromatography (TLC).

The monitoring of monoglyceride enrichment via a SFF process by capillary SFC has also been utilized in our laboratory (22). The objective of the SFF process was to produce a top product that could be used as an industrial emulsifier. A thermal-gradient fractionating column incorporating a flowing SC–CO_2 phase was utilized to enrich the monoglyceride content. Fractionation conditions were as follows: pressure of 31 MPa (4500 psi); temperatures in four consecutive zones from the bottom to top of the column were 65, 75, 85, and 95°C, respectively; CO_2 flow rate of 10 L/min (expanded CO_2 flow) with a feed flow rate of 1.2 mL/min. The monoglyceride content was increased about 20 wt. %, yielding a mixture equivalent to a commercially sold product. As shown in Fig. 9, SFC provided a total quantitative profile of the resultant top product composition that can be used to assess the results from the column fractionation process. Again, an SB-Octyl-50 capillary column (10 m × 100 µm i.d.; 0.5 µm film thickness) at 100°C was used, along with a pressure program

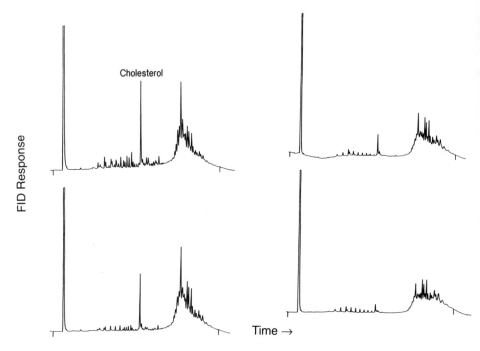

FIG. 8. SFC profiles of collected wool grease fractions obtained from SFE experiment.

from 100 atm to 300 atm at 8 atm/min (10) to effect the separation. The resultant profile is a good example of how the sequence noted by Borch–Jensen (3) can be utilized to characterize this mixed glyceride composition. The SFC profile also detects the presence of free fatty acids (FFA) and indicates that the feed material may contain (or the fractionation process may produce) the hydrolysis product (FFA) as a side product, rather than the desired mixed glyceride compositions.

Analysis of Minor Constituents in Lipid Mixtures

SFC–FID can also be used for the detection and quantitation of trace components that frequently occur in lipid-rich matrices, provided the flame ionization detector has sufficient sensitivity for the target analyte. Analysts who routinely employ capillary SFC occasionally forget that the FID has a large dynamic detection range, and this potential is frequently suppressed when one is exploiting capillary SFC for the analysis of major components in lipid-containing mixtures. A good example of this principle is shown in Figs. 10 and 11, where capillary SFC has been performed on a peanut oil sample using the standard SB-Octyl column analysis described earlier in this chapter.

At the detector signal amplification levels normally used for the FID when capillary SFC is being used for triglyceride analysis (Fig. 10), the trace components in the peanut oil are hardly detectable, relative to the main triglyceride profile.

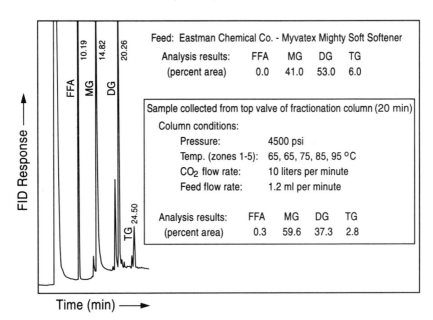

FIG. 9. Capillary SFC separation of glyceride mixture obtained from a SFF column experiment.

However, by increasing the sensitivity of the FID (Fig. 11), trace species become apparent in the chromatographic profile. For the peanut oil sample under consideration, there appears to be a series of peaks, perhaps associated with a homologous or oligomeric series of compounds, between the large solvent peak and the enlarged triglyceride profile. Mass spectrometric identification of these compounds indicated that they were a series of polysiloxane oligomers. This indicated that the peanut oil had been treated with agents to suppress its volatility and foaming as well as to suppress its deterioration via oxidation.

Another example of using capillary SFC for trace component analysis is to assess the cholesterol content of foodstuffs. SFC methods have been used for the analysis of cholesterol and cholesteryl esters in milk fat (23), egg (24), and human serum (24,25) and yielded accurate determinations. Our approach was to use analytical SFE in the off-line mode in a two-step sequence, in which the interfering lipid moieties (i.e., triglycerides) were initially removed from the sample matrix using $SC-CO_2$ an extraction cell packed with various sorbent media that would adsorb the cholesterol selectively. Then, after the removal of most of the interfering fat, we used the same extraction conditions, except for a small quantity of organic solvent added to the $SC-CO_2$ as a cosolvent, to effect removal of the cholesterol from the sorbent surface. To test the effectiveness of this scheme, we monitored the $SC-CO_2$ extracts both before and after the addition of cosolvent to the $SC-CO_2$ (Figs. 12A and B) using capillary SFC. The pressure program and column temperature for the analysis

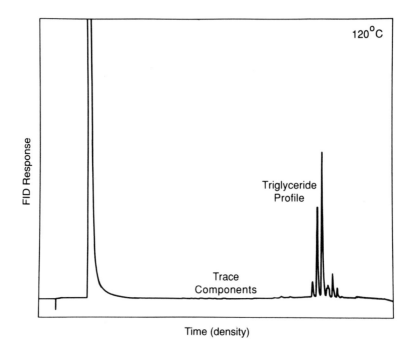

FIG. 10. Capillary SFC of a peanut oil sample using a low amplification of the FID signal.

are noted on Fig. 12. An SB-Octyl-50 column (100 μm i.d., 10 m in length, 0.5-μm film thickness) was used to effect the separation shown in Fig. 12.

Figure 12A shows that the cholesterol was effectively retained on the sorbent surface during SFE of the sample (the sample was the same fish oil from the capsule cited in Fig. 1). The large unresolved band in Fig. 12A is part of the triglyceride content contained in the fish oil sample. Figure 12B shows that addition of a cosolvent (methanol) to the SC–CO_2 released the cholesterol from the matrix, along with some of the excess fat also adsorbed on the sorbent surface. Note that the preliminary sample cleanup step, termed "inverse SFE" by the authors, has reduced the interfering fat band substantially—so much so that the interfering lipid species can be programmed out of the capillary SFC column after elution of the cholesterol. Had the fat level in the sample matrix not been reduced beforehand, it would probably have interfered with the cholesterol peak shown in Fig. 12B.

As shown in Fig. 13, the calibration linearity of the FID for cholesterol in this SFC application was excellent. Using the method just described, we have been able to recover cholesterol from a variety of fat-containing food matrices at recovery levels exceeding 75%. This provides, then, a technique that uses very low levels of organic solvent in the analytical laboratory on both the extraction and chromatography steps and can be used for cholesterol determination as mandated by the Nutritional Labeling and Education Act (NLEA).

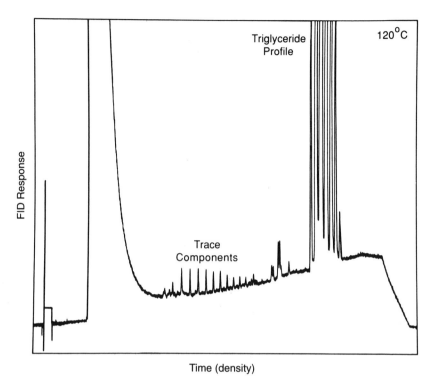

FIG. 11. Capillary SFC of a peanut oil sample using a high amplification of the FID signal.

Deformulation of Commercial Products

We have found in our laboratory that capillary SFC is an excellent technique for deformulating complex commercial products because of the mobile-phase programming capabilities inherent in it and hence its ability to chromatograph a wide molecular weight range of nonpolar to moderately polar compounds. Control of the mobile-phase pressure or density with respect to time is very precise in commercial capillary SFC instrumentation, allowing a repeatability of retention time that is very precise. This feature of capillary SFC can be used in conjunction with analytical standards to identify many of the components that occur in commercial products that contain lipophilic ingredients.

An example of this precision in retention and how it can be utilized is shown in Fig. 14, where a supposed lanolin sample has been chromatographed by SFC. The density program utilized was the standard one developed for use with an SB-Octyl-50 column and mentioned earlier in this chapter. However, upon chromatographing the lanolin sample, it was found that the classic signature compound in lanolin samples, cholesterol, did not have the same retention time as provided by the injection of a cholesterol standard. However, the isolated peak that appears in Fig. 14 did

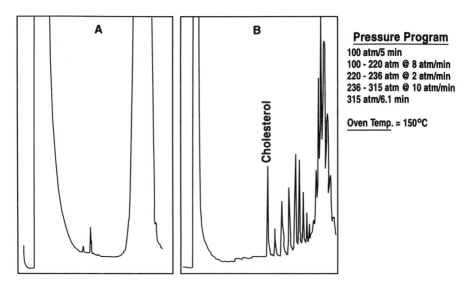

FIG. 12. SFC chromatograms of two fractions collected from an analytical SFE experiment to determine the cholesterol content in a fish oil capsule.

match the retention time of the cholesterol ester, cholesteryl acetate. This indicated that the sample we were chromatographing was not lanolin but acetylated lanolin, a derivative of lanolin utilized in cosmetic preparations because of its superior penetrating properties.

We have also used capillary SFC to deformulate certain types of cosmetic products, such as lipsticks and lip balms, and certain health aid products (4). In this case, approximately a 1% by weight solution of the product is dissolved in *n*-hexane and

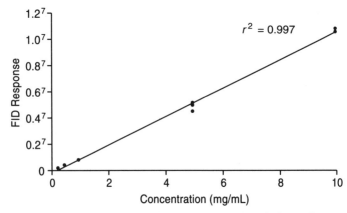

FIG. 13. Calibration curve for FID response to cholesterol using capillary SFC for the analysis.

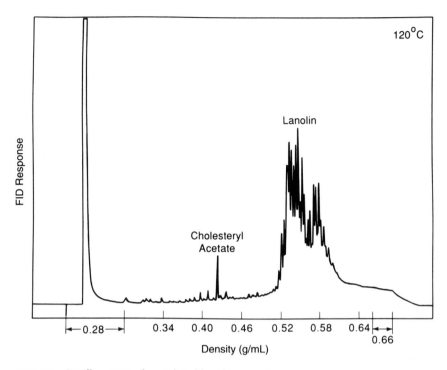

FIG. 14. Capillary SFC of acetylated lanolin sample.

the solution filtered if turbidity persists. An injection of this preparation is then made into the capillary SFC instrument, and then a lengthy density or pressure program is used to separate the dissolved components. Ingredients in the cosmetic formulation are identified by matching their retention times with known compounds or mixtures described in the formulation. In some cases, the ingredients have distinct patterns under SFC programming conditions (e.g., lanolin, petrolatum, or waxes), which aid in their identification. Using this approach, distinctly different chromatographic profiles are produced by different commercial products, thus aiding a formulator in matching a competitor's product.

An example of such a profile generated by the SFC method is shown in Fig. 15 for a decongestant ointment. Note that two volatile medicinal ingredients, camphor and menthol, elute at a mobile-phase density of approximately 0.20 g/mL or lower, according to the density program described on the horizontal axis in Fig. 3. Likewise, one of the ointment base ingredients, petrolatum, elutes between 0.35 and 0.55 g/mL consistently. Similarly, using the same chromatographic conditions as in Fig. 15, one can identify some of the major ingredients in a quencher lipstick formulation, as shown in Fig. 16. We believe that this method has considerable utility in deformulating products, particularly when lipophilic ingredients are present in sample matrix that can be solubilized ultimately in the $SC-CO_2$ mobile phase.

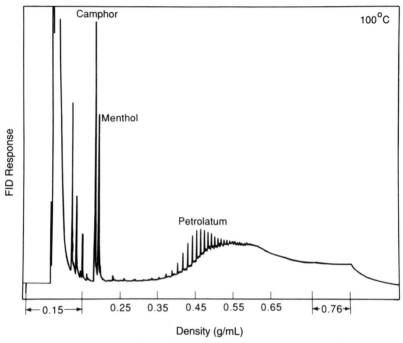

FIG. 15. Capillary SFC of lipophilic ingredients found in a decongestant ointment.

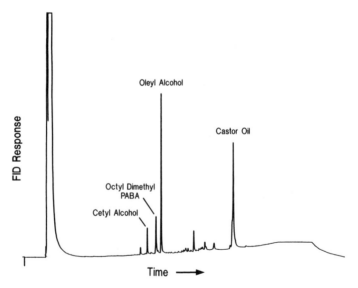

FIG. 16. Capillary SFC of components in quencher lipstick formulation.

Monitoring of Reactions

The relative ease at which SFC can analyze lipid-containing matrices with little or no prior sample preparation has proven invaluable when monitoring reactions of interest in our research program. Others, such as Staby et al. (26), have applied SFC to the analysis of the reaction products from the transesterification of fish oils and noted the superiority of SFC over other means of analysis. In our case, the separation of mixtures of fatty acids (FA), fatty acid methyl esters (FAMEs), monoglycerides (MG), diglycerides (DG), and triglycerides (TG) by SFC has been utilized to determine the conversion of TGs to FAMEs (27). For example, Fig. 17 shows the conversion of soybean oil triglycerides to the corresponding FAMEs, as catalyzed by a supported lipase in the presence of flowing SC–CO_2, which serves as a reaction medium because it contains dissolved soybean oil and methyl alcohol. The SFC chromatogram in Fig. 17 shows that over a 97% conversion of the oil to FAMEs has been achieved. The capillary SFC separation in Fig. 17 was accomplished using an SB-Octyl-50 column (100 µm i.d., 10 m long, 0.5 µm film thickness) programmed from 120 atm (5 min) to 300 atm at a rate of 8 atm/min and a temperature gradient from 100°C (5 min) to 190°C at a rate of 8°C/min. Similarly, capillary SFC has been used to measure completeness of reaction during the development of an SFE/lipase-catalyzed reaction (SFE/SFR) method, used for the determination of fat food matrices (28).

Capillary SFC separation has also been achieved on the reaction products resulting from the enzymatically-catalyzed tranesterification of soybean oil with ethylene glycol to produce a mixture of mono- and diesters, which can be used in the lubri-

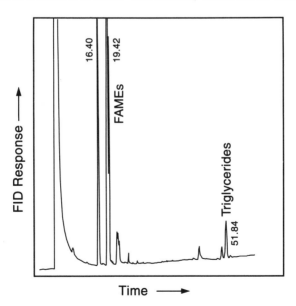

FIG. 17. Separation of FAMEs from soybean oil triglycerides using capillary SFC.

cation and the cosmetic industries. The reaction products, shown in Fig. 18, were produced by using a lipase (Novozyme 435, Novo Nordisk, Danbury, CT) in conjunction with SC–CO_2 to solubilize the starting reactants at 4000 psi (27.6 MPa) and 70°C, eventually yielding a product with 54.9% monoester and 16.7% diester content. Note that capillary SFC analysis also allows separation of the starting material (soybean oil TAG) and intermediate species such as di- and monoglycerides.

Similarly, SFC profiles can be generated on a native beeswax before and after it is enzymatically-transesterified, as shown in Fig. 19. In Fig. 19A, which is the capillary SFC of the native beeswax, the first cluster of peaks represents the odd-carbon number alkanes, C_{25}–C_{33}, whereas the second cluster of peaks comprises the monoesters of palmitic acid with C_{24}–C_{34} diols. The third cluster of peaks represents the diesters formed between the C_{24}–C_{34} diols and palmitic acid, which elute last, based on their corresponding higher molecular weights. Transesterification with methyl acetate randomizes the alcohol chains, producing methyl esters and acetates

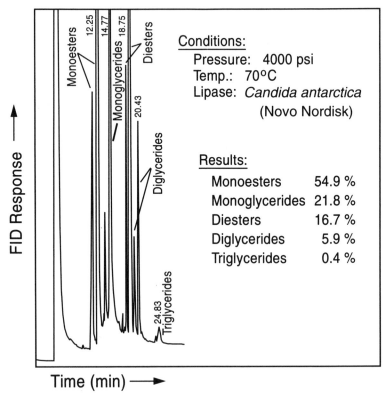

FIG. 18. Capillary SFC analysis of the product mixture obtained from the enzymatically catalyzed transesterification of soybean oil with ethylene glycol.

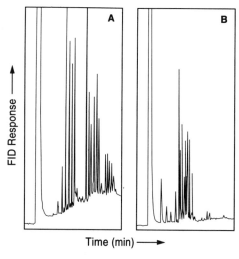

FIG. 19. Beeswax profile by capillary SFC(A) before and (B) after transesterification over an enzyme in flowing SC–CO_2.

of alcohols of lower molecular weight, leading to products having shorter retention times during the SFC run (Fig. 19B). Some of these aforementioned compounds, however, do overlap with the unreactive n-alkane cluster of peaks that were in the starting, native beeswax.

Another reaction sequence that we have studied by capillary SFC is a glycerolysis reaction run in the presence of SC–CO_2 in a stirred reactor (29). The SC–CO_2 in this case is apparently acting as an autocatalyst to initiate the production of mono- and diglycerides in the presence of low levels of water. The conditions for the SFC analysis were the same as the ones used for the aforementioned tranesterification reaction studies. Table 4 shows product compositions for the glycerolysis of five different vegetable oils from capillary SFC analysis. Note that free fatty acids are also detected by the capillary SFC runs, indicating that the competing hydrolysis reaction is also present in the stirred autoclave. Such data shows that glycerolysis conducted in an SC–CO_2 atmosphere can eliminate the use of alkali salts, traditionally used as glycerolysis catalysts, thereby avoiding the need to separate the catalyst from the final product.

TABLE 4
Composition (wt%) of Glycerolysis Products from Different Vegetable Oils[a]

Type of Oil	MG[b]	DG[b]	TG[b]	FFA[b]
Soybean	49.2	26.6	10.1	14.0
Peanut	46.6	32.1	12.5	8.8
Cottonseed	41.1	35.0	12.6	11.3
Corn	45.6	32.3	13.0	9.2
Canola	41.7	33.0	16.0	9.3

[a]Obtained at 250°C, 20.7 MPa, glycerol/oil ratio of 25, and addition of 4% water, 4-h reaction time.
[b]MG = monoglyceride; DG = diglyceride; TG = triglyceride; FFA = free fatty acid.

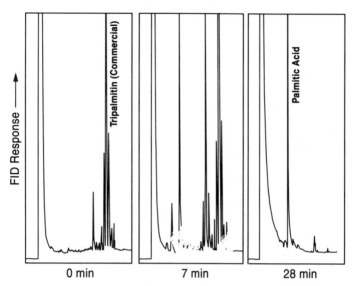

FIG. 20. Hydrolysis of tripalmitin in subcritical water as monitored by capillary SFC.

More recently, we have used capillary SFC to monitor the results from vegetable oil hydrolysis, conducted in subcritical water media to produce mixtures of fatty acids (30). The reaction temperature in this case is usually in the region of 250 to 300°C. As shown by the SFC trace (Fig. 20), a commercial tripalmitin sample can be converted to an intermediate series of diglycerides and palmitic acid after 7 min reaction time. SFC analysis also showed that total conversion to the acid could be achieved in 28 min (30). Although the resultant product mixtures had to be separated from an aqueous emulsion by solvent partition, SFC could be directly applied to the organic solvent layer for the assay of the lipophilic reaction products. The SFC analysis was accomplished on an SB-Phenyl-50 capillary column (50 µm i.d., 10 m long, 0.25 µm film thickness), using a pressure program that consisted of holding the pressure at 100 atm for 5 min, followed by a pressure ramp of 4 atm/min to 240 atm, and finally a ramp to 320 atm at a rate of 10 atm/min. The oven temperature was held isothermally at 100°C.

Conclusions

In summary, we believe we have shown that SFC can be a vital technique for both basic and applied research, as illustrated by the numerous examples cited in this chapter. We have noted in this chapter only results obtained via capillary SFC using $SC-CO_2$ as a mobile phase with detection by FID. However, by using capillary SFC with modifiers (31) or other supercritical fluids, as well as packed column technology along with other supercritical fluid media, the SFC of very polar lipid species can probably be achieved. Many of the separations noted in this chapter were accomplished in less than one hour of analysis time, suggesting that further optimization of the analysis time for specific target analytes may make these methods amenable for on-line process control. It should be noted that SFC is claiming new niche applications, particularly in the chromatography of chiral compounds and operation on a preparative scale.

References

1. Salisbury, C.L., Wilson, H.O., and Priznar, F.J. (1992) Source Reduction in the Lab, *Environmental Testing and Analysis, 1*(2), 48–52.
2. King, J.W. (1997) Capillary Supercritical Fluid Chromatography of Cosmetic Ingredients and Formulations, *J. Microcol. Sep.* 9, in press.
3. Borch-Jensen, C., Magnussen, M.P., and Mollerup, J. (1995) Supercritical Fluid Chromatography of Shark Liver Oils, *INFORM 6*, 465.
4. King, J.W. (1990) Applications of Capillary Supercritical Fluid Chromatography— Supercritical Fluid Extraction to Natural Products, *J. Chromatogr. Sci. 28*, 9–14.
5. King, J.W., Hill, Jr. H.H., and Lee, M.L. (1993) in *Physical Methods of Chemistry Series* (Rossiter, B.W., and Baetzold, R.C., eds.), 2nd edn., Vol. X, John Wiley & Sons, Inc., New York, pp. 1–83.
6. Markides, K.E., Lee, M.L., and Later, S.W. (1989) in *Microbore Column Chromatography, A Unified Approach to Chromatography* (Yang, F.J., ed.), Marcel Dekker, Inc., New York, pp. 239–266.
7. Laakso, P. (1992) in *Advances in Lipid Methodology* (Christie, W.W., ed.), Vol. 1, The Oily Press, Alloway, Scotland, pp. 81–118.
8. Sandra, P., and David, F. (1996) in *Supercritical Fluid Technology in Oil and Lipid Chemistry* (King, J.W., and List, G.R., eds.), AOCS Press, Champaign, Illinois, pp. 321–347.
9. King, J.W. (1993) Analysis of Fats and Oils by SFE and SFC, *INFORM 4*, 1089–1098.
10. Chester, T.L., Pinkston, J.D., and Raynie, D.E. (1996) Supercritical Fluid Chromatography and Extraction, *Anal. Chem. 12*, 487R–514R.
11. Mossoba, M.M., and Firestone, D. (1996) New Methods for Fat Analysis in Foods, *Food Test. & Anal. 2*(2), 24–32.
12. King, J.W. (1988) in *SFC Applications* (Markides, K.E., and Lee, M.L., eds.), Brigham Young University Press, Provo, Utah, p. 334.
13. Fjeldsted, J.C., Jackson, W.P., Peaden, P.A., and Lee, M.L. (1983) Density Programming in Capillary Supercritical Fluid Chromatography, *J. Chromatogr. Sci. 21*, 222–225.
14. Chester, T.L., and Innis, D.P. (1995) Quantitative Open-Tubular Supercritical Fluid Chromatography Using Direct Injection onto a Retention Gap, *Anal. Chem. 67*, 3057–3063.
15. White, C.M., and Houck, R.K. (1985) Analysis of Mono-, Di-, and Triglycerides by Capillary Supercritical Fluid Chromatography, *J. High Resolut. Chromatogr. Chromatogr. Commun. 8*, 293–296.
16. Hayes, D.G., and Kleiman, R. (1996) *Supercritical Fluid Chromatography Analysis of New Crop Seed Oils and Their Reactions 73*, 1691–1697.

17. Knowles, D. (1988) in *SFC Applications* (Markides, K.L., and Lee, M.L., eds.), Brigham Young University Press, Provo, Utah, p. 127.
18. Snyder, J.M., Taylor, S.L., and King, J.W. (1993) Analysis of Tocopherols by Capillary Supercritical Fluid Chromatography and Mass Spectrometry, *J. Am. Oil Chem. Soc. 70*, 349–354.
19. Lee, T.W., Bobik, E., and Malone, W. (1991) Quantitative Determination of Mono- and Diglycerides with and without Derivatization by Capillary Supercritical Fluid Chromatography, *J. Assoc. Off. Anal. Chem. 74*, 533–537.
20. Favati, F., King, J.W., and Mazzanti, M. (1991) Supercritical Carbon Dioxide Extraction of Evening Primrose Oil, *J. Am. Oil Chem. Soc. 68*, 422–427.
21. Cygnarowicz-Provost, M., King, J.W., Marmer, W.N., and Magidman, P. (1994) Extraction of Woolgrease with Supercritical Carbon Dioxide, *J. Am. Oil Chem. Soc. 71*, 223–225.
22. King, J.W., Sahle-Demessie, E., Temelli, F., and Teel, J.A. (1997) Thermal Gradient Fractionation of Glyceride Mixtures under Supercritical Fluid Conditions, *J. Supercrit. Fluid 10*, in press.
23. Huber, W., Molero, A., Pereyra, C., and Martinez de la Ossa, E. (1995) Determination of Cholesterol in Milk Fat by Supercritical Fluid Chromatography, *J. Chromatogr. A 715*, 333–336.
24. Ong, C.P, Ong, H.M, Li, S.F.Y., and Lee, H.K. (1990) The Extraction of Cholesterol from Solid and Liquid Matrices Using Supercritical CO_2, *J. Microcol. Sep. 2*, 69–73.
25. Nomura, A., Yamada, J., Takatsu, A., Horimoto, Y., and Yarita, T. (1993) Supercritical Fluid Chromatographic Determination of Cholesterol and Cholesteryl Esters in Serum on ODS–Silica Gel Column, *Anal. Chem. 65*, 1994–1997.
26. Staby, A., Borch-Jensen, C., Balchen, S., and Mollerup, J. (1994) Supercritical Fluid Chromatographic Analysis of Fish Oils, *J. Am. Oil Chem. Soc. 71*, 355–359.
27. Jackson, M.A., and King, J.W. (1996) Methanolysis of Seed Oils in Flowing Supercritical Carbon Dioxide, *J. Am. Oil Chem. Soc. 73*, 353–356.
28. Snyder, J.M., King, J.W., and Jackson, M.A. (1996) Fat Content for Nutritional Labeling by Supercritical Fluid Extraction and an On-Line Lipase Catalyzed Reaction, *J. Chromatogr. A 750*, 201–207.
29. Temelli, F., King, J.W., and List, G.R. (1996) Conversion of Oils to Monoglycerides by Glycerolysis in Supercritical Carbon Dioxide Media, *J. Am. Oil Chem. Soc. 73*, 699–706.
30. Holiday, R.L., King, J.W., and List, G.R. (1997) Hydrolysis of Vegetable Oils in Sub- and Supercritical Water, *Ind. Eng. Chem. Res. 36*, in press.
31. Raynie, D.P., Fields, S.M., Djordevic, N.M., Markides, K.E., and Lee, M.L. (1989) A Method for the Preparation of Binary Mobile Phase Mixtures for Capillary Supercritical Fluid Chromatography, *J. High Resolut. Chromatogr. 12*, 51–52.

Chapter 8
Analysis of Unusual Triglycerides and Lipids Using Supercritical Fluid Chromatography

D.G. Hayes

Department of Chemical Engineering, The University of Alabama in Huntsville, 307 Material Sciences Building, Huntsville, AL 35899

Introduction

The employment of supercritical fluid chromatography (SFC) in analytical chemistry has grown rapidly in the past decade. Its use has been applied recently to pesticides, polymers, pharmaceutics [including chiral separations (1,2)], agrochemicals, hydrocarbons, and carbohydrates (3–6). Perhaps the largest application area for SFC is lipids (7–13).

Supercritical Fluids

The uniqueness of SFC in comparison to alternative lipid analytical techniques, such as gas and liquid chromatographies, centers upon the properties of supercritical fluids (SCFs). A substance enters the supercritical state when its pressure and temperature are raised above its critical pressure (P_c) and temperature (T_c), respectively. The properties of SCFs (e.g., density, viscosity, diffusion coefficients, and solvent strength/polarity) are intermediate between those of gases and liquids and can be varied within a wide range of values through minor changes in pressure and temperature. Moreover, properties such as solvent strength can be easily "tuned" by changing the SCF density, which in turn is varied by adjusting the system's pressure and temperature. Most SFC applications employ a gradient of density (or equivalently, pressure and/or temperature) to separate analytes in the same fashion that temperature programming is employed for gas chromatography (GC) and gradient elution for reversed-phase high-performance liquid chromatography (HPLC). Most commercially available SFC equipment allows pressure to be varied between 0.1 and 40.5 MPa and temperature between 25 and 250°C, although reports from various research laboratories demonstrate that pressures higher than 40.5 MPa are obtainable [e.g., (14)]. (Most commercially-available SFC open tubular columns degrade at temperatures higher than 250°C.) Equivalently, the carrier fluid density (of CO_2) can be varied between 0.1 and 0.7 g/mL.

The most commonly used SCF for the SFC of lipids is carbon dioxide, because of its low and obtainable values of P_c and T_c (7.39 MPa and 304.2 K, respectively); its abundance and low cost, its low toxicity and corrosivity, and its compatibility with a variety of detectors. The only cases in which CO_2 alone cannot be used as carrier fluid are those involving polar analytes. In these cases, a modifier is added to CO_2 to increase the solubility of the analyte. Common modifiers include polar alcohols (e.g., methanol, ethanol, propanols), polar organic solvents (e.g., water, tetrahydrofuran, dimethylsulfoxide, 1,4'-dioxane, and acetonitrile), and chlorinated solvents (e.g., methylene chloride). However, with the exception of water, the presence of modifiers limits the choice of detectors (see below).

SFC Equipment

The equipment used in SFC is similar to that employed in other chromatographic techniques, such as GC and HPLC. Moreover, carrier fluid (CO_2 and perhaps a modifier) is transported from a source (usually a gas cylinder) to a column using a pump. A sample is injected at the front of the column and transported though the column by the carrier fluid. Separation of analytes occurs in the column as a result of differences in relative volatility (the degree of which is controlled by the carrier fluid pressure and the column temperature) and through molecular interactions with the column's stationary phase. The carrier fluid transports the separated analytes to a detector, which is interfaced with a data acquisition system.

In contrast to GC and HPLC, SFC equipment possesses unique features. First, the pump and tubing fittings must withstand high pressures. In addition, the pump must accurately and precisely control the carrier fluid pressure. The most common type of pump used to fulfill these requirements is a syringe pump (5); these pumps are especially common with open tubular columns. For packed columns, HPLC-like reciprocating or diaphragm pumps are becoming common (3); these pumps are less expensive alternatives to syringe pumps.

Second, the introduction of sample to the column is more difficult because of the presence of high pressure in the system. Direct on-column injection can be applied to packed columns because of their large diameters; however, split injection techniques must be applied for open tubular type columns. The two most common methods of split injection are the use of a splitter assembly with microvolume valve and timed-split injection. The latter technique controls the amount of sample injected by controlling the duration that a pneumatic valve is allowed in an "inject" position (typically 0.01 to 1 s).

Third, adjustments must be made to the connection between the column outlet and the detector inlet. This is because of the combination of low mass flow rate and high fluid velocity and pressure existing in the column outlet stream, particularly for open tubular columns (5,6,15). In addition, in the presence of pressure or density programming, the changing fluid velocity and pressure result in significant changes in mass flow rate during the run. A frit restrictor is applied in the connection between column outlet and detector inlet to reduce the fluid pressure and change the fluid flow characteristics to those acceptable for detection. Importantly, the frit restrictor helps control the column pressure drop, hence the fluid linear velocity. A detailed and comprehensive study of restrictors, as well as all other SFC equipment, is provided elsewhere (3,5,6,8,11).

Two categories of SFC columns exist: open tubular and packed column. Open tubular (OT) columns, also referred to as capillary columns, are typically 50 to 100 µm in inner diameter and 5 to 25 m in length. They typically consist of tubing (fused silica is the most common tubing material) with a stationary phase coated on the inner walls (5). The columns are similar to the capillary columns used in GC, except that SFC OT columns are usually smaller in diameter. In addition, due to the higher pressures involved, the stationary phase used in SFC columns must be more strongly cross-linked and more tightly bound to the inner walls of the column. Like capillary GC, separation of analytes using open tubular SFC is based mostly on differences in relative volatility (except at high pressures, where the carrier fluid is more liquidlike) and partly on the selective adsorption of analytes on to the column's stationary phases. Typical stationary phases for OT SFC columns and their selectivity are similar to those encountered with capillary GC.

Packed SFC columns are very similar to HPLC columns. Most SFC columns consist of normal- or reversed-phase silica packings, with particle diameter on the order of microns and column diameter on the order of millimeters (for analytical-scale separations). Chiral packed SFC columns are becoming popular (1–3). Separation using a packed SFC column is similar to HPLC separation. In both cases, analytes are separated by the relative strength of analyte–stationary phase–carrier fluid interactions, in the simplest of terms. One problem encountered with packed-column SFC is the very poor desorption of polar analytes from the column due to their poor solubility in CO_2. This problem has been overcome by using a polar cosolvent modifier (discussed above) and by deactivating some of the packing material's adsorption sites. Silver-impregnated packed (silica) columns, which have been demonstrated to separate lipids by the number of double bonds, double bond position, and *cis* versus *trans* configuration in HPLC, have also been applied recently for SFC (16). However, these columns usually have short lifetimes (7).

One of the major advantages of SFC over GC and HPLC is its compatibility with several different detectors. SFC can employ detectors commonly used for GC—flame ionization detection (FID), electron capture (ECD), nitrogen-phosphorus (NPD), and mass spectrometry (MS)—as well as those used for HPLC—ultraviolet/visible (UV/Vis), fluorescence, and evaporative light scattering (ELSD). The most common detector used for lipids, particularly neutral lipids, is the FID, because of the strong signal it generates and the linearity of its response with concentration. Supercritical CO_2 is compatible with all of these detector types; however, the inclusion of most cosolvents (with water being an exception) prevents the employment of the FID. (Refractive index and infrared spectroscopy detectors are not compatible with CO_2 under most circumstances; however, the latter has been used successfully when supercritical xenon was employed [5].) The use of NMR detection for SFC is being developed (3). More detail on the detector types and their employment in SFC analysis is reviewed elsewhere (3,5,6).

Comparison Between SFC and Other Chromatography Techniques

SFC is being more frequently employed for lipid separations. However, many of the separations encountered in the analytical chemistry of lipids can be equally or better performed using GC, HPLC, or thin-layer chromatography (TLC), which are less expensive than SFC and have been used in lipid separations for decades. Fig. 1 compares SFC, GC, HPLC, and TLC based on several criteria. Although TLC is quite accurate, rapid, and versatile, it is also quite labor-intensive, and quantitation is not as straightforward with TLC than with other chromatographic techniques. Therefore, TLC is not as universally applied as GC or HPLC for analytical-scale separations.

GC is the method of choice for separating simple mixtures of neutral lipids (glycerides). Triglycerides and glyceride classes are readily separated by molecular weight (MW) or carbon number using high-temperature capillary GC. (This mode of separation, which is due to differences in relative volatility, is controlled through programming of the column temperature). The resolution achieved for such separations in unequaled by other chromatographic techniques. In addition, use of an FID simplifies the quantitation. (GC–MS is quite helpful in identifying unknown analytes.) The disadvantages of GC for lipid separations are as follows.

1. GC analysis of several lipid types (e.g., partial glycerides and free fatty acids) often requires tedious and expensive derivatization procedures.
2. For many analytes the high temperatures required for elution promote thermal degradation. In addition, the limited thermal stability of most GC columns (*ca.* 350–450°C temperature limit) reduces the range of analyte MW permitted.
3. Polar lipids are often degraded or do not produce a signal by the FID.

Moreover, GC cannot be used successfully for analysis of complex lipid mixtures without employing several different sets of chromatographic conditions and derivatization methods.

HPLC is also commonly employed for lipid separations. In the simplest of terms, separations occur because of the relative interactions between analytes, column, and mobile phase. Due to the variety of column types and choices of mobile phase, HPLC is the most versatile and selective of the chromatographic techniques. However, for lipid separations, use of HPLC presents significant difficulties:

1. The choice of usable detectors is limited. Since most lipids are not very UV- or visible light–sensitive, absorbance detectors, the most common HPLC detectors, are frequently employed (unless analytes are derivatized with absorbance-sensitive labels). Refractive index (RI) detectors historically have been the most commonly employed detectors for lipid analysis using HPLC; however, RI detectors cannot be employed with gradient elution techniques that must often accompany reverse-phase HPLC. The recent development of the ELSD has been very useful for HPLC lipid analysis, but this detector is still relatively expensive; in addition, its response is logarithmically proportional to the analyte concentration, so quantitation is not as straightforward as when the FID is employed.

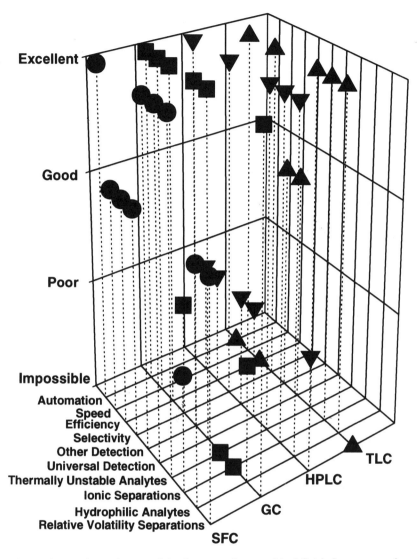

Fig. 1. Comparison of some of the features of supercritical fluid chromatography (SFC), gas chromatography (GC), high-performance liquid chromatography (HPLC), and thin-layer chromatography (TLC). Redrawn from Ref. (8), with permission of author and publisher.

2. Run times for HPLC analyses are often significantly longer than those encountered for GC or SFC.
3. The development of HPLC conditions for a given separation often takes more time and effort relative to GC and SFC.

4. The increasing costs of highly pure (HPLC-grade) solvents and the solvent waste removal costs involved with HPLC are becoming important issues in the selection of analytical chemistry equipment.

SFC has many advantages over its competitors for several types of separations, but not for all separations. SFC can separate analytes based on relative volatility, as does GC, but without the use of high temperatures. In addition, for the separation of a given analyte mixture SFC yields a smoother baseline than GC when the FID detector is employed, based on the author's experience. However, SFC has yet to match the degree of resolution achieved using GC (17). In addition, packed-column SFC can separate analytes based on analyte–stationary phase–carrier fluid interactions as with HPLC, but with shorter run times (*ca.* four times faster using SFC [7]) and straightforward quantitation *via* the FID. However, SFC has yet to match the degree of selectivity achieved using HPLC (even with a modifier present, which may also prevent the use of an FID).

There are two analyte matrices for which SFC analysis is preferred over GC and HPLC analysis:

1. Lipids that are susceptible to degradation at high temperatures, such as glycerides that contain multiple double bonds or multiple oxygen atoms, or that have high MW.
2. Complex lipid mixtures and lipid reaction mixtures.

These application areas will now be discussed in detail.

SFC of Unusual Glycerides

Glycerides Containing Oxygenated Fatty Acyl Groups

Table 1 lists SFC analyses of glycerides (mostly triglycerides, or TG) containing either epoxy or hydroxy free fatty acids (FFA). In most cases, open tubular (OT) columns have been employed along with carbon dioxide as carrier fluid and a FID (18–24). When a nonpolar-stationary-phase column is used under these conditions, separation occurs based on relative volatility, or equivalently, molecular weight (MW). Using nonpolar stationary phases, the MW-based separation of TG and glycerides using either pressure or density programming has been well established (17,18,21,25–33). Use of nonpolar columns does not permit significant separation based on double bond type, configuration, or position, but it partially separates saturated and unsaturated lipids; also, analytes that contain acyl groups with double bond positions within six carbons of a carbonyl group are partially separated from those without such acyl groups (18,34,35). Moreover, through comparison of an analyte's retention time with those of standards, its MW can be estimated, as will be discussed below.

The proper selection of temperature can improve the resolution of overlapping peaks. When nonpolar OT-columns are employed, an increase of temperature up to

TABLE 1
SFC Analysis of Glycerides Containing Hydroxy or Epoxy Fatty Acids[a]

FFA Type	Glyceride type	SFC conditions	Reference No.
Vernolic, $18:1^{9c}$-epoxy^{12,13c}	*Vernonia galamensis* and *Euphorbia lagascae* oils (TG)	OT-nonpolar column CO_2, *P*- and *T*- ramping	(18)
Vernolic	*E. lagascae* oil TG and FFA	OT-polar column CO_2, ρ-ramping (isothermal)	(19)
Dimorphecolic, $18:2^{10t,12t}$-OH^9	*Dimorphotheca pluvialis* oil (TG)	OT-nonpolar column CO_2, *P*-ramping (isothermal)	(20)
Ricinoleic $18:1^{9c}$-OH^{12}	Castor oil (TG) *Ricinus communis* oil	OT-nonpolar column CO_2, *P*-ramping	(21)
C_{18}, C_{20} HFA (e.g., lesquerolic)	Oils from genus *Lesquerella*	OT-nonpolar column CO_2, *P*-, and *T*-ramping	(22,23)
2-Hydroxy acids	FFA (HFA)	OT-nonpolar column CO_2, constant *P* and *T*	(24)

[a]Abbreviations: FFA, free fatty acid; HFA, hydroxy fatty acid; OT, open tubular; *P*, carrier fluid pressure; *T*, carrier fluid temperature; ρ, carrier fluid density.

150°C leads to an increase in resolution; however, further increases lead to only small improvements in resolution (18,21,31). This is shown for the separation of the TG species of vernonia oil (Fig. 2) (18). The MW-based separation of vernonia oil's TGs is challenging, because of the small difference in MW between species. For example, TG containing 0, 1, 2, or 3 vernolic ($18:1^{9c}$-epoxy^{12,13c}) acyl groups per TG (with the other acyl groups being C_{18} unsaturates) have molecular weight differences of only 14 Da; equivalently, when assigning vernolic acid a carbon number of 19, these TG species differ in carbon number by only one unit. When operated at 100°C, very little separation was achieved; however, at 150 to 200°C resolution was greatly improved (Fig. 2). To prevent the degradation of the analytes, temperature was not further increased. Note that the elution density decreased as temperature was increased. Simultaneous pressure (or density) and temperature programming further improved the resolution of vernonia oil TG (data not shown) (18). (A positive temperature ramp led to a greater improvement in resolution than did a negative temperature ramp.) Others have also improved MW-based resolution by applying simultaneous pressure and temperature programming (32,36,37). Vernonia oil was also examined using reversed-phase HPLC, where TG species were separated by the degree of polarity (38). Although trivernolin was isolated by reversed-phase HPLC (demonstrated through ^{1}H-NMR), other TG species were not isolated to the degree achieved using SFC, because the HPLC technique does not provide MW-based separation (18,19,38).

The other separations listed in Table 1 are similar to those displayed for vernonia oil. An exception is the work of Borch-Jensen and Mollerup, who resolved the TG peaks of *Euphorbia lagascae* oil to a similar degree as that displayed in Fig.

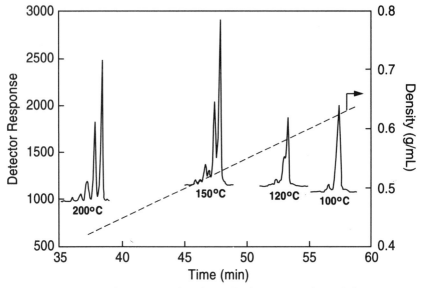

Fig. 2. Separation of vernonia oil triglycerides by SFC. Isothermal density programming of the carrier fluid, CO_2, was employed [0.15 g/mL hold for 9 min.; density increased at 0.01 g/(mL min) until all analytes eluted] using a SB-methyl-100 open tubular column (10 m × 50 μm i.d.) (Dionex, Salt Lake City, UT) and a flame ionization detector at 350°C. From Ref. (18), with permission of author and publisher.

2 for vernonia oil, where a nonpolar OT column was used (19). Both vernonia and *E. lagascae* oils are rich in vernolic acid content and share very similar TG structure (18). The researchers employed an OT polar column with isothermal (170°C) density programming (19). Thus, the separation achieved by Borch-Jensen and Mollerup is based in part on analyte-column interactions, and therefore on the relative polarity of the TG species (19).

Hydroxy FA–containing analytes have also been examined using GC and HPLC. Carlson et al. employed high-temperature, short-column GC to resolve the TG species of lesquerella oils based on their differences in MW (39). The MW-based separation of lesquerella oils achieved using SFC and simultaneous pressure and temperature programming was superior in its degree of separation and the degree of smoothness of its baseline (23). However, the GC analysis was shorter than that by SFC (23,39), and it is not clear whether the GC conditions were optimized to yield the greatest possible resolution of peaks (39). In addition, it may be possible that the hydroxy acid–containing TG may have degraded in part during the GC analysis because of the high temperatures employed (*ca.* 350°C) (39). HPLC is superior to SFC with OT columns for separating hydroxy and epoxy FFA species, because the latter analytical technique often yields skewed FFA peaks, particularly for hydroxy FFA (discussed below) (40,41).

Estolide-Containing Lipids and Polymeric Lipids

In addition to hydroxy and epoxy fatty acid–containing glycerides, lipids composed of estolides are also susceptible to thermal degradation at higher temperatures such as those commonly employed in GC. Estolides—hydroxy fatty acid oligomers attached by ester bonds between the –OH moiety of one acyl group and the carbonyl moiety of a second acyl group—have potential applications in printing inks, cosmetic formulations, and lubricants. Examples of estolide separations are given in Table 2 and Fig. 3 (18,20,22,42,43). Estolides with free –COOH groups were well resolved for species with up to 5 monomeric units; however, the esterification of these groups with oleyl alcohol increased the resolution range up to 6 monomeric units and reduced the peak width (Fig. 3). The results for the SFC of free estolides are comparable to (but not quite at the same level of resolution as) that yielded by reversed-phase HPLC; however, the latter method cannot employ the FID (44). An increase in column temperature improved the resolution of low-MW estolides but shortened the elution range (18). The increase of peak width with retention time is a well-known phenomenon in all chromatographies (due to translational diffusion). However, column efficiency is further decreased at higher carrier fluid pressure (or equivalently, density) in systems that feature a fixed frit restrictor (e.g., the system employed to generate Fig. 3), because of the increase in carrier fluid velocity and the decrease in the diffusion coefficient (5,6). The degree of change in linear velocity and diffusion coefficients with pressure is a strong function of the column design and the device used to connect the column outlet to the detector (5,6,15,45–47).

SFC has also been applied to analyze other polymeric lipids, such as polyglycerol and polyol esters (48–51), and glycerol tetraether lipids (52), as well as other surfactants and waxes. SFC analysis of polymers is reviewed elsewhere (3,53).

Polyunsaturated Lipids

Separations involving highly unsaturated lipids are traditionally operated using reversed-phase HPLC, since GC can lead to thermal degradation (16,54–57). SFC separations of polyunsaturated lipids are listed in Table 3 (34,35,58–66). Blomberg

TABLE 2
SFC Analysis of Estolide-Containing Samples[a]

Sample	Largest MW resolved	Reference No.
12-hydroxystearic acid estolides	Not available	(42)
Estolides derived from oleic acid	1685 (pentamer)	(18,43)
Oleic acid-derived estolide oleyl esters	2213 (hexamer oleyl ester)	(18,43)
Estolide-containing TG[b]	1490 (five acyl groups)	(20,22)

[a]All results discussed here employed similar SFC conditions: open tubular SFC column, CO_2 as the carrier fluid, and isothermal pressure- or density-programming.

[b]TGs were from seed oils of processed *D. pluvialis* oil (20), species from the genus *Lesquerella* (22), or *Heliophilia amplexicaulis* (22). Triglyceride molecules contained estolides consisting of only two repeat units (monoestolides), except for *H. amplexicaulis* oil, which also contained a small amount of diestolide.

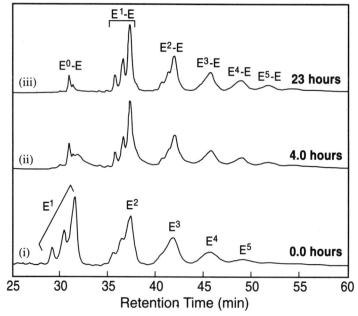

Fig. 3. SFC of samples from the lipase-catalyzed esterification of oleyl alcohol and polyestolide derived from oleic acid. Reaction times are listed in the figure. E^i refers to estolide containing i estolide bonds ($i + 1$ acyl groups). E^i - E refer to estolide oleyl esters. The analysis occurred using the same column and detector described in Fig. 2. The carrier fluid (CO_2) pressure was held for 9 min at 12.7 MPa, then increased from 12.7 MPa to 40.53 MPa at a rate of 0.507 MPa/min. The column temperature was held isothermally at 100°C. From Ref. (18), with permission of author and publisher.

and coworkers have applied fused silica columns (0.25 mm i.d.) micropacked with Nucleosil gel (Macherey Nagel, Düren, Germany) and impregnated with silver (or permanganate) ions for the separation of various polyunsaturated lipids (i.e., argentation chromatography) (58–63). Cosolvents (acetonitrile and isopropanol) were required to improve the solvent power of CO_2. As demonstrated in Fig. 4, silver-ion SFC separated TG based on their molecular weight, number of double bonds, and double bond position (58). The latter trend can be realized by comparing peaks 15 and 16. According to a recent review by Dobson et al. (16), silver-ion SFC has several advantages over silver-ion HPLC:

1. Significant MW-based separation
2. Improved separation of TGs containing the same number of double bonds but differing in acyl composition and double bond position, such as the T54:6

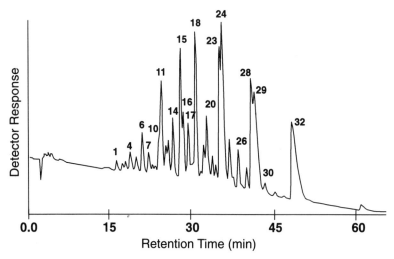

Fig. 4. SFC of borage oil. Column: fused silica, 300 mm × 25 mm, packed with Nucleosil 4 SA (Macherey Nagel, Düren, Germany) and impregnated with $AgNO_3$. Conditions: 2-min hold at 115°C and 26.35 MPa, followed by a ramp of –0.5°C/min and 0.101 MPa/min to 85°C and 33.24 MPa. Mobile phase: CO_2/acetonitrile/isopropanol at 92.8:6.5:0.7 mole %. Detector: ultraviolet (210 nm). Peaks: 1: PPL; 4: PPG; 6: POL; 7: SOL; 10: POG; 11:PLL; 14: OOL; 15: PLG; 16: PLLn; 17: OOG; 18: OLL; 20: SGG; 23: OLG; 24: LLL, 26: GLEr; 28: GGO; 29: GLL; 30: GGGo; 32: LGG, where P = palmitic (16:0), S = stearic (18:0), O = oleic ($18:1^{9c}$), L = linoleic ($18:2^{9c,12c}$), Ln = α-linolenic ($18:3^{9c,12c,15c}$), G = γ-linolenic ($18:3^{6c,9c,12c}$), Go = gondoic ($20:1^{11c}$), and Er = erucic ($22:1^{13c}$) acids. Redrawn from Ref. (58), with permission.

isomers trilinolein (LLL) and oleic–linoleic–α-linolenic TG (OLLn) in linseed oil (58).

In addition, packed-column SFC has been applied to separate glyceride positional isomers as well as 1,3- and 1,2-dioleylglycerol (67).

Open tubular columns with polar stationary phases (e.g., carbowax and cyanopropyl phases) have also been successfully applied to separate lipids by degree of unsaturation (Table 3) (34,35,66). Manninen and coworkers were able to separate a variety of TGs and FAMEs using polar open tubular columns based on MW, number of double bonds, and double bond position (e.g., molecules containing α- versus γ-linoleic acid) (34,35). However, unlike silver-ion packed-column SFC, SFC with OT columns could not separate positional isomers (e.g., LLL and OLLn) (34,35). Depending on the depth of analysis required, the use of OT columns may be adequate. Such separations also have the advantage of FID usage.

TABLE 3
SFC Analysis of Polyunsaturated Lipids[a]

Sample	SFC conditions	Reference No.
Linseed oil (TG)	Micropacked silver (or MnO_4^-) ion column CO_2/ACN/IPA; P- and T-ramping	(58–60)
Borage oil	Micropacked silver-ion column[b]	(58,60)
Columbine[c]	Micropacked silver-ion column[b]	(61)
FAMEs	Micropacked silver-ion column[b]	(62)
FAMEs	Micropacked column (CO_2)	(64)
Fish oil	Micropacked silver (or MnO_4^-) ion column[b]	(58,63)
Fish oil	OT polar column CO_2; isothermal r-ramping	(34,65)
Sea buckthorne oil and pulp[d]	OT-polar column[e]	(34)
Cloudberry seed oil[f]	OT-polar column[e]	(34)
Alpine and black currant oils[g]	OT-polar column[e]	(35)
Fish oil FAEEs	OT-polar column[e]	(66)

[a]Abbreviations: FAME, fatty acid methyl ester, FAEE, fatty acid ethyl ester; ACN, acetonitrile; IPA, isopropyl alcohol (2-propanol); P, carrier fluid pressure; T, carrier fluid temperature, r, carrier fluid density; OT, open tubular.
[b]Similar chromatographic conditions to those listed above for linseed oil.
[c]*Aquilegia vulgaris* lipids.
[d]*Hippophae rhamnoides*.
[e]Similar conditions to those listed above for the OT column separation of fish oil.
[f]*Rubus shamaemorus*.
[g]*Ribes alpinum* and *R. nigrum*, respectively; FAME also examined.

SFC of Complex Lipid Mixtures and Lipid Reaction Mixtures

Trace Components in Complex Lipid Mixtures

SFC has been very useful in quantifying trace components in complex lipid samples (Table 4) (19,65,68–76). Such trace components include FFA, partial glycerides, vitamins, sterols, cholesterol, tocopherols, squalene, and hydroperoxides. The advantages of using SFC for these analyses include simplified sample preparation (i.e., derivatization is not required) and the employment of the FID for

TABLE 4
SFC Analysis of Trace Components in Lipid Samples[a]

Trace component	Lipid sample	Reference No.
FFA	Abused vegetable oil	(68)
FFA	*Euphorbia lagascae* oil	(19)
FFA, VIT, CHOL, TOC	Onion seed lipids	(69)
FFA, VIT, CHOL, TOC, squalene	Fish oils	(65,70–73)
CHOL	Milk fat[b]	(74)
Sterols	Fungi lipid	(75)
Hydroperoxides	Peanut oil	(76)

[a]Abbreviations: FFA, free fatty acids; VIT, vitamins; CHOL, cholesterol; TOC, tocopherols.
[b]Glycerides were removed from the milk fat by saponification before SFC analysis.

quantification. For example, an isolation step is usually employed before analyzing tocopherols via HPLC (77). Huber et al. found that SFC analysis gave a more accurate and precise measurement of cholesterol in milk fat than GC did (74). A more complete discussion of the SFC conditions used in separating minor components from each other is contained elsewhere (10).

Lipid Reaction Mixtures

SFC has been found useful for analyzing lipid reactions, especially those involving the enzyme lipase (18,20,22,43,72,78–80). One application of SFC is the on-line analysis of an enzyme-catalyzed reaction occurring in supercritical CO_2 (79). Another use of SFC is for analyses where standards of the products formed are not available. Moreover, through use of a MW–retention time calibration plot, unknown peaks can be identified (14,18,33,81). These plots are useful for analyses performed using nonpolar OT columns, where separation by degree of saturation does not occur; hence, the only mode of separation is by MW. Figure 5

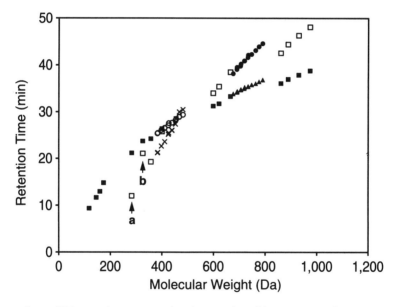

Fig. 5. SFC retention time–molecular weight calibration curve for reactants and products appearing during the lipase-catalyzed alcoholysis of lesquerella oil by α, ω-diols. The column, detector, and pressure programming are the same as those described in Fig. 3. The column temperature was held at either 100°C or 200°C. (100°C, 200°C)—(■,□) lesquerella oil free fatty acid (FFA), monoglycerides, diglycerides, and triglycerides; (○,X) diol-lesquerella monoester; (▲,●) diester. Peaks a and b refer to C_{18} and 20-OH FFA, respectively. From Ref. (18), with permission of author and publisher.

illustrates a such a plot for the lipase-catalyzed alcoholysis of lesquerella (*Lesquerella fendleri*) oil by α,ω-diols (18). The slope of the calibration curve is initially steep, then gradually decreases. The initial steepness occurs because separation in the low-MW region is based on the relative volatility of the analytes, as occurs for GC. In the higher-MW portion of the curve, where a higher (hence more liquidlike) carrier fluid density is required, separation is based less on relative volatility and more on interactions among solute, solvent, and stationary phase, as occurs for HPLC, hence the smaller slope. When column temperature was increased to 200°C, both resolution (proportional to the slope of the calibration curve) and retention times decreased for low-MW (<200 Da) analytes and increased for higher-MW analytes (Fig. 5). The low-MW analytes, in fact, were poorly resolved from the solvent peak. When elution density replaces retention time as the ordinate, two separate but nearly parallel curves result, with the curve for the analysis at 200°C being below the 100°C curve (Fig. 6). The use of a density program with similar parameters as the pressure program led to only minor differences in the calibration curve (Fig. 6).

Data points *a* and *b* in Figs. 5 and 6 refer to C_{18} and C_{20} FFA species present in lesquerella oil. The retention behavior of these species are not well described by the calibration curves, because their peaks at 200°C are quite wide and the peak

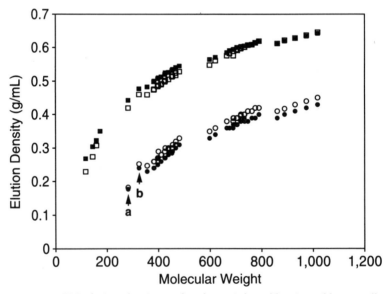

FIG. 6. SFC elution density–molecular weight calibration of lesquerella oil glycerides and diol esters by using pressure (■,□) or density (●,○) programming at 100°C or 200°C, respectively. SFC column, carrier fluid, and detector are described in Fig. 2. Density and pressure programs are given in Figs. 2 and 3, respectively. From Ref. (18), with permission of author and publisher.

shapes are non-Gaussian (18). Others have also encountered tailing of FFA peaks and found that the resolution of FFA decreased at higher temperatures during SFC (24,68,82–84). The shape of FFA peaks improves with modification of the SFC column stationary phase (85) or by the addition of cosolvent to CO_2 (68,83,84).

Conclusions

Supercritical fluid chromatography is becoming a common tool for the analysis of lipids, alongside GC and HPLC. For analyzing highly oxygenated and polyunsaturated lipids and complex lipid mixtures, SFC may be the tool of choice.

References

1. Petersson, P., and Markides, K.E. (1994) Chiral Separations Performed by Supercritical Fluid Chromatography, *J. Chrom. 666*, 381–394.
2. Kot, A., Sandra, P., and Venema, A. (1994) Sub- and Supercritical Fluid Chromatography on Packed Columns: A Versatile Tool for the Enantioselective Separation of Basic and Acidic Drugs, *J. Chrom. Sci. 32*, 439–448.
3. Chester, T.L., Pinkston, J.D., and Raynie, D.E. (1996) Supercritical Fluid Chromatography and Extraction, *Anal. Chem. 68*, 487R–514R.
4. Lafosse, M., Herbreteau, B., and Morin-Allory, L. (1996) Supercritical Fluid Chromatography of Carbohydrates, *J. Chrom A 720*, 61–73.
5. Lee, M.L., and Markides, K.E. (1990) *Analytical Supercritical Fluid Chromatography and Extraction*. Chromatography Conferences, Inc., Provo, Utah.
6. King, J., Hill, H.H., Jr., and Lee, M.L. (1993) Analytical Supercritical Fluid Chromatography and Extraction, *Physical Methods of Chemistry Series*, Rossiter, B.W., and Baetzold, R.C., 2nd edn., *Vol. X*, John Wiley and Sons, New York, pp. 1–83.
7. Sandra, P., and David, F. (1996) Basic Principles and the Role of Supercritical Fluid Chromatography in Lipid Analysis, *Supercritical Fluid Technology in Oil and Lipid Chemistry*, King, J.W., and List, G.R., American Oil Chemists' Society, Champaign, Illinois, pp. 321–347.
8. Chester, T.L. (1996) Supercritical Fluid Chromatography for the Analysis of Oleochemicals, *Supercritical Fluid Technology in Oil and Lipid Chemistry*, King, J.W., and List, G.R., American Oil Chemists' Society, Champaign, Illinois, pp. 348–375.
9. Chester, T.L., Pinkston, J.D., and Raynie, D.E. (1994) Supercritical Fluid Chromatography and Extraction, *Anal. Chem. 66*, 106R–130R.
10. Artz, W.E. (1996) Supercritical Fluid Chromatography of Trace Components in Oils and Fats, *Supercritical Fluid Technology in Oil and Lipid Chemistry*, King, J.W., and List, G.R., American Oil Chemists' Society, Champaign, Illinois, pp. 376–385.
11. Laakso, P. (1993) Supercritical Fluid Chromatography of Lipids, *Advances in Lipid Methodology—One*, Christie, W.W., The Oily Press, Ayr, Scotland, pp. 81–119.
12. Bartle, K.D., and Clifford, T.A. (1992) Supercritical Fluid Extraction and Chromatography of Lipids and Related Compounds, *Advances in Applied Lipid Research*, Padley, F.D., JAI Press, London, pp. 217–264.
13. Artz, W.E. (1993) Analysis of Lipids by Supercritical Fluid Chromatography, *Analyses of Fats, Oils and Derivatives*, Perkins, E.G., American Oil Chemists' Society, Champaign, Illinois, pp. 270–300.

14. Chester, T.L., and Innis, D.P. (1993) Investigation of Retention and Selectivity in High-Temperature, High-Pressure, Open-Tubular Supercritical Fluid Chromatography with CO_2 Mobile Phase, *J. Microcol. Sep. 5*, 441–449.
15. Smith, R.D., Fulton, J.L., Petersen, R.C., Kopriva, A.J., and Wright, B.W. (1986) Performance of Capillary Restrictors in Supercritical Fluid Chromatography, *Anal. Chem. 58*, 2057–2064.
16. Dobson, G., Christie, W.W., and Nikolova-Damyanova, B. (1995) Silver ion chromatography of lipids and fatty acids. *J. Chrom. B 671*, 197–222.
17. Sandra, P., David, F., Munari, F., Mapelli, G., and Trestianu, S. (1988) HR-CGC and CSFC for the Analysis of Relatively High Molecular Weight Compounds, *Supercritical Fluid Chromatography*, Smith, R.M., Royal Society of Chemistry, London, pp. 137–158.
18. Hayes, D.G., and Kleiman, R. (1996) Supercritical Fluid Chromatographic Analysis of New Crop Seed Oils and their Reactions, *J. Am. Oil Chem. Soc., 73*, 1691–1697.
19. Borch-Jensen, C., and Mollerup, J. (1996) Determination of Vernolic Acid Content in the Oil of *Euphorbia lagascae* by Gas and Supercritical Fluid Chromatography, *J. Am. Oil Chem. Soc. 73*, 1161–1164.
20. Hayes, D.G., Kleiman, R., Weisleder, D., Adlof, R.O., Cuperus, F.P., and Derksen, J.T.P. (1995) Occurrence of Estolides in Processed *Dimorphotheca pluvialis* Seed Oil, *Ind. Crops Prod. 4*, 295–301.
21. Baiocchi, C., Saini, G., Cocito, C., Giacosa, D., Roggero, M.A., Marengo, E., and Favale, M. (1993) Analysis of Vegetable and Fish Oils by Capillary Supercritical Fluid Chromatography with Flame Ionization Detection, *Chromatographia 37*, 525–533.
22. Hayes, D.G., and Kleiman, R. (1995) The Triglyceride Composition, Structure, and Presence of Estolides in the Oils of *Lesquerella* and Related Species, *J. Am. Oil Chem. Soc. 72*, 559–569.
23. Hayes, D.G., and Kleiman, R. (1996) A Detailed Triglyceride Analysis of *Lesquerella fendleri* Oil: Column Chromatographic Fractionation Followed by Supercritical Fluid Chromatography, *J. Am. Oil Chem. Soc. 73*, 267–269.
24. Doehl, J., Farbrot, A., Greibrokk, T., and Iversen, B. (1987) Supercritical Fluid Chromatographic Separation of Low- to Medium-Polarity Compounds that are Difficult to Elute from Packed Columns, *J. Chrom. 392*, 175–184.
25. Anton, K., Bach, M., Chalackal, M., Eppinger, J., Frederiksen, L., Pericles, N., Muringer, F., and Volkart, M. (1993) SFC in Pharmaceutical Drug Analysis. Possibilities and Limitations, *The Second European Symposium on Analytical Supercritical Fluid Chromatography and Extraction,* Sandra, P., Markids, K., and Devos, G., Dr. Alfred Huethig Verlag, Heidelberg, Germany, pp. 182–184.
26. Bhaskar, A.R., Rizvi, S.S.H., and Harriott, P. (1993) Performance of a Packed Column for Continuous Supercritical Carbon Dioxide Processing of Anhydrous Milk Fat, *Biotechnol. Prog. 9*, 70–74.
27. Huopalahti, R., Laakso, P., Saaristo, J., Linko, R., and Kallio, H. (1988) Preliminary Studies of Triacylglycerols of Fats and Oils by Capillary SFC, *J. High Resol. Chrom. Chrom. Comm. 11*, 899–901.
28. Kallio, H., Laakso, P., Houpalahti, R., and Linko, R.R. (1989) Analysis of Butterfat Triacylglycerols by Supercritical Fluid Chromatography/Electron Impact Mass Spectrometry, *Anal. Chem. 61*, 698–700.

29. Kallio, H., and Laakso, P. (1990) Effect of Carbon Dioxide Flow-Rate on the Separation of Triacylglycerols by Capillary Supercritical Fluid Chromatography, *J. Chrom. 518*, 69–75.
30. Kallio, H., Vauhkonen, T., and Linko, R.R. (1991) Thin-Layer Silver ion Chromatography and Supercritical Fluid Chromatography of Baltic Herring (*Clupea harengus membras*) triacylglycerols, *J. Agric. Food Chem. 39*, 1573–1577.
31. Proot, M., Sandra, P., and Geeraert, E. (1986) Resolution of Triglycerides in Capillary SFC as a Function of Column Temperature, *J. High Res. Chrom. Chrom. Comm. 9*, 189–192.
32. Richter, B.E., Anderson, M.R., Knowles, D.E., Campbell, E.R., Porter, N.L., Nixon, L., and Later, D.W. (1988) Capillary Supercritical Fluid Chromatography: Use for the Analysis of Food Components and Contaminants, *Supercritical Fluid Extraction and Chromatography: Techniques and Applications (ACS Symposium Series 366)*, Charpentier, B.A., and Sevenants, M.R., American Chemical Society, Washington, DC, pp. 179–190.
33. White, C.M., and Houck, R.K. (1985) Analysis of Mono-, Di-, and Triglycerides by Capillary Supercritical Fluid Chromatography, *J. High Resol. Chrom. Chrom. Comm. 8*, 293–296.
34. Manninen, P., Laakso, P., and Kallio, H. (1995) Method for Characterization of Triacylglycerols and Fat-Soluble Vitamins in Edible Oils and Fats by Supercritical Fluid Chromatography, *J. Am. Oil Chem. Soc. 72*, 1001–1008.
35. Manninen, P., Laakso, P., and Kallio, H. (1995) Separation of γ- and α-Linolenic Acid Containing Triacylglycerols by Capillary Supercritical Fluid Chromatography, *Lipids 30*, 665–671.
36. Schmitz, F.P., Leyendecker, D., Leyendecker, D., and Gemmel, B. (1987) Possibilities for Optimization of Oligomer Separation with Supercritical Fluid Chromatography, *J. Chrom. 395*, 111–123.
37. Leyendecker, D., Leyendecker, D., Schmitz, F.B., and Klesper, E. (1986) Multiple Gradients in Supercritical Fluid Chromatography (SFC), *J. High Res. Chrom. Chrom. Comm. 9*, 525–527.
38. Neff, W.E., Adlof, R.O., Konishi, H., and Weisleder, D. (1993) High-Performance Liquid Chromatography of the Triacylglycerols of *Veronia glamensis* and *Crepis alpina* Seed Oils, *J. Am. Oil Chem. Soc. 70*, 449–453.
39. Carlson, K.D., Chaudhry, A., Peterson, R.E., and Bagby, M.O. (1990) Preparative Chromatographic Isolation of Hydroxy Acids from *Lesquerella fendleri* and *L. gordonii* Seed Oils, *J. Am. Oil Chem. Soc. 67*, 495–498.
40. Gérard, H.C., Moreau, R.A., Fett, W.F., and Osman, S.F. (1992) Separation and Quantitation of Hydroxy and Epoxy Fatty Acids by High-Performance Liquid Chromatography with an Evaporative Light-Scattering Detector, *J. Am. Oil Chem. Soc. 69*, 301–304.
41. Akasaka, K., Akama, T., Ohrui, H., and Meguro, H. (1993) Measurement of Hydroxy and Hydroperoxy Fatty Acids by a High-Pressure Liquid Chromatography with a Column-Switching System, *Biosci. Biotech. Biochem. 57*, 2016–2019.
42. Raynor, M.W., Bartle, K.D., and Clifford, A.A. (1990) Analysis of Aliphatic and Phenolic Carboxylic Acids by Capillary Supercritical Fluid Chromatography–Fourier Transform Infrared Microspectrometry, *J. Chrom. 505*, 179–190.

43. Hayes, D.G., and Kleiman, R. (1995) Lipase-Catalyzed Synthesis and Properties of Estolides and their Esters, *J. Am. Oil Chem. Soc. 72*, 1309–1316.
44. Isbell, T.A., and Kleiman, R. (1994) Characterization of Estolides Produced from the Acid-Catalyzed Condensation of Oleic Acid, *J. Am. Oil Chem. Soc. 71*, 379–383.
45. Köhler, U., Biermanns, P., and Klesper, E. (1994) Influence of Linear Velocity, Column Length, and Pressure Drop in SFC. I. Capacity Ratios, Selectivities, and Theoretical Plate Heights, *J. Chrom. Sci. 32*, 461–470.
46. Klesper, E., and Schmitz, F.P. (1988) Gradient Methods in Supercritical Fluid Chromatography, *J. Supercrit. Fluids 1*, 45–69.
47. Wright, B.W., and Smith, R.D. (1986) Rapid Analysis Using Capillary Supercritical Fluid Chromatography, *J. High Res. Chrom. Chrom. Comm. 9*, 73–77.
48. Artz, W.E., and Myers, M.R. (1995) Supercritical Fluid Extraction and Chromatography of Emulsifiers, *J. Am. Oil Chem. Soc. 72*, 219–224.
49. Carey, J.M., and Sutton, G.P. (1995) Polyol Ester Synthetic Refrigeration Lubricant Analysis by Supercritical Fluid Chromatography, *Anal. Chem. 67*, 1632–1636.
50. Chester, T.L., and Innis, D.P. (1986) Effect of Free Hydroxyl Groups in the Separation of Polyglycerol Esters by Capillary Supercritical Fluid Chromatography, *J. High Res. Chrom. Chrom. Comm. 9*, 178–181.
51. Ye, M.Y., Hill, K.D., and Walkup, R.G. (1994) Separation of T-MAZ Ethoxylated Sorbitan Fatty Acid Esters by Supercritical Fluid Chromatography, *J. Chrom. A. 662*, 323–327.
52. DeLuca, S.J., Voorhees, K.J., Langworthy, T.A., and Holzer, G. (1986) Capillary Supercritical Fluid Chromatography of Archaebacterial Glycerol Tetraether Lipids, *J. High Res. Chrom. Chrom. Comm. 9*, 182–185.
53. Anton, K., Bach, M., Berger, C., Walch, F., Jaccard, G., and Carlier, Y. (1994) From Potential to Practice: Relevant Industrial Applications to Packed-Column Supercritical Fluid Chromatography, *J. Chrom. Sci. 32*, 430–438.
54. Perrin, J.L., Prevot, A., Traitler, H., and Bracco, U. (1987) Analysis of Triglyceride Species of Blackcurrant Seed Oil by HPLC via a Laser LIght Scattering Detector, *Rev. Franç Corps Gras 34*, 221–223.
55. Aitzetmüller, K., and Grönheim, M. (1992) Separation of Highly Unsaturated Triacylglycerols by Reversed Phase HPLC with Short Wavelength UV Detection, *J. High Res. Chrom. 15*, 219–226.
56. Chang, M.K., Conkerton, E.J., Chapital, D., and Wan, J.P. (1994) Behavior of Diglycerides and Conjugated Fatty Acid Triglycerides in Reverse-Phase Chromatography, *J. Am. Oil Chem. Soc. 71*, 1173–1175.
57. Christie, W.W. (1988) Separation of Molecular Species of Triacylglycerols by High-Performance Liquid Chromatography with a Silver Ion Column, *J. Chrom. 454*, 273–284.
58. Blomberg, L.G., Demirbürker, M., and Andersson, P.E. (1993) Argentation Supercritical Fluid Chromatography for Quantitative Analysis of Triacylglycerols, *J. Am. Oil Chem. Soc. 70*, 939–946.
59. Demirbürker, M., and Blomberg, L.G. (1991) Separation of Triacylglycerols by Supercritical-Fluid Argentation Chromatography, *J. Chrom. 550*, 765–774.
60. Demirbürker, M., and Blomberg, L.G. (1991) Permanganate-Impregnated Packed Capillary Columns for Group Separation of Triacylglycerols Using Supercritical Media as Mobile Phases, *J. Chrom. 600*, 358–363.

61. Demirbürker, M., Blomberg, L.G., Olsson, N.U., Bergqvist, M., Herslöf, B.G., and Jacobs, F.A. (1992) Characterization of Triacylglycerols in the Seeds of *Aquilegia valgaris* by Chromatographic and Mass Spectrometric Methods, *Lipids 27*, 436–441.
62. Demirbürker, M., Hägglund, I., and Blomberg, L.G. (1992) Separation of Unsaturated Fatty Acid Methyl Esters by Packed Capillary Supercritical Fluid Chromatography, *J. Chrom. 605*, 263–267.
63. Demirbürker, M., and Blomberg, L.G. (1990) Group Separation of Triacylglycerols on Micropacked Argentation Columns Using Supercritical Media as Mobile Phases, *J. Chrom. Sci. 28*, 67–72.
64. Sakaki, K. (1993) Supercritical Fluid Chromatographic Separation of Fatty Acid Methyl Esters on Aminopropyl-Bonded Silica Stationary Phase, *J. Chrom. 648*, 451–457.
65. Borch-Jensen, C., Staby, A., and Mollerup, J. (1993) Supercritical Fluid Chromatographic Analysis of a Fish Oil of the Sand Eel *Amodytes* sp., *J. High Res. Chrom. 16*, 621–623.
66. Staby, A., Borch-Jensen, C., Mollerup, J., and Jensen, B. (1993) Flame Ionization Detector Responses to Ethyl Esters of Sand Eel (*Amodytes lancea*) Fish Oil Compared for Different Gas and Supercritical Fluid Chromatographic Systems, *J. Chrom. 648*, 221–232.
67. Schmeer, K., Nicholson, G., Zhang, S., Bayer, E., and Bohning-Gaese, K. (1996) Identification of the Lipids and the Ant Attractant 1,2-Dioleoylglycerol in the Arils of *Commiphora guillaumini* Perr. (Burseraceae) by Supercritical Fluid Chromatography–Atmospheric Pressure Chemical Ionization Mass Spectrometry, *J. Chrom. A. 727*, 139–146.
68. France, J.E., Snyder, J.M., and King, J.W. (1991) Packed-Microbore Supercritical Fluid Chromatography with Flame Ionization Detection of Abused Vegetable Oils, *J. Chrom 540*, 271–278.
69. Hannan, R.M., and Hill, H.H., Jr. (1991) Analysis of Lipids in Aging Seed Using Capillary Supercritical Fluid Chromatography, *J. Chrom. 547*, 393–401.
70. King, J.W. (1990) Application of Capillary Supercritical Fluid Chromatography–Supercritical Fluid Extraction to Natural Products, *J. Chrom. Sci. 28*, 9–14.
71. Staby, A., Borch-Jensen, C., Balchen, S., and Mollerup, J. (1994) Quantitative Analysis of Marine Oils by Capillary Supercritical Fluid Chromatography, *Chromatographia 39*, 697–705.
72. Staby, A., Borch-Jensen, C., Balchen, S., and Mollerup, J. (1994) Supercritical Fluid Chromatographic Analysis of Fish Oils, *J. Am. Oil Chem. Soc. 71*, 355–359.
73. Borch-Jensen, C., and Mollerup, J. (1996) Supercritical Fluid Chromatography of Fish, Shark, and Seal Oils, *Chromatographia 42*, 252–258.
74. Huber, W., Molero, A., Pereyra, C., and Martinez de la Ossa, E. (1995) Determination of Cholesterol in Milk Fat by Supercritical Fluid Chromatography, *J. Chrom. A. 715*, 333–336.
75. Sakaki, K., Sako, T., Yokochi, T., Suzuki, O., and Hakuta, T. (1988) Studies on Production of Lipids in Fungi. XIX. Separation of Neutral Lipids in Fungi by Using Supercritical Fluid Chromatography, *Yukugaku 37*, 54–56.
76. Sugiyama, K., Shiokawa, T., and Moriya, T. (1990) Application of Supercritical Fluid Chromatography and Supercritical Fluid Extraction to the Measurement of Hydroperoxides in Foods, *J. Chrom. 515*, 555–562.

77. Chase, G.W., Jr., Akoh, C.C., and Eitenmiller, R.R. (1994) Analysis of Tocopherols in Vegetable Oils by High-Performance Liquid Chromatography: Comparison of Fluorescence and Evaporative Light-Scattering Detection, *J. Am. Oil Chem. Soc. 71*, 877–886.
78. Miller, D.A., Blanch, H.W., and Prausnitz, J.M. (1991) Enzyme-Catalyzed Interesterification of Triglycerides in Supercritical Carbon Dioxide, *Ind. Eng. Chem. Res. 30*, 939–946.
79. Berg, B.E., Hansen, E.M., Gjørven, S., and Greibrokk, T. (1993) On-Line Enzymatic Reaction, Extraction, and Chromatography of Fatty Acids and Triglycerides with Supercritical Carbon Dioxide, *J. High Res. Chrom. 16*, 358–363.
80. Hayes, D.G., and Kleiman, R. (1996) Lipase-Catalyzed Synthesis and Properties of Lesquerolic Acid Wax and Diol Esters, *J. Am. Oil Chem. Soc., 73,* 1385–1392.
81. Jones, B.A., Shaw, T.J., and Clark, J. (1992) The Effect of Temperature on Selectivity in Capillary SFC, *J. Microcol. Sep. 4*, 215–220.
82. Hellgeth, J.W., Jordan, J.W., Taylor, L.T., and Khorassani, M.A. (1986) Supercritical Fluid Chromatography of Free Fatty Acids with On-Line FTIR Detection, *J. Chrom. Sci. 24*, 183–188.
83. Geiser, F.O., Yocklovich, S.G., Lurcott, S.M., Guthrie, J.W., and Levy, E.J. (1988) Water as a Stationary Phase Modifier in Packed-Column Supercritical Fluid Chromatography. I. Separation of Free Fatty Acids, *J. Chrom. 459*, 173–181.
84. Ibañez, E., Li, W., Malik, A., and Lee, M.L. (1995) Low Flow Rate Modifier Addition in Packed Capillary Column Supercritical Fluid Chromatography, *J. High Resol. Chrom. 18*, 559–563.
85. Nomura, A., Yamada, J., Tsunoda, K., Sakaki, K., and Yokochi, T. (1989) Supercritical Fluid Chromatographic Determination of Fatty Acids and their Esters on an ODS–Silica Gel Column, *Anal. Chem. 61*, 2076–2078.

Chapter 9

Oxidation Products of Conjugated Linoleic Acid and Furan Fatty Acids

M.P. Yurawecz, N. Sehat, M.M. Mossoba, J.A.G. Roach, and Y. Ku

Center for Food Safety and Applied Nutrition, U.S. Food and Drug Administration, Washington, DC 20204

Introduction

"Conjugated linoleic acid" or "CLA" is an imprecise term for a mixture of octadecadienoic fatty acid moieties containing two conjugated double bonds. In general, CLA refers to the octadecadienoic fatty acid moieties that are formed when the double bonds in linoleic acid, [9(Z),12(Z)-octadecadienoic acid] (Fig. 1, compound I), are shifted in their positions between carbons with or without a change in their *cis/trans* configurations. The imprecision, even with this broad definition, results from the use of the term "CLA" for many matrices that contain conjugated octadecadienoates, such as partially purified beef extract (1,2), alkali-isomerized vegetable oil (3), hydrogenated soybean oil (4), dietary supplements, and synthetic materials of varying composition.

Increasing interest in conjugated octadecadienes began in the 1980s when CLA was found to be anticarcinogenic in animal model systems (2) and when a new mechanism for *in vivo* formation of the so-called "diene conjugation" (DC) in humans was proposed (5–10). A simplified schematic for two different oxidation paths from linoleic acid to DC is shown in Fig. 1. Path A indicates the hydroperoxides (LOOHs) that are known to form (11–15) during the *in vitro* autoxidation of linoleic acid. The increase in UV absorption at 234 nm in a system containing oxidized linoleic acid has been generally accepted (15) to result from conjugation of the rearranged double bonds. In 1983, Cawood et al. (5) reported that linoleic acid, when exposed to UV irradiation in the presence of human albumin or gamma globulin, gives rise to a series of products that contain no additional oxygen but that undergo double bond isomerization with or without conjugation, and that one of these fatty acids had the same HPLC retention volume as the component that gives rise to DC in humans. In 1984, the identity of this fatty acid was reported to be octadeca-9,11-dienoic acid (6). In 1991, the configuration of this compound was reported to be 9Z,11E-octadecadienoic acid (compound II in Fig. 1) (7). On the basis of these results, these investigators (5–10) have suggested that the nonclassical path B shown in Fig. 1 is more indicative of *in vivo* oxidation, whereas path A is indicative of *in vitro* peroxidation (12). However, these studies do not address why only compound II is a major component of DC in humans.

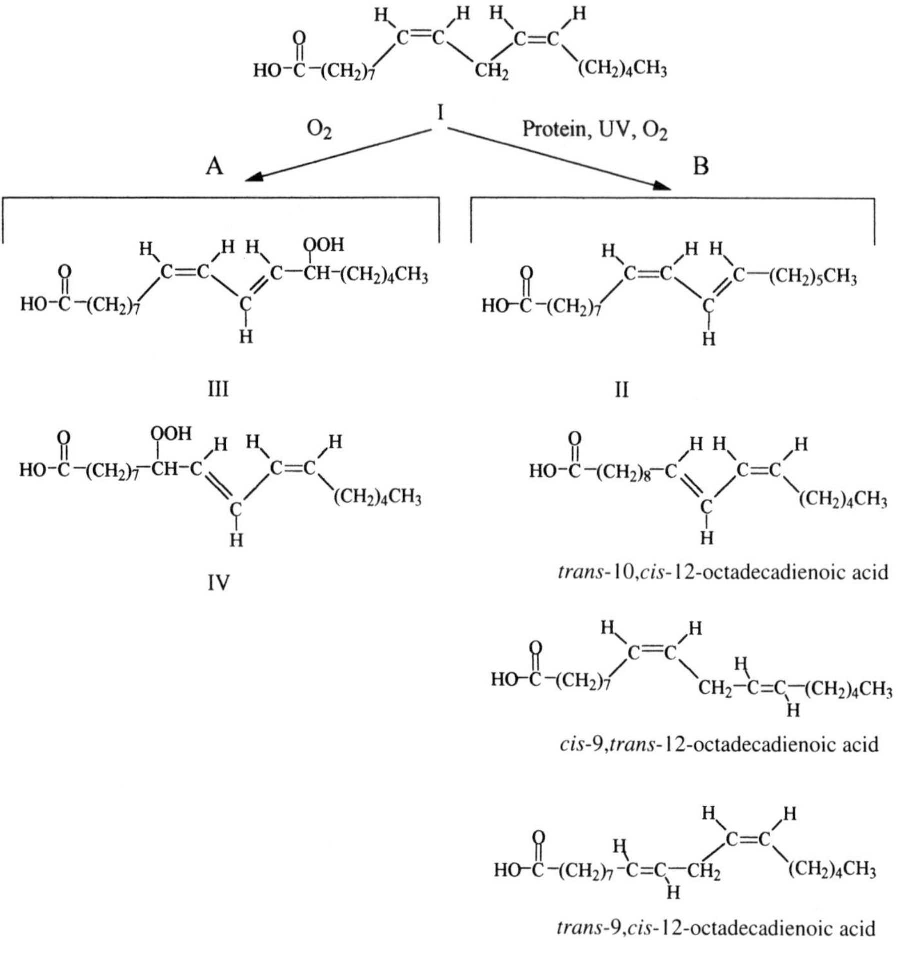

FIG. 1. Linoleic acid oxidation.

Dormandy and Wickens (10) have summarized the results of studies of the experimental and clinical pathology of DC as follows: Over 95% of DC in human serum, tissue fluids, and tissues (abnormal as well as normal) is due to the presence of a single fatty acid, compound II. Measurement of the ratio of compound II to linoleic acid may be a diagnostic tool for the detection of premalignant changes in the cervix. The level of compound II is elevated in the phospholipid-esterified serum fraction of chronic alcoholics. Phospholipid-esterified compound II is found at elevated levels in the bile and duodenal juice of patients with pancreatic disease.

Other research confirmed the presence of compound II in human serum and the presence of compound II following oxidation of linoleic acid in the presence of protein. It was also reported that compound II did not increase when human or rat blood was treated with UV irradiation or with phenylhydrazine, and it does not increase in the livers of rats treated with phenylhydrazine or bromotrichloromethane (16). These findings showed that the increase in compound II did not result from induced increases in the peroxidation process in rats.

In another study, an increase in DC was detected in the livers of rats fed a diet that included hydrogenated soybean oil (4). The increase in absorbance in this case was due to the presence of small amounts of CLA in the hydrogenated soybean oil. Many foods other than hydrogenated soybean oil contain CLA (17–21). The principal dietary sources of CLA are food products derived from ruminant animals: milk, beef, and cheese (18). As discussed in the chemistry section, CLA may also be derived from purified oleic acid.

In summary, the origins of compound II in humans are unknown. One should be cautious in interpreting increased UV absorption or changes in the ratio of compound II to linoleic acid without identifying the source (e.g., lipid oxidation or diet) of the conjugation (16,22).

Compound II was also suspected to be the active isomer in a series of experiments conducted primarily by Pariza and coworkers (1–3,23–31). Working independently of issues involving the origin of DC, Pariza found antimutagenic activity in an extract of fried ground beef (23). In subsequent work (2), examination of the extracts indicated that the material giving rise to the antimutagenic activity was a mixture of octadecadienoic acids. The isomers found in the fried ground beef extract could be synthesized by conjugating the double bonds in linoleic acid. Thus, the term "CLA" was coined. The authors speculated that compound II may be the active factor because more than 95% of the increase in conjugated octadecadienoate in phospholipids was due to the presence of compound II (1,25).

CLA has been reported to have anticarcinogenic activity in mouse epidermal tissue treated with 7,12-dimethylbenz[a]anthracene (DMBA) (2,24), inhibitory activity against benzp[a]pyrene-induced forestomach neoplasia in ICR mice (25), and inhibitory activity against the development of DMBA-induced mammary tumors in rats (26,27). In addition, CLA has been reported to modulate immune-induced growth suppression in chicks (28), to overcome catabolic responses due to injection of endotoxin (29), to reduce atherosclerosis and total cholesterol and to alter the LDL/HDL ratio in rabbits fed a hypercholesteremic diet (30). Male F344 rats treated with 2-amino-3-methylimidazo[4,5-*f*]quinoline (IQ) had significantly fewer aberrant colonic crypt foci and lower IQ-DNA adducts in the colon when treated with CLA by gavage (3). A more recent report shows that CLA inhibits the growth and metastases of human breast adenocarcinoma cells implanted in mice with severe combined immunodeficiency syndrome (SCID) (31). The authors postulated that dietary CLA blocked the growth and metastases of human breast cancer cells implanted in the

SCID mice *via* mechanisms independent of the host's immune system. The chemoprotective effects of CLA have been reviewed by Belury (32).

Furan fatty acids (F-acids), like CLA and DC, are fatty acids containing a conjugated diene system, although in the case of F-acids the system is a cyclic one. The link between F-acids and CLA was made recently with the recognition that F-acids are secondary oxidation products of CLA (33). Previously, researchers had used conjugated dienes as synthetic precursors to F-acids (34) and suggested that formation of a conjugated system was required for their *in vivo* formation (35). F-acids were first reported in *Santalaceae* seed oil in 1966 (36). Since that time, considerable efforts have been expended in determining the occurrence (37–40), function (41–45), synthesis (34,35,46–49) and origin of F-acids (50).

Methodology

Lipid chemistry and cyclodiene chemistry, in particular, are replete with examples of compounds that are artifacts of analytical procedures (38–40,42,43,51–53). CLA has been produced by injecting allylic hydroxy oleates (AHOs) into a gas chromatograph. F-acids have been detected from the GC injection of fatty acid esters containing either a 1,4-dioxo-ene or a cyclic peroxide functionality (34,38,52). Many compounds that contain these functionalities are present in foods, either because they occur naturally or because they are the result of secondary products of lipid autoxidation (11,15,51).

We determined by HPLC/UV, GC/MS, and GC/matrix isolation/Fourier transform IR that currently used acid-catalyzed methylation procedures convert AHO to CLA and other products, potentially resulting in misleading values for analysis of CLA in foods. A mixture of AHOs containing mainly (8- and 11-) hydroxy-9-octadecenoates was synthesized (51) and tested by using methylation procedures with the following catalysts: BF_3, HCl, NaOMe, and tetramethylguanidine (TMG). Both the BF_3 and HCl procedures converted AHO to CLA. The base-catalyzed procedures did not convert AHO to CLA and were therefore recommended for CLA analysis.

There is no *a priori* way to know whether AHO or other labile precursors are present in a sample. Of particular concern are samples that have undergone oxidative stress (e.g., frying, aging) or are derived from target organs of animals with induced pathologies. Methylation or extraction of these samples with strong acids may result in a low percentage of compound II relative to other CLA isomers that does not reflect the actual composition of the samples.

Procedures

Methyl esters of conjugated linoleic acid were obtained from Nu-Chek-Prep, Inc., Elysian MN. The precise isomeric distribution is unknown. The manufacturer's assay gave the following results: *c-9,t-11/t-9,c-11* 41%; *t-10,c-12* 44%; *c-10,c-12* 9.5%; *t-9,t-11/t-10,t-12* 1.3%; *c-9,c-11* 1%; *c-9,c-12* 0.7%.

CLA Oxidation (33,51,54)

Stearic acid ($C_{18:0}$), heptadecanoic acid ($C_{17:0}$), tridecanoic acid (C_{13}), or the methyl esters of these fatty acids were used as the internal reference for several of the oxidations described below. CLA acids were exposed to varying amounts of air, water, and methanol. The effects of the following variations were studied:

Oxidation 1 (Ox$_1$): Fixed quantities (75 mg CLA and 3.38 mg $C_{17:0}$) were dissolved in 1 mL methanol in test tubes. An aliquot of 0.0 mL, 0.1 mL, 0.2 mL, 0.5 mL, 1.0 mL, or 2 mL H_2O was added to each tube. The individual tubes, which contained air, were sealed and immersed in an oil bath and held at 45°C for 46 h.

Oxidation 2 (Ox$_2$): 0.5 g CLA was dissolved in 50 mL methanol in a 250-mL Erlenmeyer flask with a stirring bar and with air flowing through a pipet under the liquid surface at 100–200 mL/min. Then 100 mL water was added, and the suspension was heated on a stirrer-hot plate at temperatures varying from 48 to 69°C. Heating was stopped after 7.5 h and started again 16 h later. Methanol/water (1:2) was added as needed to maintain a volume of 100–150 mL. Aliquots (5 mL) were taken hourly and examined after methylation.

Oxidation 3 (Ox$_3$): Fixed quantities (44.2 mg CLA and 3.22 mg $C_{18:0}$) were dissolved in 2 mL methanol in test tubes. A micro stirring bar and 4 mL H_2O were added to each tube. The tubes were placed in an oil bath on a stirrer-hot plate and held at 40°C. The solutions were continuously stirred and aerated (2–5 mL/min). Methanol/water (1:2) was added as needed to maintain a volume between 4 and 6 mL. Individual solutions were removed from the oil bath at different times and examined after methylation.

Oxidation 4 (Ox$_4$): CLA methyl esters (109 mg) and methyl tridecanoate (10 mg) were dissolved in 10 mL of petroleum ether. One mL of this solution was added to each of 8 test tubes (110 mL volume), and the petroleum ether was removed from the tubes by a stream of argon. The tubes, containing air, were capped and placed in an oil bath at 50°C. Individual tubes were removed from the bath at different times and stored for at least 1 hr at −20°C prior to GC analysis. Samples were dissolved in 2 mL of isooctane for injection. A similar experiment was performed with the methyl F-acid, 9,12-epoxy-9,11-octadecadienoate ($F_{9,12}$), in place of CLA.

Oxidation 5 (Ox$_5$): Performed as in **Ox$_4$** except that the tubes were not capped and were held at room temperature, 50°C and 75°C.

Methylation Procedures

Internal standards (0.1 or 1.0 mg) were added to each sample before methylation. In the methylation procedures described below, the final solvents were dried with Na_2SO_4; hexane was used for UV analyses; isooctane was used for GC analyses.

Boron trifluoride (BF$_3$). A procedure for methylation of marine oils was used (55). Briefly, the oil was treated with 0.5 N NaOH/methanol for 7 min at 100°C, followed by BF$_3$/methanol for 5 min at 100°C. Excess BF$_3$ was reacted with saturated NaCl/H$_2$O, and fatty acid methyl esters (FAME) were extracted into isooctane.

Methanolic HCl. The procedure was based on a procedure recommended by Supelco, Inc. (Bellefonte, PA). A mixture of 7 mL ethyl ether, 0.5 mL 2,2-dimethoxypropane and 2.5 mL 3 N methanolic HCl was added to up to 100 mg oil in a 15 mL test tube with a Teflon liner. The tube was purged with argon and placed in the dark overnight at room temperature. The reaction mixture was added to 100 mL water and 10 mL hexane and shaken for 1 minute in a 250 mL separatory funnel to extract the methyl esters. The aqueous layer was removed and the hexane layer was washed with water and dried over anhydrous Na$_2$SO$_4$. The solvent was evaporated under argon and the residue redissolved in isooctane.

Tetramethylguanidine (TMG). The procedure used was based on that of Schuchardt and Lopes (56). The oil was added to a reaction tube containing 2 mL of 20% TMG/methanol. The tube was sealed and heated to 100°C for 2 min. The reaction product was transferred to a tube containing 20 mL of saturated NaCl/H$_2$O and vortexed for 1 min. Five mL of hexane or isooctane was added to the tube; then the tube was vortexed for 1 min.

Sodium methoxide. 10 mL 0.5 N NaOCH$_3$-methanol was added to up to 100 mg oil in a 15 mL test tube with a Teflon liner. The tube was vortexed and briefly immersed in hot water to dissolve the oil, then purged with argon and placed in the dark overnight at room temperature. The methyl esters were extracted with two 5-mL portions of isooctane.

Diazomethane. Five mL saturated NaCl/H$_2$O and 5 mL ethyl ether were added to each test portion in individual test tubes. Each tube was vortexed for 30–60 s. The ether fractions were dried over Na$_2$SO$_4$. One mL diazomethane in ether solution was carefully and slowly added to the dried ether portion of each sample. After 20 min, 1–2 drops of acetic acid were added to react with the remaining diazomethane. The solutions were injected into the gas chromatograph without further cleanup.

Instrumentation

Gas chromatography was performed using a Hewlett-Packard 5890A instrument under the following conditions: column, 50 m × 0.25 mm i.d. CP Sil 88 (Chrompack, Raritan, NJ) capillary; helium carrier gas; flame ionization detector; injector 220°C; detector 280°C; column 75°C for 2 min, then raised 20°C/min to 185°C and held for 33 min, then raised at 4°C/min to 225°C. Samples were run in both split and splitless modes.

Low- and high-resolution EI GC/MS analyses were obtained with a Hewlett-Packard 5890 series II gas chromatograph coupled to a Micromass (Manchester, UK) Autospec Q mass spectrometer and OPUS 4000 data system. The GC/MS system utilized version 2.1BX software.

The same capillary GC column used to obtain FID data was also used to obtain GC/MS data. Adjusting the capillary GC column head pressure to 10 psi gave chromatography comparable to that used to obtain the GC FID data. The GC/MS conditions were: splitless injection with helium sweep restored 1 min after injection; injector and transfer lines 250°C; oven 75°C for 2 min after injection, then 20°C/min to 185°C, hold 185°C 15 min, 4°C/min to 225°C, hold 225°C 5 min.

The mass spectrometer was tuned to a resolution of 1000 (5% valley) by observing m/z 305 in the EI mass spectrum of perfluorokerosene (PFK). The mass scale was calibrated with PFK for magnet scans from 440 to 44 daltons at 1 s per decade. Filament emission was 200 µA at 70 eV. Ion source temperature was 250°C. High-resolution MS data were obtained by GC/MS with the MS operated in voltage scan mode at 11,000 resolution.

A Mattson Instruments Model Sirius 100 FTIR spectrometer equipped with a matrix isolation (MI) Cryolect interface operated at 12% K under vacuum, and a Bio-Rad direct-deposition (DD) Tracer FTS-60A system were used with gas chromatography to obtain Fourier transform infrared (FTIR) spectra. These systems were used with a CP-Sil-88 capillary column and have been described in detail (57).

Analytical Chemistry

Detection of CLA as fatty acid methyl esters (FAME) is usually accomplished by gas chromatography (GC) (18) using a flame ionization or MS detector or by high-performance liquid chromatography (HPLC) using a UV detector (18). Isolation of lipid materials from various matrices is the same for CLA as any other lipid analysis except that the use of strong acids should be avoided.

At least two major considerations are important for successful analyses: First, high-purity reference materials are generally not available for isomers other than the *cis*-9, *trans*-11 or *trans*-9, *trans*-11. The second consideration is that derivatization of fatty acid to FAME using strong acid catalysts (e.g., BF_3/methanol or HCL/methanol) may yield misleading results. These results include reduced yield of CLA isomers relative to other methylation procedures (53), creation of CLA artifacts (51), and drastic changes in the relative amounts of CLA isomers. This is illustrated in Figs. 2 and 3.

In Fig. 2, chromatogram A shows a test portion of olive oil methylated with BF_3/methanol, and chromatogram B shows another test portion from the same oil source that has been methylated with TMG/methanol. The CLA isomers indicated in Fig. 2A, as well as later eluting methoxy octadecenoates, are artifacts of the procedure. The amount of artifactual *t,t*-CLA that is shown amounts to 0.2%. Much lower amounts of *c,t*-CLA are generated by the BF_3 methylation procedure. The chro-

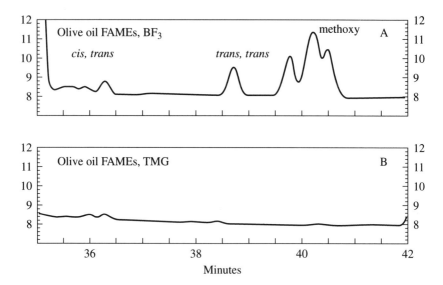

FIG. 2. Partial chromatograms (CP Sil 88) of FAME obtained from the same olive oil methylated using either (A) BF_3; or (B) TMG.

matogram in Fig. 2B demonstrates that no detectable CLA isomers are present in this sample of olive oil. A high ratio of $t,t/c,t$ isomers in naturally occurring samples containing CLA may be indicative of an unacceptable methylation procedure (18,51,53). Figure 3 shows chromatograms obtained using a 100 M SP 2560 capillary column of milk lipids methylated with NaOH/methanol (A) and with HCl/methanol (B). In addition to the obvious change in *cis/trans* isomerization, the total CLA (relative to internal standard, not shown) quantitated in A is 20 % higher than that in B. These chromatograms were generously provided by John K.G. Kramer, Agriculture and Agri-Food Canada.

Compounds such as AHOs can lead to formation of CLA artifacts. High ratios of $t,t/c,t$ isomers may be indicative of an unacceptable methylation procedure. However, the presence of CLA in foods such as hydrogenated (4) or unhydrogenated (58) soybean oil or purified oleic acid should not be dismissed. Figure 4 shows the UV spectra of a highly purified reference standard of oleic acid. The absorbance maximum shown was confirmed by GC to be due to CLA. CLA was detected (typically 0.1–0.2%) in all the oleic acid and methyl oleate standards that we have examined in our laboratory. Diene conjugation, probably due to the presence of CLA, has also been reported in triolein (58). Formation of hydroxides at allylic positions to double bonds can occur naturally or as the result of reduction of hydroperoxides. Bleaching, distilling and other processing of purified fatty acid products leads to dehydration of the allylic hydroxide, producing an extra double bond conjugated with the bond to which the -OH was allyic. This results in the presence of a small amount of a fatty acid containing one extra double bond (conjugated) in purified

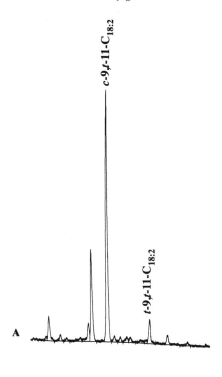
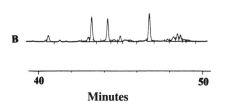

FIG. 3. Partial chromatograms (SP 2560) of FAME obtained from the same milk lipids methylated using either (A) sodium methoxide; or (B) methanolic HCl.

fatty acid reference materials. Thus, CLA is a low-level contaminant of oleic acid; conjugated triene is a low-level contaminant of linoleic acid, and so forth. The origin of compound II in humans has not been established, so it is premature to eliminate the possibility that it originates *in vivo* via an octadecenoate (e.g., oleate or vaccinate) precursor.

The availability of derivatives other than FAME offer certain advantages. Of special note are 4,4-dimethyloxazoline (DMOX) derivatives, which, when examined by GC/EI/MS, can indicate the position of the double bonds in the chain (59,60). Reaction of the methyl esters of fatty acids with 2-amino-2-methyl-1-propanol results in a DMOX derivative that strongly directs EI remote site fragmentation of the carbon chain.

The determination of double bond location by EI GC/MS is illustrated in Fig. 5. The EI mass spectra in Fig. 5 were recorded for the DMOX derivative of 9,11-CLA

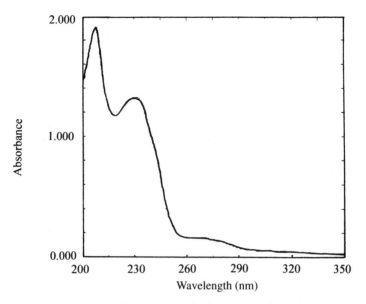

FIG. 4. UV spectra of a commercial reference standard of oleic acid, 1 mg/mL in hexane.

(Fig. 5A) and the DMOX derivative of a compound extracted from cooked hamburger (Fig. 5B). The 9,11-CLA was obtained from the reaction of allylic hydroxy oleate (AHO) with BF_3. The cooked-hamburger compound was obtained via TMG methylation of the fat extracted from cooked hamburger (51). The two DMOX derivatives have the same GC retention time. Their EI mass spectra are characterized by an ion series consisting of a loss of 15 from the molecular ion followed by successive losses of 14 until both carbons of a double bond in the carbon chain are lost with the neutral species. At this point, the observed loss in the ion series is 12, rather than 14. The difference of 12 in the ion series is a clear indication of the location of a double bond in the carbon chain.

Confirmation of *trans/trans* and *cis/trans* isomers by GC/MI/FTIR is shown in Figs. 6 and 7. The GC/MI/FTIR of *trans,trans* CLA is shown in Fig. 6. The spectra at the retention time of *trans,trans* CLA are presented for (A) reference 9,11-CLA obtained commercially, (B) the reaction of AHO with BF_3, and, (C) FAME derived from an olive oil *via* the BF_3 procedure. Resolution is 4 cm^{-1} for the spectra shown. Bands at wavenumbers 3005 and 3022 are for =C-H stretch, and the band at 990 is for the =C-H out-of-plane deformation. Confirmation of *cis/trans* CLA isomers are shown in Fig. 7 at the same resolution for the same injected portion. Bands near wavenumbers 3011 and 3032 are for =C-H stretch, and those near 985 and 950 are for out-of-plane deformation of =C-H (51).

The IR spectra of F-acids have been previously reported (36). The increased resolution of the GC/MI/FTIR technique is demonstrated in Fig. 8. The GC/MI/FTIR

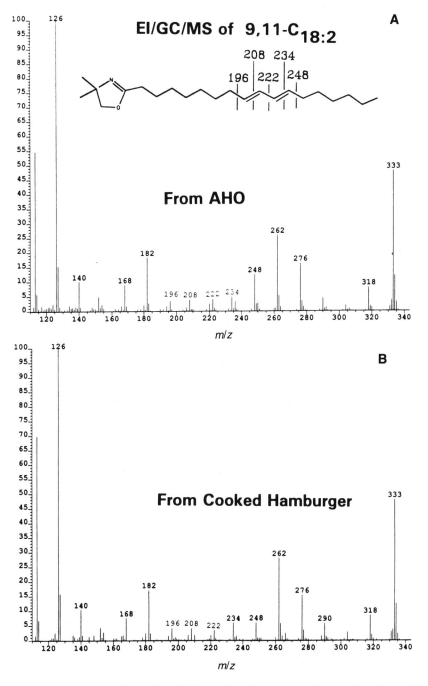

FIG. 5. EI GC/MS of 4,4-dimethyloxazoline derivatives of 9,11-C$_{18:2}$ from (A) reaction of AHO with BF$_3$; (B) product obtained from TMG methylation of fat from cooked hamburger.

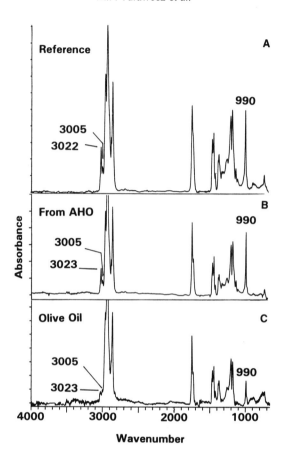

FIG. 6. GC/MI/FTIR spectra, resolution 4 cm^{-1}, of conjugated t,t-$C_{18:2}$ compounds from (A) a commercial reference material; (B) products of the reaction of AHO with BF_3; (C) olive oil FAME derived with BF_3.

of the methyl F-acid, $F_{9,12}$, is shown in Fig. 8A. The labeled bands are indicative of a FAME. The spectrum of an oxidation product of methyl CLA at the same retention time as $F_{9,12}$ is shown in Fig. 7B. The labeled bands are indicative of a methyl F-acid ester. Tables 1 and 2 list the corresponding functional groups that yielded the characteristic frequencies in Fig. 8. The GC/MI/FTIR spectra of other methylated 2,5 disubstituted F-acids are very similar to that observed for $F_{9,12}$ (33).

The analysis of 2,5 disubstituted F-acids by GC/MS has also been previously reported in detail (35,36). Figure 9 shows the mass spectrum of an unusual F-acid, methyl 11,14-epoxy-11,13-octadecadienoate ($F_{11,14}$) identified as an oxidation product of a commercial CLA mixture (33). A series of low-mass ions in the spectrum (m/z 67, 81, and 95) would be consistent with the mass spectra of dienes or

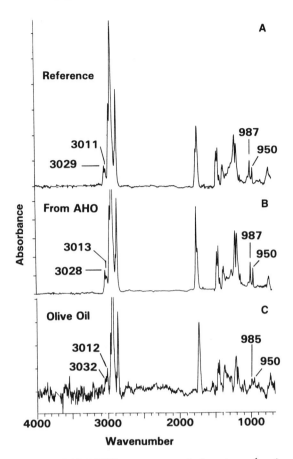

FIG. 7. GC/MI/FTIR spectra, resolution 4 cm^{-1}, of conjugated c,t-$C_{18:2}$ compounds from (A) a commercial reference material; (B) products of the reaction of AHO with BF_3; (C) olive oil FAME derived from BF_3.

cyclic oxygen compounds. High-resolution MS confirmed that the elemental composition of m/z 95 was C_6H_7O. Cleavage of the carbon chain beta to the furan ring gives rise to m/z 137 and 265. These proposed fragmentations for the principal ions in Figure 9 are consistent with their experimentally determined elemental compositions and published spectra.

Oxidation Products

Primary oxidation products of conjugated dienes are not similar to those of methylene-interrupted systems (such as linoleic acid). LOOHs, such as III and IV (see Fig. 1), are not stable primary products of CLA or F-acids. Conjugated double bond

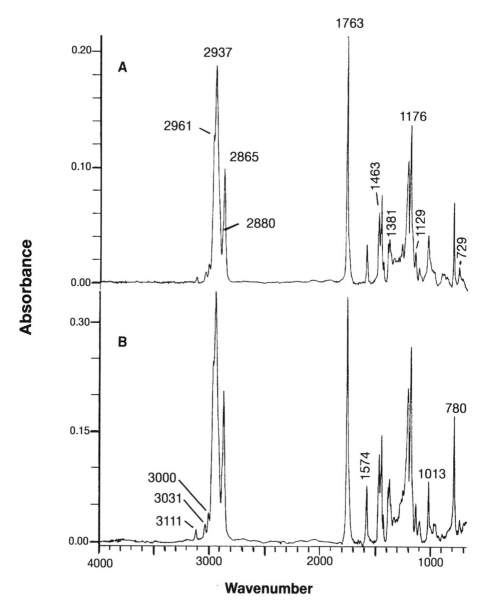

FIG. 8. GC/MI/FTIR spectra of (A) synthesized $F_{9,12}$; and (B) a compound at identical retention time in the methylated oxidation mixture. The wavenumbers shown on A refer to those bands indicative of a FAME compound; and B refer to those bands indicative of a 2,5 disubstituted furan.

TABLE 1
FAME MI-FTIR Bands at 4 cm^{-1} Resolution

Position (cm^{-1})	Assignment
2961	**CH$_3$** asym. str.
2937	**CH$_2$** asym. str.
2880	**CH$_3$** sym. str.
2865	**CH$_2$** sym. str.
1753	Ester **C=O** str.
1463	**CH$_2$** scissors
1381	**CH$_3$** sym. scissors
1176	Ester sym. **C–O** str.
1129	**CH$_3$** in-plane rock
729	**CH$_2$** rock

TABLE 2
Furan Ring Bands by MI-FTIR at 4 cm^{-1} Resolution

Position (cm^{-1})	Assignment
3111, 3031, 3000	**=C–H** str.
1574	C=C str.
1013	**C–O** str.
780	**=C–H** out-of-plane bend

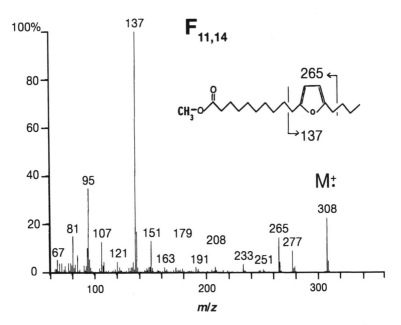

FIG. 9. EI mass spectrum attributed to methyl 11,14-epoxy-11,13-octadecadienoic acid (F$_{11,14}$).

systems are classically thought to be very unstable in the presence of O_2 (61). CLA and F-acids are very stable in triacylglycerols and in methanol (CLA) or borate buffer (F-acids) solutions, even with applied heat and aeration. However, they will oxidize when exposed to air on the laboratory bench (33,44,54).

Singlet Oxygen ($^1O_2$)

The singlet oxygen oxidation of conjugated diene systems is well established (34,62–64) and has been reported for the octadecadienoate FAME system (34). Fig. 10 shows that the *t,t* double bond configuration (V) is most suitable for the $^1O_2$ synthesis of the cyclic peroxide (VI). The identification of compound VI is based on its proton NMR spectrum after isolation by TLC. Analysis of VI following its synthesis showed that VI had the same retention time and mass spectrum as the FAME of the 2,5-disubstituted F-acid (VII) (34). This is consistent with the scheme in Fig. 11, in which VI is converted by heat or acid to [methyl 9,12-epoxy-9,11-octadecadienoate ($F_{9,12}$)] VII (34,64). Path A presumes the next logical Diels-Alder addition product. In Path B, it is shown that by using $^1O_2$, lipoxygenase-1 (LOX) [LOOH is necessary for this reaction, as reported elsewhere (43,44)] or meta-

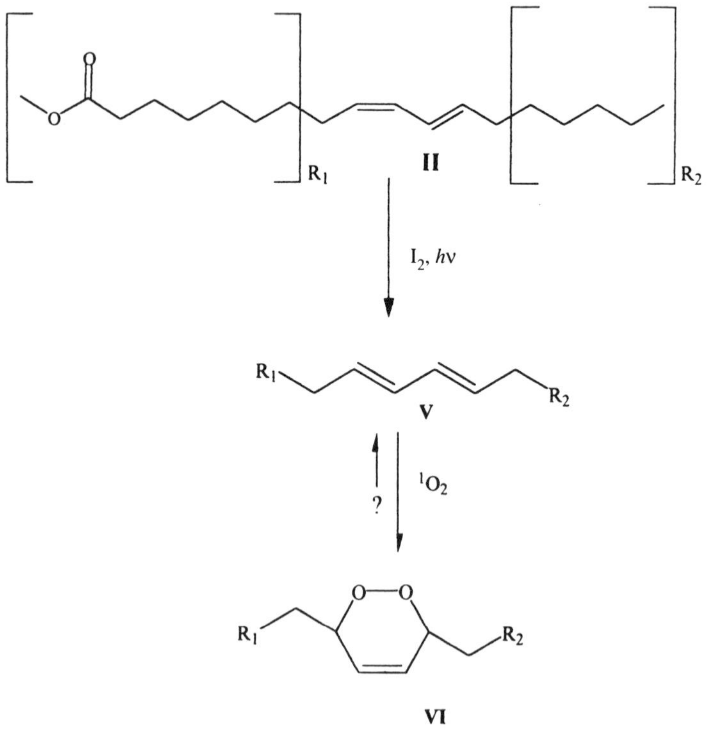

FIG. 10. Singlet oxygen oxidation of CLA.

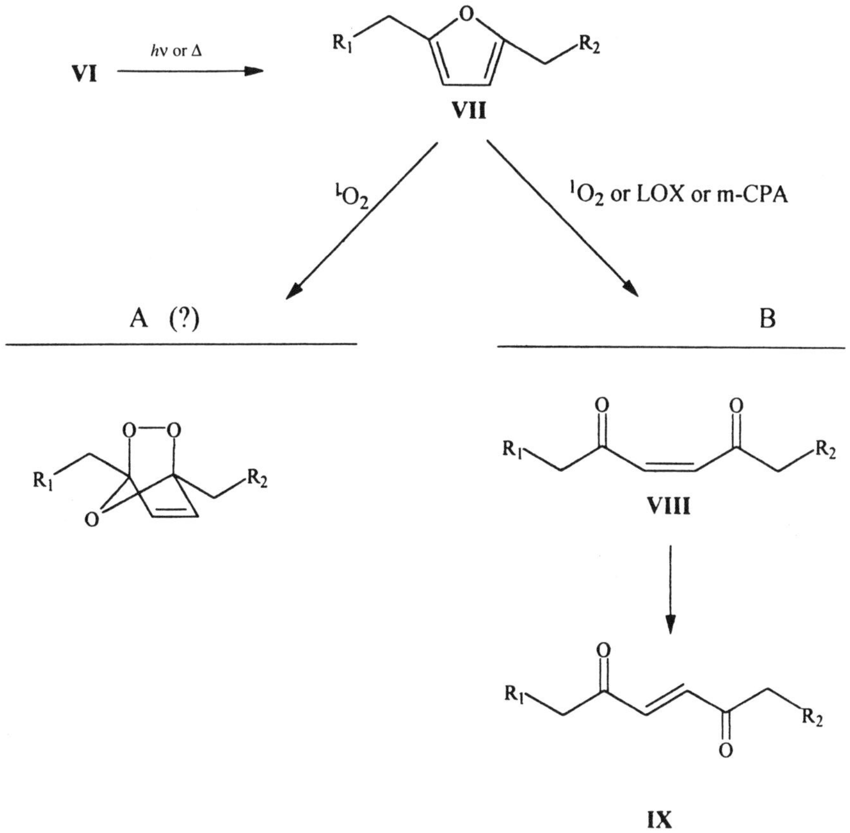

FIG. 11. Singlet oxygen oxidation of methyl F-acid.

chloroperoxybenzoic acid (m-CPA), VII can be oxidized to methyl 9,12-dioxo-*cis*,10-octadecenoate (VIII), which can isomerize to methyl 9,12-dioxo-*trans*,10-octadecenoate (IX) [Ref. (45) reports the reaction for methyl 10,13-epoxy-10,12-octadecadienoate ($F_{10,13}$), whereas VII corresponds to $F_{9,12}$].

Triplet Oxygen ($^3O_2$)

CLA was very stable in methanol solutions (33). Experiment Ox_1 (see the Procedures section) was performed to determine whether addition of H_2O would increase the rate of oxidation. No F-acids or other volatile compounds were detected when methanol or 10% H_2O/methanol was used as the solvent. The F-acids produced with increasing percent H_2O were 1.9 mg FFA with 20% H_2O; 2.8 mg FFA with 33% H_2O; 3.0 mg FFA with 50% H_2O; and 2.3 mg FFA with 67% H_2O. Higher aqueous solvent mixtures resulted in the CLA coming out of solution to form a suspension. The yields of F-acids increased in inverse proportion to the amount of CLA in solution.

Open air oxidations (54) were performed at room temperature, 50 and 75°C. Half lives (see **Ox$_5$**, in Procedures) for the CLA FAME were: 14 d at ambient temperature, 20 h at 50°C, and 7 h at 75°C. Figure 12 shows a GC/MS chromatogram of the oxidation products of methyl CLA at 75°C that includes the peaks of the methyl F-acids. The identities of the numerically indicated peaks as determined by GC/MS and GC/MI/FTIR are given in Table 3. The primary oxidation products are not known, but they consist of at least two types. Type 1 (Path A in Fig. 13) is similar in result to a 1,4 addition of O_2 because the resulting F-acids are bounded by oxygen at the 1,4 positions of the conjugated diene system. The other types (Path B in Fig. 13) appear to result from reaction pathways involving attack by triplet oxygen on the π system. The primary autoxidation products of CLA shown in Fig. 13 have not been identified (33). In Path A the primary Type 1 oxidation shows parenthetically, $VI_{m,n}$. This has not yet been confirmed in the autoxidation reaction. However, a series of compounds were detected in the autoxidized matrix with molecular weights of 326 and mass spectra that were similar to methyl F-acids.

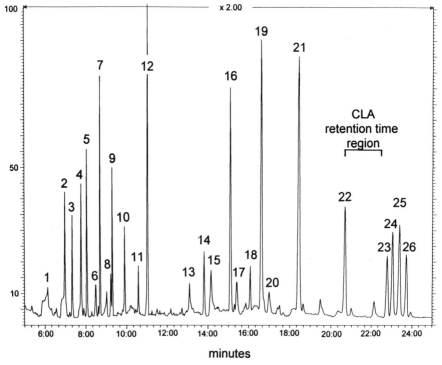

FIG. 12. Total ion current chromatogram (CP Sil 88) of identified methyl CLA oxidation products (**Ox$_5$**) that elute at or before the retention time of methyl F-acids.

TABLE 3
Volatile Oxidation Products of CLA Eluting Before F-Acids

Peak #	Name
1	Hexanal
2	Heptanal
3	Methyl heptanoate
4	Octanal
5	Methyl octanoate
6	trans-2-heptanal
7	Methyl nonanoate
8	trans-2-octanal
9	Methyl decanoate
10	trans-2-nonenal
11	trans-2-decanal
12	Methyl tridecanoate (internal reference)
13	Lactone, R-CHO, α, β-unsaturated
14	Methyl 7-oxo-heptanoate
15	Lactone, R-CHO, α, β-unsaturated
16	Methyl 8-oxo-octanoate
17	Lactone, R-CHO, α, β-unsaturated
18	Methyl octadecenoate
19	Methyl 9-oxo-nonanoate
20	Lactone, R-CHO, α, β-unsaturated
21	Methyl 10-oxo-decanoate
22	Methyl 11-oxo-undecanoate
23	Methyl 8,11-epoxy-8,10-octadecadienoate ($F_{8,11}$)
24	Methyl 9,12-epoxy-9,11-octadecadienoate ($F_{9,12}$)
25	Methyl 10,13-epoxy-10,12-octadecadienoate ($F_{10,13}$)
26	Methyl 11,14-epoxy-11,13-octadecadienoate ($F_{11,14}$)

This is consistent with the presence of a series of compounds similar to VI (34). In our laboratory (54), we found that BF_3/methanol treatment of the oxidized extracts typically doubles the amount of methyl F-acids such as VII. This is consistent with an intermediate such as VI (34).

In other oxidations (see the Procedures section above) of CLA and 2, 5-disubstituted methyl F-acids, we measured an increase in 2,5-disubstituted methyl F-acids, even when the remaining amount of CLA was negligible and even though 2,5-disubstituted methyl F-acids have a shorter half-life [ca 35 h in the Ox_4 reaction (at 50°C the half-life of $F_{9,12}$ using a restricted amount of O_2 is ca. 35 h)] than CLA [ca. 40 h in Ox_4 (at 50°C the half-life of CLA FAME using a restricted amount of O_2 is ca. 40 h; with unrestricted O_2 the half-life is ca. 20 h)] under similar conditions. In the experiment used to construct Fig. 14, tubes containing 110 mL O_2 and 11 mg CLA were sealed airtight to exclude ambient singlet oxygen because it is known to interfere in open air experiments (65). At 93 h, CLA was at its detection limit because of interference with other compounds (e.g., methyl 11-oxo-undecanoate). Methyl F-acids, shown at 10 times the levels found in Fig. 14, were still detectable after 140

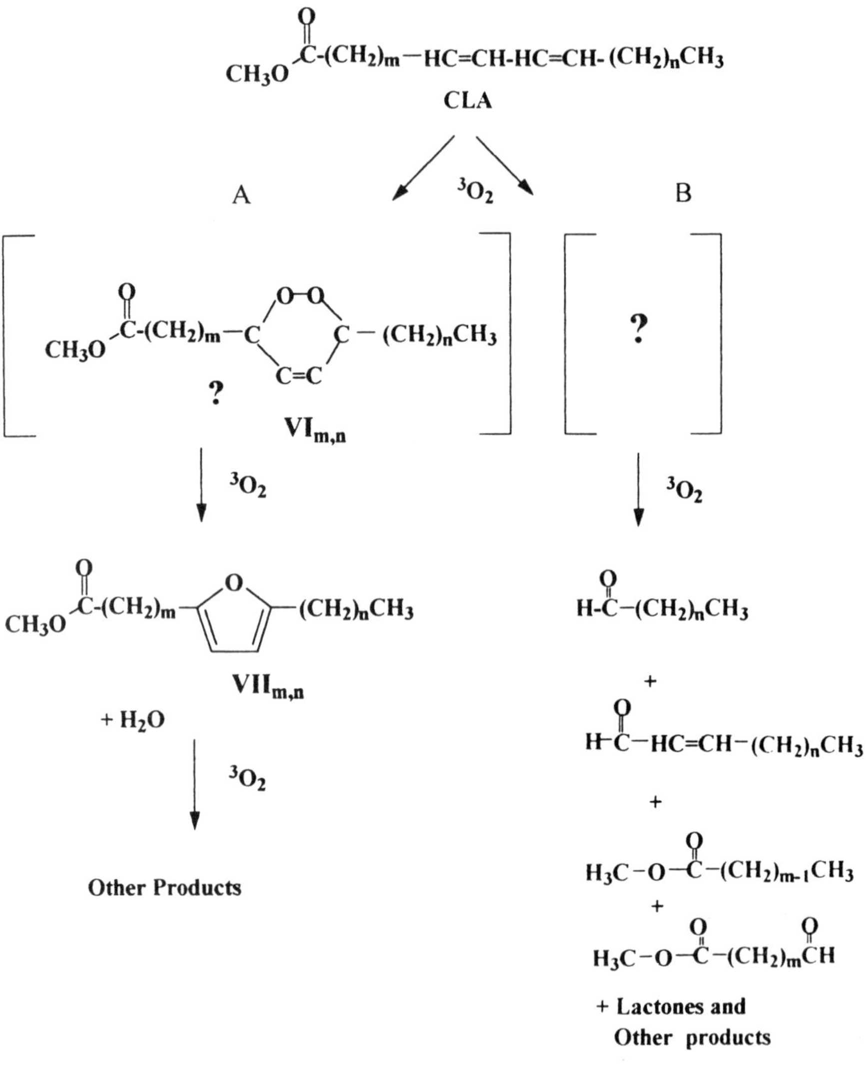

FIG. 13. Methyl CLA autoxidation.

h. Figure 15 shows the oxidation of VII ($F_{9,12}$) under oxidation conditions (Ox_4) similar to those shown in Fig. 14. The half life of VII ($F_{9,12}$) was ca 35 h. The compounds X and XI in Fig. 15 are oxidation products of VII and are discussed below. The increase and decrease in levels of methyl octanoate are shown in Fig. 16 for open-air oxidation experiments at 50 and 75°C. These curves are typical for all the products (aldehydes, medium chain FAME, 2-enals, al-esters) indicated in path B of the scheme shown in Fig. 13. After a given amount of time, which is tempera-

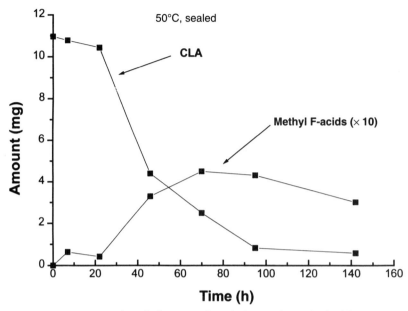

FIG. 14. Responses of methyl CLA and methyl F-acids (multiplied by 10) vs. oxidation time at 75°C (**Ox₄**).

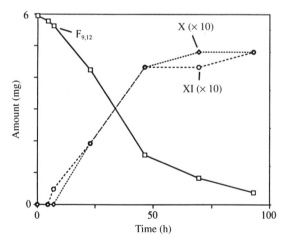

FIG. 15. Responses (adjusted by area of methyl $C_{18:0}$) of $F_{9,12}$ (VII), and oxidation products X and XI (multiplied by 10) (see Fig. 17 for structures) vs. **Ox₄** oxidation time.

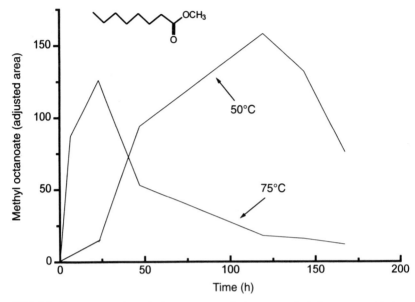

FIG. 16. Responses of methyl octanoate relative to methyl $C_{13:0}$ vs. oxidation (Ox_5) time at 50 and 75°C.

ture-dependent, the levels of all GC-determined compounds decline, even those as chemically stable as octanoic FAME, presumably because of free radical-induced polymerization. Identification of the specific lactones (Table 3) identified generically by GC/MS and GC/MI/FTIR may yield more definitive clues to the mechanism of CLA autoxidation. If CLA oxidizes to VI during autoxidation, as it does during singlet oxygen oxidation, the reaction might be reversible (see Fig. 11). Reversible reactions have been reported for other primary oxidation products (13). In this case, it would be a reverse Diels–Alder reaction. However, the expected diene product of the reverse reaction may be in the *trans, trans* configuration. This may partially explain why the *trans, trans* isomer is usually found accompanying compound II in sample analyses.

The oxidation of VII ($F_{9,12}$) is illustrated in Fig. 17. Path A shows that VII can be converted to VIII in the presence of a hydroperoxide, $^1O_2$, or radicals and that VIII is readily converted to IX in solution [(43–45), where Ref. 45 refers to $F_{10,13}$; Ref. 43 and 44 refer to $F_{9,12}$ (see Table 3 for structures)]. The identification (66) of two isomeric compounds—methyl 8-oxo-9,12-epoxy-9,11-octadecadienoate (X) and methyl 13-oxo-9,12-epoxy-9,11-octadecadienoate (XI)—from the autoxidation of VII is shown in path B. Path C shows a series of representative compounds produced by oxidation of VII in the presence of silica (42). The products 9,12-dioxo-10,13-octadecadienoate (XII) and 9,12-dioxo-7,10-octadecadienoate (XIII) in Path C contain a 1,4-dioxo-2-ene function that may be responsible for inhibition of

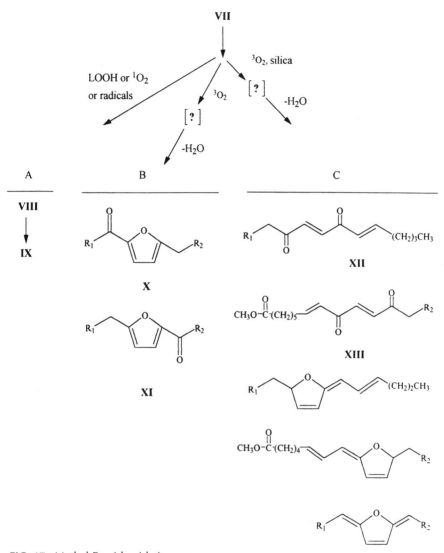

FIG. 17. Methyl F-acid oxidation.

bacterial urease (42). Compounds X and XI are of identical composition as XII and XIII ($C_{19}H_{30}O_4$). The circumstances by which these groups of compounds may undergo interconversion with each other are under investigation by our laboratory.

Figure 18 shows a FID chromatogram from the GC analysis of the oxidation products of VII ($F_{9,12}$) after 93 h at 50°C in a sealed tube (the **Ox$_4$** reaction). The retention time of compound VII was 35 min, whereas those of compounds X and XI

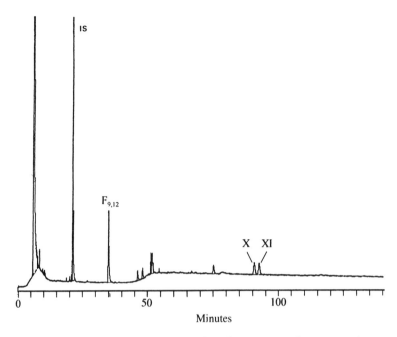

FIG. 18. Chromatogram (CP Sil 88) of oxidation (Ox_4) of $F_{9,12}$ at 93 h.

were at 90–95 min. Figure 19 shows the EI mass spectrum of compound X. The presence of keto and furan functional groups in compounds X and XI suppresses many of the usual ester ions. Loss of the elements of water (m/z 304) from the molecular ion (m/z 322) and the presence of the acylium ion (m/z 291) are consistent with a methyl ester structure. A characteristic methyl ester ion series, 87, 101, 115, 129, 143, 157, ... , is not evident in the mass spectra of either compound.

The abundant ions in the mass spectra of X and XI arise from fragmentation directed by the keto and furan functions. The most readily understood processes that may occur are gamma hydrogen rearrangement, gamma cleavage, cleavage beta to the oxygen of the carbonyl, and cleavage beta to the ring. The structures of these two compounds are such that postulated competing ionization processes frequently produce isobaric products, so the task of distinguishing between the two molecules requires rationalizing the fragmentations that are unique for each structure.

Gamma hydrogen rearrangement to a carbonyl can occur at the ester or keto function. This process is the origin of m/z 74 in the spectra of methyl esters. Gamma hydrogen rearrangement at the keto function will give rise to an abundant ion at m/z 194, if the keto is at C8, or 266, if the keto is at C13.

Gamma cleavage may proceed through the enol form of the ketone to produce a stable 4-membered ring as the product ion. For the 8-oxo compound this gives rise to m/z 207, and for 13-oxo the product ion is m/z 279.

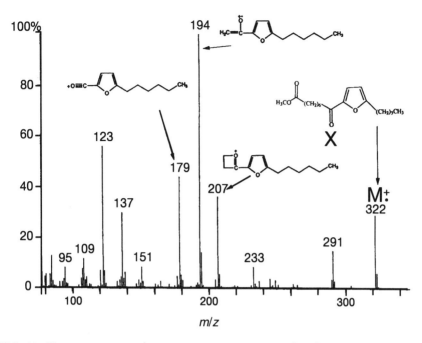

FIG. 19. EI mass spectrum of 8-oxo-9,12-epoxy-9,11-octadecadienoate (X).

Cleavage beta to the oxygen of the carbonyl and cleavage beta to the ring can give rise to isobaric species from the two compounds. The low relative abundance of ions attributed only to cleavage beta to the ring suggests that cleavage beta to the oxygen of the carbonyl is the preferred pathway. Thus, the abundant ion at m/z 179 in the mass spectrum of GC peak X is suggested to arise from cleavage adjacent to the carbonyl. These data are consistent with an interpretation that peak × is the 8-oxo compound.

Additional GC/MS and GC/DD/FTIR data facilitated this interpretation. High-resolution MS was used to determine the elemental compositions of the ions in the EI mass spectra of compounds × and XI. GC/DD/FTIR revealed bands at 1665 cm^{-1} (C=O stretch) and 1588 and 1517 cm^{-1} (C=C stretch) indicative of conjugation of a carbonyl group with a C=C double bond.

A significant difference in the oxidation of conjugated dienes vs. methylene-interrupted fatty acid moieties is that the consumed oxygen results in products that have very different properties. This is illustrated for conjugated dienes in the following stoichiometric sequence:

CLA + $^3O_2$ → [?] – H_2O → VII + $^3O_2$ → [?] – H_2O → X or XI (or XII and XIII).

The oxygen consumed is converted to H_2O and more unsaturated compounds or unidentified compounds that contain additional oxygen. This contrasts with the propagation step in autoxidation schemes for methylene-interrupted polyunsaturated fatty acids, in which oxygen consumed in the reaction is primarily converted to LOOHs (11–15).

Summary

More detailed knowledge of the *in vivo* oxidation and other pertinent reactions of compounds I and II is needed to account for the variety of physiological responses reported (see the Introduction section). It is possible that more than one mechanism is responsible for such a broad range of physiological activities. It should be clear from the issues covered in this review that one must be very careful in validating methodology for findings of CLA or DC during sample analysis. Artifacts, *cis/trans* isomerizations, and low recoveries may occur with the use of strong acids in methylation or extraction procedures. Intake of various foods can result in apparent DC when no lipid oxidation is indicated (4). The oxidation of conjugated diene systems, including CLA and F-acids, with singlet oxygen has been well studied (34). Autoxidation of CLA and F-acids is less well understood from the point of view of the mechanisms involved.

How does the oxidation of CLA and possibly F-acids relate to physiological effects of compound II in human serum and tissues or to CLA in the human diet? Is CLA an antioxidant? It has been reported that CLA inhibits the production of linoleic acid LOOHs more effectively than α-tocopherol and in a manner comparable to that of BHT in a system consisting of phosphate buffer (pH 8), water, and ethanol (25). In other experiments, it was shown that CLA failed to protect linoleic acid either as a free acid or as a phospholipid (1-palmitoyl-2-linoleoyl phosphotidylcholine) under conditions of metal ion-dependent and independent oxidative stress (67). In this system, BHT and α- tocopherol were effective antioxidants. It is widely recognized that a test compound may or may not serve the function of an antioxidant, depending on the relative amount of the test compound used and many other variables. It is clear that at this time, the definitive experiment that reveals how the potential antioxidant effect(s) of CLA relate to chemoprotective mechanism(s) has not been performed.

Does CLA have a role in basic cellular repair mechanisms? Compounds that contain the *trans* dioxo-ene functionality, such as compounds IX, XII, and XIII, are reported to be strong inhibitors of aggregation induced by thrombin and arachidonic acid (68). Batna and Spiteller have hypothesized a general scheme by which I is isomerized to II and transformed to compounds with the *trans* dioxo-ene functionality (43,44). They further hypothesize that these types of compounds covalently bond with proteins containing thiol groups to change the nature of plant cells in a manner not yet understood (43,44). As discussed above, the *in vitro* oxidation of CLA proceeds through an F-acid intermediate before products are produced that are stable to further oxidation under a given set of external conditions. We noted above that both CLA and F-acids are remarkably stable under certain conditions (e.g., in methanolic solutions) (33–34,43–44). Experiments have been reported that show that the F-acid 10,13-epoxyoctadeca-10,12-dienoic acid (methyl ester or Na^+ salt), dissolved in borate buffer and vigorously aerated with O_2, does not oxidize over a 3-h period while being irradiated with fluorescent lamps (44). Addition of LOX did not initiate the oxidation reaction. However, if LOX and linoleic acid are added, or if

LOOH is added, VII will oxidize to VIII. Compound II has been reported in phospholipids (PL) (25). The experimental conditions under which the phospholipid-containing compound II (II_{PL}) should react have not yet been determined. The scheme in Fig. 20 is based on the scheme previously proposed by Batna and Spiteller (43), except that it begins with II_{PL} instead of a linoleate phospholipid. It is not currently known why compound II is selectively found in phospholipids as

II_{PL}

VI_{PL}

VII_{PL}

IX_{PL}

XIV

XV

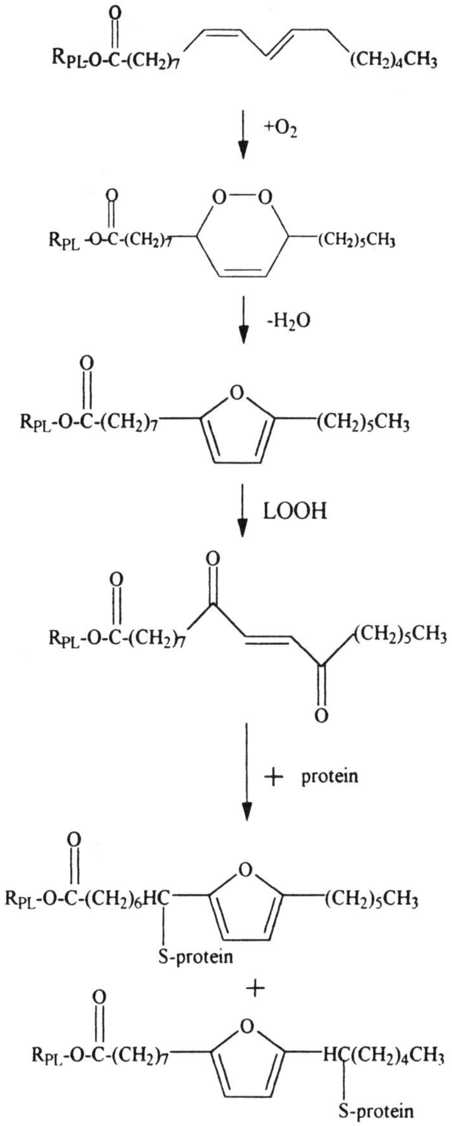

FIG. 20. Speculative CLA mechanism.

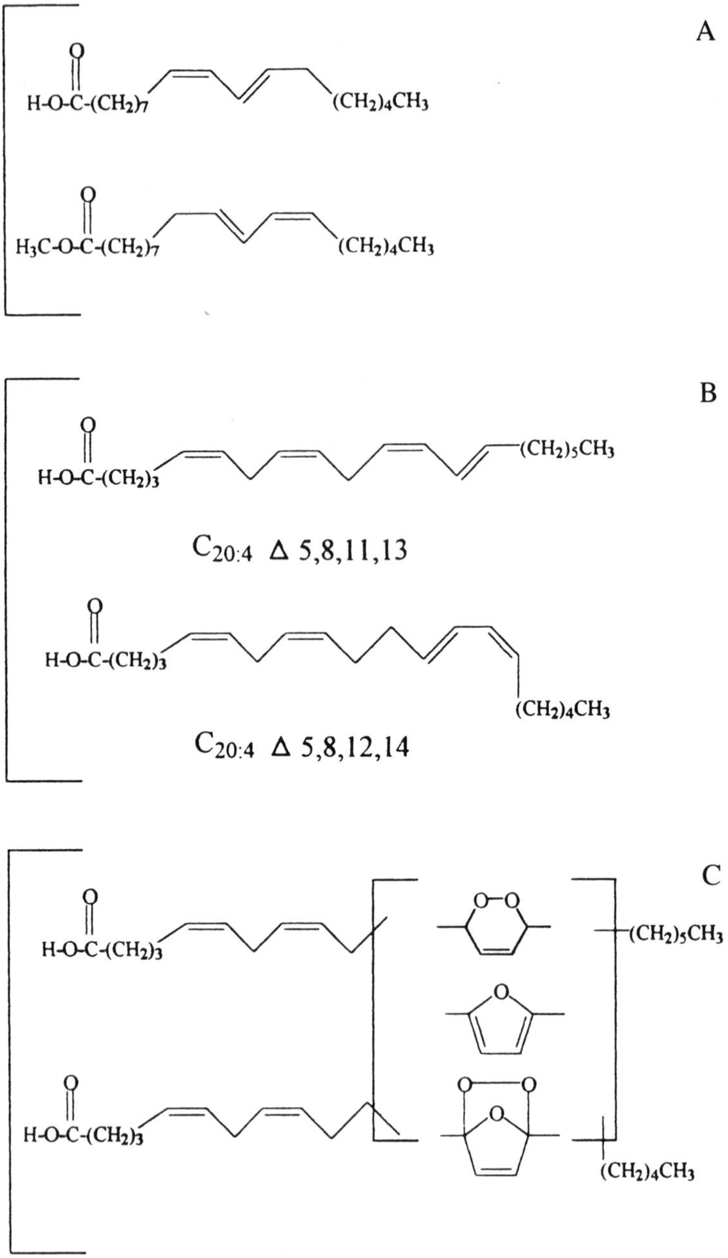

FIG. 21. Desaturation and elongation of CLA.

compared to other CLA isomers. The possibility of the formation of compound II_{PL} in situ from linoleic acid or from some other fatty acid moiety (e.g., oleate), as a result of tissue damage, has not been eliminated. If compound II_{PL} oxidizes to VI and VII in vivo as occurs in vitro, then the remainder of the scheme in Fig. 20 may proceed as reported (43). In the presence of LOOH, VII_{PL} would oxidize to IX_{PL} (or XII_{PL}, XIII_{PL}, or another component that contains the dioxo-ene functionality), and IX_{PL} reacts with the thiol group contained in selected proteins to form the covalently bonded compounds XIV and XV. If compounds such as XIV or XV are formed, their presence would indicate that the protein component of the cell has been altered and this may mark the cell for repair or destruction.

Does CLA mediate prostaglandin function? It has been reported that isomers of dietary CLA are desaturated and elongated in the livers of rats (4,69). Commonly reported CLA isomers in the diet—compound II and trans-10,cis-12-CLA—are shown in Fig. 21A. The elongated desaturated compounds metabolized from CLA are shown in Fig. 21B. (Note: The cis/trans configurations were not reported in the references cited). Fig. 21C shows some of the plausible oxidation products of the desaturated, elongated compound indicated in Part B. The compounds proposed in Part C are not at this time known to exist, nor are their relationship(s) to the chemoprotective effects of CLA understood.

Many substances, nominally described as CLA, are being studied in chemical and biological laboratories, and, as mentioned above, are being used as dietary supplements. Other compounds such as conjugated trienes (70,71) and cyclic fatty acid monomers (53) are also very physiologically active and are potentially included in some "CLA."

References

1. Pariza, M.W., and Ha, Y.L. (1990) Newly Recognized Anticarcinogenic Fatty Acids, in *Antimutagenesis and Anticarcinogenesis Mechanism II* (Kuroda, Y., Shankel, D., and Waters, M.D., eds.), Plenum Press, New York and London, pp. 167–170.
2. Ha, Y.L., Grimm, N.K., and Pariza, M.W. (1987) Anticarcinogens from Fried Ground Beef: Heat-Altered Derivatives of Linoleic Acid, *Carcinogenesis 8*, 1881–1887.
3. Liew, C., Schut, H.A.J., Chin, S.F., Pariza, M.W., and Dashwood, R.H. (1995) Protection of Conjugated Linoleic Acids Against 2-Amino-3-Methylimidazo[4,5-*f*]Quinoline-Induced Colon Carcinogenesis in the F344 Rat: A Study of Inhibitory Mechanisms, *Carcinogenesis 16*, 3037–3043.
4. Banni, S., Day, B.W., Evans, R.W., Corongiu, F.P., and Lombardi, B. (1995) Detection of Conjugated Diene Isomers of Linoleic Acid in Liver Lipids of Rats Fed a Choline-Devoid Diet Indicates that the Diet Does Not Cause Lipoperoxidation, *Nutritional Biochemistry 6*, 281–289.
5. Cawood, P., Wickens, D.G., Iversen, S.A., Braganza, J.M., and Dormandy, T.L. (1983) The Nature of Diene Conjugation in Human Serum, Bile and Duodenal Juice, *FEBS Lett. 162*, 239–243.
6. Iversen, S.A., Cawood, P., Madigan, M.J., Lawson, A.M., and Dormandy, T.L. (1984) Identification of a Diene Conjugated Component of Human Lipid as Octadeca-9,11-Dienoic Acid, *FEBS Lett. 171*, 320–324.

7. Smith, G.N, Taj, M., and Braganza, J.M. (1991) On the Identification of a Conjugated Diene Component of Duodenal Bile as 9Z,11E-Octadecadienoic Acid, *Free Rad. Biol. Med. 10*, 13–21.
8. Harrison, K., Cawood, P., Iversen, A., and Dormandy, T.L. (1985) Diene Conjugation Patterns in Norman Human Serum, *Life Chem. Rep. 3*, 41–44.
9. Guyan, P.M.,Uden, S., and Braganza, J.M. (1990) Heightened Free Radical Activity in Pancreatitis, *Free Rad. Biol. Med. 8*, 347–354.
10. Dormandy, T.L., and Wickens, D.G. (1987) The Experimental and Clinical Pathology of Diene Conjugation, *Chem. Phys. Lipids 45*, 353–364.
11. Frankel, E.N. (1985) Chemistry of Free Radical and Singlet Oxidation of Lipids, *Prog. Lipid Res. 23*, 197–221.
12. Frankel, E.N. (1980) Lipid Oxidation, *Prog. Lipid Res. 19*, 1–22.
13. Paquette, G., Kupranycz, D.B., and van de Voort, F.R. (1985) The Mechanisms of Lipid Oxidation I. Primary Oxidation Products, *Can . Inst. Food Sci. Technol. J. 18*, 112–118.
14. Porter, N.A., Lehman, L.S., Weber, B.A., and Smith, K.J. (1981) Unified Mechanism for Polyunsaturated Fatty Acid Autoxidation. Competition of Peroxy Radical Hydrogen Atom Abstraction, β-Scission, and Cyclization, *J. Am. Chem. Soc. 103*, 6447–6455.
15. Gray, J.I. (1978) Measurement of Lipid Oxidation: A Review, *J. Am. Chem. Soc. 55*, 539–546.
16. Thompson, S., and Smith, M.T. (1985) Measurement of the Diene Conjugation Form of Linoleic Acid in Plasma by High-Performance Liquid Chromatography: A Questionable Non-Invasive Assay of Free Radical Activity?, *Chem.-Biol. Inter. 55*, 357–366.
17. Parody, P.W. (1977) Conjugated Octadecadienoic Acids of Milk Fat, *J. Dairy Sci. 60*, 1550–1553.
18. Chin, S.F., Liu, W., Storkson, J.M., Ha, Y.L., and Pariza, M.W. (1992) Dietary Sources of Conjugated Dienoic Isomers of Linoleic Acid, A Newly Recognized Class of Anticarcinogens, *J. Food Comp. Anal. 5*, 185–197.
19. Ha, Y.L., Grimm, N.K., and Pariza, M.W. (1989) Newly Recognized Anticarcinogenic Fatty Acids: Identification and Quantification in Natural and Processed Cheeses, *J. Agric. Food Chem. 37*, 75–81.
20. Shantha, N.C., Decker, E.A., and Ustunol, Z. (1992) Conjugated Linoleic Acid Concentration in Processed Cheese, *J. Am. Oil Chem. Soc. 69*, 425–428.
21. Werner, S.A., Luedecke, L.O. and Shultz, T.D. (1992) Determination of Conjugated Linoleic Acid Content and Isomer Distribution in Three Cheddar-Type Cheeses: Effects of Cheese Cultures, Processing, and Aging, *J. Agric. Food Chem. 40*, 1817–1821.
22. Halliwell, B., and Chirico, S. (1993) Lipid Peroxidation: Its Mechanism, Measurement, and Significance, *Am. J. Clin. Nutr. 57(supp.)*, 715s–725s.
23. Pariza, M.W., Loretz, L.J., Storkson, J.M., and Holland, N.C. (1983) Mutagens and Modulator of Mutagenesis in Fried Ground Beef, *Cancer Research (suppl.) 43*, 2444s–2446s.
24. Pariza, M.W., and Hargraves, W.A. (1985) A Beef Derived Mutagenesis Modulator Inhibits Initiation of Mouse Epidermal Tumors by 7,12-Dimethylbenz[A]anthracene, *Carcinogenesis 6*, 591–593.
25. Ha, Y.L., Storkson, J., and Pariza, M.W. (1990) Inhibition of Benzo(*a*)pyrene-Induced Mouse Forestomach Neoplasia by Conjugated Dienoic Derivatives of Linoleic Acid, *Cancer Res. 50*, 1097–1101.

26. Ip, C., Chin, S.F., Scimeca, J.A. and Pariza, M.W. (1991) Mammary Cancer Prevention by Conjugated Dienoic Derivatives of Linoleic Acid, *Cancer Res. 51*, 6118–6124.
27. Ip, C., Scimeca, J.A., and Thompson, H. (1995) Effects of Timing and Duration of Dietary Conjugated Linoleic Acid on Mammary Cancer Prevention, *Nutr. Cancer 24*, 241–247.
28. Cook, M.E., Miller, C.C., Park, Y., and Pariza M. (1993) Immune Modulation by Altered Nutrient Metabolism: Nutritional Control of Immune-Induced Growth Depression, *Poultry Sci. 72*, 1301–1305.
29. Miller, C.C., Park, Y., Pariza, M.W., and Cook, M.E. (1994) Feeding Conjugated Linoleic Acid to Animals Partially Overcomes Catabolic Responses Due to Endotoxin Injection, *Biochem. Biophys. Res. Commun. 198*,1107–1112.
30. Lee. K.N., Kritchevsky, D., and Pariza, M.W. (1994) Conjugated Linoleic Acid and Atherosclerosis in Rabbits, *Atherosclerosis 108*, 19–25.
31. Visonneau, S., Cesano, A., Tepper, S.A., Scimeca, J.A., Santoli, D., and Kritchevsky, D. (1997) Conjugated Linoleic Acid Suppresses the Growth of Human Breast Adenocarcinoma Cells in SCID Mice, *Anticancer Res. 17*, 1–5, in press.
32. Belury, M.A. (1995) Conjugated Dienoic Linoleate: A Polyunsaturated Fatty Acid with Unique Chemoprotective Properties, *Nutr. Rev. 53*, 83–89.
33. Yurawecz, M.P., Hood, J.K., Mossoba, M.M., Roach, J.A.G., and Ku, Y. (1995) Furan Fatty Acids Determined as Oxidation Products of Conjugated Octadecadienoic Acid, *Lipids 30*, 595–598.
34. Bascetta, E., Gunstone, F.D., and Scrimgeour, C.M. (1984) Synthesis, Characterization, and Transformations of a Lipid Cyclic Peroxide, *J. Chem. Soc. Perkin Trans. I* 2199–2205.
35. Gunstone, F.D., and Wijesundera, R.C. (1979) Fatty Acids, Part 54. Some Reactions of Long-Chain Oxygenated Acids With Special Reference to Those Furnishing Furanoid Acids, *Chem. Phys. Lipids 24*, 193–208.
36. Morris, L.J., Marshall, M.O., and Kelly, W. (1966) A Unique Furanoid Fatty Acid From *Exocarpus* Seed Oil, *Tetrahed. Letters 36*, 4249–4253.
37. Hannemann, K., Puchta, V., Simon, E., Ziegler, H., Ziegler, G., and Spiteller, G. (1989) The Common Occurrence of Furan Fatty Acids in Plants, *Lipids 24,* 296–298.
38. Ishii, K., Okajima, H., Koyamatsu, T., Okada, Y., and Watanabe, H. (1988) The Composition of Furan Fatty Acids in the Crayfish, *Lipids 23,* 694–700.
39. Ishii, K., Okajima, H., Okada, Y., and Watanabe, H. (1988) Studies of Furan Fatty Acids of Salmon Roe Phospholipids, *J. Biochem. 103*, 836–839.
40. Schödel, R., and Spiteller, G. (1986) Über das Vorkommen von F-Säuren in Rinderleber und deren enzymatischen Abbau bei Gewebeverletzung (The Occurrence of F-Acids in Cattle Liver and Their Enzymatic Degradation during Tissue Damage), *Liebigs Ann. Chem. 5,* 459–462.
41. Okada, Y., Okajima, H., Konishi, H., Terauchi, M., Ishii, K., Liu, I.-M., and Watanabe, H., (1990) Antioxidant Effect of Naturally Occurring Furan Fatty Acids on Oxidation of Linoleic Acid in Aqueous Dispersion, *J. Am. Oil Chem. Soc. 67*, 858–862.
42. Rosenblat, G., Tabak, M., Lie Ken Jie, M.S.F., and Neeman, I. (1993) Inhibition of Bacterial Urease by Autoxidation of Furan C-18 Fatty Acid Methyl Ester Products, *J. Am. Oil Chem. Soc. 70*, 501–505.
43. Batna, A., and G. Spiteller (1994) Effects of Soybean Lipoxygenase-1 on Phosphatidylcholines Containing Furan Fatty Acids, *Lipids 29,* 397–403.

44. Batna, A., and Spiteller, G. (1994) Oxidation of Furan Fatty Acids by Soybean Lipoxygenase-1 in the Presence of Linoleic Acid, *Chem. Phys. Lipids 70*, 179–185.
45. Boyer, R.F., Litts, D., Kostishak, J., Wijesundera, R.C., and Gunstone, F.D. (1979) The Action of Lipoxygenase-1 on Furan Derivatives, *Chem. Phys. Lipids 25*, 237–246.
46. Lie Ken Jie, M.S.F., Chan, H.W.M., Wai, J.S.M., and Sinha, S. (1981) Fatty Acids: XXIII. A Rapid Method for the Preparation of C_{18} Furanoid Fatty Ester Involving Dry-Column Chromatography, *J. Am. Oil Chem. Soc. 50*, 705–706.
47. Lie Ken Jie, M.S.F., and Lam, C.H. (1978) Fatty Acids, Part XVI: The Synthesis of All Isomeric C_{18} Furan-Containing Fatty Acids, *Chem. Phys. Lipids 21*, 275–287.
48. Gunstone, F.D., and Said, A.I. (1971) Fatty Acids, Part 29: Methyl 12- mesyloxyoleate as a Source of Cyclopropane Esters and of Conjugated Octadecadienoates, *Chem. Phys. Lipids 7*, 121–134.
49. Lie Ken Jie, M.S.F., and Sinha, S. (1981) Fatty Acids, Part 21: Ring Opening Reactions of Synthetic and Natural Furanoid Fatty Esters, *Chem. Phys. Lipids 28*, 99–109.
50. Gorst-Allman, C.P., Puchta, V. and Spiteller, G. (1988) Investigations of the Origin of the Furan Fatty Acids (F-Acids), *Lipids 23*, 1032–1036.
51. Yurawecz, M.P., Hood, J.K., Roach, J.A.G., Mossoba, M.M., Daniels, D., Ku, Y., Pariza, M.W., and Chin, S.F. (1994) Conversion of Allylic Hydroxy Oleate to Conjugated Linoleic Acid and Methoxy Oleate by Acid-Catalyzed Methylation Procedures, *J. Am. Oil Chem. Soc. 71*, 1149–1155.
52. Morris, L.J., Holman, R.T., and Fontell, K. (1960) Alteration of Some Long- Chain Esters During Gas–Liquid Chromatography, *J. Lipid Res. 1*, 412–420.
53. Shantha, N.C., Decker, E.A., and Hennig, B. (1993) Comparison of Methylation Methods for the Quantitation of Conjugated Linoleic Acid Isomers, *J. AOAC Int. 76*, 644–649.
54. Yurawecz, M.P., Roach, J.A.G., Mossoba, M.M., and Ku, Y. (1995) Temperature Dependence of Air Oxidation of Methyl Esters of Conjugated Linoleic Acid (CLA), (Abstract G), *INFORM 6*, 515.
55. *Official Methods and Recommended Practices of the American Oil Chemists' Society*, Method Ce 1b-89, (1990) 4th edn., American Oil Chemists' Society, Champaign, Illinois.
56. Schuchardt, U., and Lopes, O.C. (1988) Tetramethylguanidine Catalyzed Transesterification of Fats and Oils: A New Method for Rapid Determination of Their Composition, *J. Am. Oil Chem. Soc. 65*, 1940–1941.
57. Mossoba, M.M. (1993) Applications of Capillary GC–FTIR, *INFORM 4*, 854–859.
58. Brown, H.G., and Snyder, H.E. (1982) Conjugated Dienes of Crude Soy Oil: Detection by UV Spectrophotometry and Separation by HPLC, *J. Am. Oil Chem. Soc. 59*, 280–283.
59. Fay, L., and Richli, U. (1991) Location of Double Bonds in Polyunsaturated Fatty Acids by Gas Chromatography–Mass Spectrometry after 4,4-Dimethyloxazoline Derivatization, *J. Chrom. 541*, 89–98.
60. Luthria, D.L., and Sprecher, H. (1993) 2-Alkenyl-4,4-Dimethyloxazolines as Derivatives for the Structural Elucidation of Isomeric Unsaturated Fatty Acids, *Lipids 28*, 561–564.
61. Hopkins, C.Y. (1972) Fatty Acids with Conjugated Unsaturation, in *Topics in Lipid Chemistry* (Gunstone, F.D., ed.), Vol 3., John Wiley and Sons Inc., New York, pp. 68–71.

62. Foote, C.S., Wuesthoff, M.T., Wexler, S., Burstain, I.G., Denny, R., Schenck, G.O., and Schulte-Elte, K.-H. (1967) Photosensitized Oxygenation of Alkyl-Substituted Furans, *Tetrahed. Letters 23*, 2583–2599.
63. Kearns, D.R. (1969) Selection Rules for Singlet-Oxygen Reactions. Concerted Addition Reactions, *J. Am. Chem. Soc. 91*, 6554–6563.
64. Ideses, R., Shani, A., and Klug, J.T. (1982) Cyclic Peroxide: An Isolable Intermediate in Singlet Oxygen Oxidation of Pheromones to the Furan System, *Chem. Ind. 19*, 409–410.
65. Ogawa, S., Fukui, S., Hanasaki, Y., Asano, K., Uegaki, H., Fujita, S., and Shimazaki, R. (1991) Determination Method of Singlet Oxygen in the Atmosphere by Use of α-Terpinene, *Chemosphere 22*, 1211–1225.
66. Sehat, N., Yurawecz, M.P., Mossoba, M.M., Roach, J.A.G., and Ku, Y. (1996) GC–FTIR and GC–MS Analyses of Autoxidation Products of Furan Fatty Acids, *Abstracts of the 87th AOCS Annual Meeting & Expo*, April 28–May 1, Indianapolis, Indiana, p. 2.
67. Van den Berg, J.J.M., Cook, N.E., and Tribble, D.L. (1995) Reinvestigation of the Antioxidant Properties of Conjugated Linoleic Acid, *Lipids 30*, 599–605.
68. Graff, G., Gellerman, J.L., Sand, D.M., and Schlenk, H. (1984) Inhibition of Blood Platelet Aggregation by Dioxoene Compounds, *Biochim. Biophys. Acta 799*, 143–150.
69. Ramilison, I., Sebedio, J.L., Juanéda, P., Dobson, G., and Christie, W.W. (1996) Conversion of Conjugated Linoleic Acid (CLA) Isomers in C20:3 and C20:4 Metabolites, *Oil Processing & Biochemistry of Lipids,* in the 1st Congress of the European Section of AOCS, September 19–20, 1996, University of Burgundy Dijon, France, Abstract D8.
70. Nugteren, D.H., and Christ-Hazelhof, E. (1987) Naturally Occurring Conjugated Octadecatrienoic Acids are Strong Inhibitors of Prostaglandin Biosynthesis, *Prostaglandins 33*, 403–417.
71. Yurawecz, M.P., Molina, A.A., Mossoba, M.M., and Ku ,Y. (1993) Estimation of Conjugated Octadecatrienes in Edible Fats and Oils, *J. Am. Oil Chem. Soc. 70*, 1093–1099.

Chapter 10
Analysis of Lipid Oxidation Products by Combination of Chromatographic Techniques

G. Márquez-Ruiz and M.C. Dobarganes

Instituto de la Grasa (CSIC), Avda. Padre García Tejero, 4, 41012 Sevilla, SPAIN

Introduction

The oxidation of lipids is of great concern, because it affects the wholesomeness of foods through the development of rancidity (1–3) and the decay of nutritional value and food safety (4–6). The autoxidation of unsaturated lipids is a catalytic process that proceeds via free radicals and involves the reaction of oxygen with the unsaturated fatty acid moieties of triglycerides or any other lipid molecule. As hydroperoxides form and further oxidation, decomposition, and polymerization reactions take place, a wide range of oxidation intermediates and end compounds is produced (7–9). Hence, great difficulties are encountered in attempting to evaluate oxidative alteration, and new analytical procedures of general application are demanded.

Peroxide value, UV absorption, and assay of thiobarbituric acid–reactive substances probably stand out among the classical methods used to evaluate oxidation. These methods all focus on partial and distinct aspects of the oxidation process—formation of hydroperoxides in the first case, secondary products in the others. Analyses of specific oxidation compounds can be attained through application of chromatographic techniques (10–12). Given the complexity of the reactions involved, the chromatographic methods most commonly applied are headspace analyses of volatiles (particularly hexanal and pentane) obtained after hydroperoxide scission (13,14).

Each of the methods cited is, however, applicable to particular stages of the process, and the information obtained depends on the type of method chosen (15). The methods currently used have been reviewed by several authors, and it is generally concluded that, in spite of the multitude of assays available, there is no universal method that allows evaluation of the extent of oxidation throughout the entire process (15–20). Therefore, the need arises for improving methodologies to reevaluate aspects of particular concern, such as the effectiveness of antioxidants and, in general, the influence of the different variables that modify the rate of oxidative reactions.

Defining the actual levels of oxidized compounds resulting from oxidation at high temperature (such as during frying) is also of great interest because of the increasing concern of consumers and health administrations about the nutritional effects of these compounds. Under such conditions, some initial oxidation products typical of low temperature exist only transiently, and polymerization reactions are accelerated; ultimately, however, many of the compounds generated contain oxygenated functions (21–24).

The analytical procedures reported in this chapter are based on combinations of chromatographic techniques and were developed to quantitate oxidation products in a variety of substrates, including model compounds, oils, and food lipids, subjected to oxidation under very different conditions, from room to frying temperatures.

Quantitation of Oxidized Triglycerides by Solid-Phase Extraction (SPE) and High-Performance Size Exclusion Chromatography (HPSEC)

During the last several years we have developed and widely applied a methodology based on a combination of adsorption and size exclusion chromatographies to study the alteration of used frying oils. Use of this methodology enables quantitation of oxidized and polymeric triglycerides as well as hydrolytic products (diglycerides and fatty acids) (25). Application of this procedure is not limited to fats and oils heated at high temperature; it was later found to be of great utility for quality evaluation of refined oils (26, 27) and quantitation of oxidation compounds in fats and fatty foods at low temperature (28–31). A recently introduced modification of this methodology is directed toward using small quantities of samples and solvents for rapid quantitation of different groups of oxidized triglycerides, using monostearin as internal standard. As illustrated in Fig. 1, polar compounds are obtained in a first step by solid-phase extraction, and quantitative results of different groups of oxidized triglycerides are subsequently attained through exclusion chromatography analysis.

Methodology

Briefly, 2 mL of the sample solution, containing 50 mg of sample and 1 mg of monostearin (used as internal standard), are placed on the silica cartridge for SPE; the solvent is passed through, while the sample is retained on the column. Next, the nonpolar fraction is eluted with 15 mL of petroleum ether:diethyl ether 90:10. A second fraction, containing polar compounds and the internal standard, is eluted with 15 mL of diethyl ether. Nonpolar and polar fractions are evaporated under reduced pressure and redissolved in tetrahydrofuran for further analyses—i.e., TLC (to check the efficiency of the separation) and HPSEC. Fractions of polar compounds are analyzed by HPSEC using a refractive-index detector and two 100 and 500 Å Ultrastyragel columns connected in series, operating at 35°C. The columns were 25 cm × 0.77 cm inner diameter, packed with a porous, highly cross-linked styrenedivinylbenzene copolymer (<10 µm). Tetrahydrofuran (high-performance liquid chromatography grade) serves as the mobile phase, with a flow of 1 mL/min. As can be observed in Fig. 1, the peaks resolved, eluting by inverse order of molecular weight, are triglyceride polymers (TGP), triglyceride dimers (TGD), oxidized

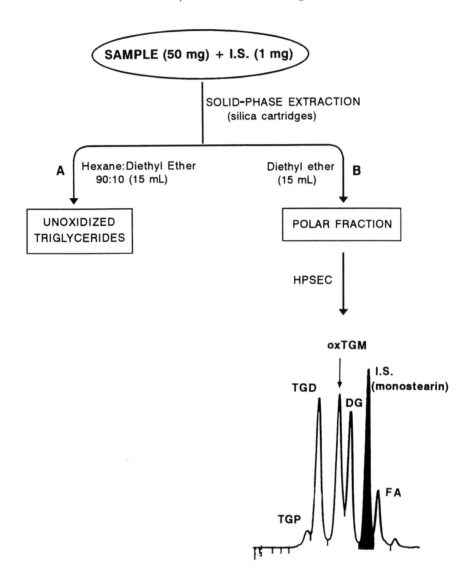

Fig. 1. Analytical procedure used for quantitation of oxidized triglycerides. *Abbreviations:* HPSEC, high-performance size exclusion chromatography; I.S., internal standard; TGP, triglyceride polymers; TGD, triglyceride dimers; oxTGM, oxidized triglyceride monomers; DG, diglycerides; FA, fatty acids.

TABLE 1
Quantitative Determination of Polar Compounds (wt %) in Fats and Oils of Different Origins

Sample	Internal standard method		Gravimetric determination	
	Mean	RSD(%)[a]	Mean	RSD(%)
1	3.7	12.1	3.5	21.1
2	4.8	7.3	5.1	12.0
3	6.7	7.3	7.1	9.0
4	7.5	4.0	7.8	10.0
5	13.0	4.2	13.6	5.1
6	18.5	3.5	18.0	4.8
7	20.1	4.6	20.7	4.0
8	24.3	3.1	25.2	1.9

[a]RSD% = (Standard deviation / Mean found) × 100
n = 3

triglyceride monomers (oxTGM), diglycerides (DG), monostearin (filled peak), and fatty acids (FA), this last peak also including polar unsaponifiable fraction. The HPSEC analysis requires a total run time of just 15 min, after which quantitation of the overall groups of compounds can be attained. This methodology was described in detail, including precision, accuracy, and recovery data, in a recent publication (32).

Table 1 shows results of the application of the procedure to eight samples with different contents of polar compounds, ranging from 3.7 to 24.3%. Polar compounds were also determined gravimetrically according to the IUPAC Standard Method, starting from 1 g of fat and using silica columns (33). There were no significant differences between the mean values obtained by the two different procedures, although the lower the polar-compound concentration, the lower the standard deviation was for the internal-standard method. In view of these results, it seems that the method proposed is useful for samples within a wide range of alteration and especially adequate for samples of low oxidation level, which otherwise give higher RSD% by gravimetric determination. Additionally, using the internal-standard method only 50-mg samples are required, and the use of silica and solvents is drastically reduced.

A large number of experiments are currently underway in our laboratory, using model systems and oils, in which this methodology is applied for studying the useful storage life and determining the efficacy of antioxidants. Some representative examples of the utility of the methodology will be shown in the following section.

Applications

Evolution of oxidation in model systems. Figure 2 presents a representative profile of the evolution of oxidation at 60°C for trilinolein (LLL) samples, without (A) and with (B) 250 mg/kg α-tocopherol added. LLL was used as a model unsaturated triglyceride in order to test the possibilities of the analytical procedure for evaluating modifications throughout oxidation, without the interference of minor

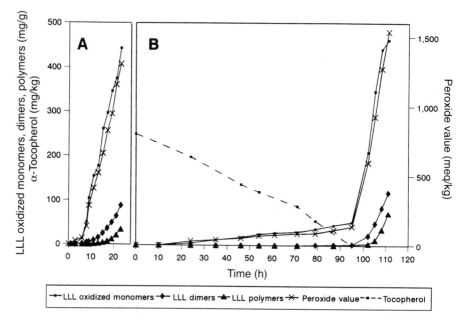

Fig. 2. Evolution of oxidized monomers, dimers, and polymers; peroxide value; and α-tocopherol content in samples of trilinolein (LLL) without (A) and with (B) α-tocopherol initially added, throughout oxidation at 60°C.

compounds, potentially having prooxidant or antioxidant effects, that could be present in oils. Accelerated oxidative tests were carried out at 60°C in an oven, because it has been reported that results at this temperature correlate well with evaluations of actual shelf lives, while a number of side reactions occurring at higher temperatures are minimized (34). Under these conditions, testing stability is not only much more rapid but also easier than it is under ambient conditions, since many variables are difficult to control over prolonged storage (15).

As can be observed, LLL oxidized monomers showed a progressive increase during the earlier stages of oxidation, which was parallel to peroxide value evolution. An excellent correlation between the two determinations was found ($Y = -15.5387 + 2.8911X$, $r = 0.9941$ for LLL (A) and $Y = -23.2369 + 3.0560X$, $r = 0.9910$ for LLL (B)). Hydroperoxides are the primary oxidation products formed; hence, peroxide value is the most common method used for early oxidation. On the other hand, the peak of oxidized triglyceride monomers comprises those monomeric triglycerides containing at least one oxidized fatty acyl (either a peroxide group or any other oxygenated function, such as epoxy, keto, hydroxy, etc.) and therefore provides a global measurement of both primary and secondary oxidation products. During the early stages of oxidation, LLL oxidized monomers would consist primarily of hydroperoxides; hence, their increase was closely parallel to that of peroxide value. However, peroxides are intermediate and labile products, readily converted into a multitude of secondary oxidation products. Among them, radical recombination gives rise to oxygenated side products of the same chain length as the parent hydroperoxide (here quantitated globally as LLL oxidized monomers) because of their similar polarity and molecular weight. Quantitation of this group of compounds can therefore be of great utility, not only to detect initially formed oxidized products—even before rancidity—but also to follow oxidation during further stages (35).

In the course of oxidation it is clearly observed that at a certain point, oxidation is accelerated, as shown by the sharp increase of LLL oxidized monomers and initiation of polymerization. This occurred at around 9 h in LLL devoid of α-tocopherol, vs. approximately 95 h in the case of LLL initially containing α-tocopherol. At that time, the α-tocopherol was practically consumed in the sample that had contained it initially, indicating that it exerted a protective effect till it was exhausted. Defining the induction period as the time interval before oxidation proceeds rapidly, presenting a sudden change in the rate of oxidation, it is observed that, under these conditions, around 50 mg/g of oxidized monomers denotes the end of the induction period and the onset of advanced oxidation. A significant rise in dimers indicates additionally that the course of oxidation has entered a second and accelerated phase. The development of oxidation was considerably delayed in LLL samples with α-tocopherol added, but, interestingly, once the tocopherol was used up, these samples showed a similar evolution pattern to their counterparts without α-tocopherol.

Figure 3 shows representative HPSEC chromatograms of polar fractions of LLL stored at moderate temperature (60°C) at the initial point (A) and three different points of the oxidation process (B,C, and D). Changes in the groups of compounds resolved are well illustrated, given that polar fractions were dissolved in equal volumes of solvent. Thus, it can be observed that the peak corresponding to monostearin, used as internal standard, is of similar magnitude in all chromatograms. Chromatogram A corresponds to the polar fraction of the starting, unoxidized LLL; it is composed exclusively of the internal standard added (1 mg), since LLL was otherwise recovered in the nonpolar fraction. The polar fraction in B was separated from a sample taken out during the early oxidation period (peroxide value = 52.6 meq/kg); it showed the presence only of oxidized triglyceride monomers, in this case LLL oxidized monomers, accounting for 2.3% on total sample. Sample C (peroxide value = 92.0 meq/kg) was withdrawn around the end of the induction period, as shown by the appearance of dimers (0.2% on total sample), thus indicating that the sample was entering the period of advanced oxidation. However, rancidity was not yet detected at this point under these conditions. In turn, LLL oxidized monomers had reached a level of 5.4% on total sample. Finally, the polar fraction in D, separated from a sample taken further along in the advanced-oxidation stage (peroxide value = 516.3 meq/kg), presented a significant increase in oxidized monomers (17.7% on total sample) and dimers (1.4% on total sample). Also, polymers could be already detected (Rt: 11.4 min).

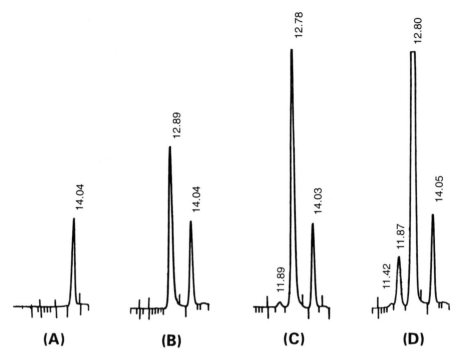

Fig. 3. High-performance size exclusion chromatograms of polar fractions isolated from starting trilinolein (A) and from trilinolein samples withdrawn at different points during oxidation at 60°C (B, C and D). Retention times (min): 11.4, LLL polymers; 11.8, LLL dimers; 12.8, LLL oxidized monomers; 14.0, monostearin (internal standard).

It is worthy to remark the great utility offered by this methodology, which allows concomitant evaluation of initial and secondary oxidation compounds, hence enabling the degree of oxidation to be determined at any point during the course of oxidation. Quantitation of oxidized triglyceride monomers provides an useful measurement for early oxidation stages, and parallel quantitation of dimers gives a good indication of the onset of advanced oxidation.

Evolution of oxidation during storage of foods. Table 2 shows evolution of total polar compounds and the different groups of oxidized triglycerides in crisps prepared with sunflower oil (SO), high-oleic sunflower oil (HOSO), and palm olein (PO), stored at room temperature up to 6 months—a period far beyond that usually established for commercialization, normally 3 months. In the course of an extensive project, these crisps were industrially prepared in a continuous 700-L fryer at 180°C for 20 h, in order to compare conventional and high-oleic sunflower oils with saturated fats in terms of frying performance and behavior of fried products during storage (36). As can be observed, values for polar compound contents in initial crisps indicated that

TABLE 2
Total Polar Compounds (wt% on sample) and Distribution in Groups of Compounds (mg/g sample) in Oils Extracted from Crisps Stored at Room Temperature

Sample	Time (weeks)	Total	Distribution				
			oxTGM	TGD	TGP	DG	FA
SO	0	5.2	15.3	15.0	1.7	12.7	6.9
	15	7.5	37.3	16.9	1.9	12.9	6.2
	25	10.2	61.5	17.5	2.0	13.7	6.8
HOSO	0	4.9	12.4	10.3	1.0	18.9	5.9
	15	4.8	12.4	10.0	1.3	18.1	5.8
	25	5.2	16.4	10.7	1.3	17.9	5.4
PO	0	8.9	11.4	12.5	1.0	60.3	3.4
	15	8.6	11.6	12.2	1.2	58.2	3.1
	25	8.9	11.9	12.2	1.0	59.2	4.2

Abbreviations: oxTGM, oxidized triglyceride monomers; TGD, triglyceride dimers; TGP, triglyceride polymers; DG, diglycerides; FA, fatty acids; SO, sunflower oil; HOSO, high-oleic sunflower oil; PO, palm olein.

frying performance had been excellent due to the high turnover period (13 h). After storage for 25 weeks, only SO crisps showed a considerable rise in polar compounds, in turn attributable specifically to the increase of oxTGM, and changes from initial values were significant even at 15 weeks. The remaining groups of compounds quantitated remained at the initial levels. Quite in contrast, HOSO and PO samples after 25 weeks presented roughly the same oxidation levels as initially, thus showing a notable shelf life at room temperature (37). Interestingly, these results were in excellent agreement with parallel sensory assessments by a panel test, showing that SO crisps were distinctly rancid from the 17th week, whereas HOSO and PO presented a rather similar behavior, maintaining fruity characteristics for odor and taste for over 6 months (38).

Quantitation of Oxidized Fatty Acid Methyl Esters by Silica Column Chromatography and HPSEC

Complementary information to that obtained by the methodology just outlined can be attained by analyzing specifically the oxidized fatty acyls included in triglyceride molecules, following the procedure shown in Fig. 4, which starts from fatty acid methyl esters (FAMEs). In this approach the range of molecular weights of the compounds separated decreases, and it is not easy to find a suitable internal standard. Hence, quantitative data are obtained here on the basis of gravimetric determinations.

Methodology

Briefly, FAMEs are obtained by transesterification of 1 g of sample with sodium methoxide and hydrochloric acid/methanol and subsequent recovery of methyl

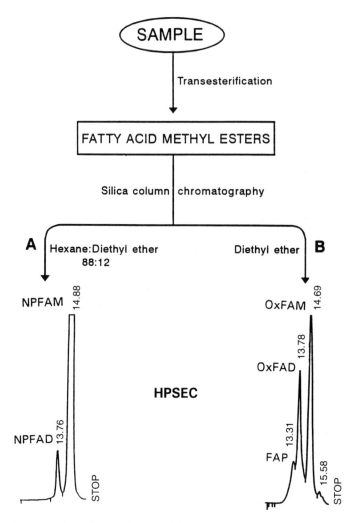

Fig. 4. Analytical procedure used for quantitation of oxidized fatty acid methyl esters. *Abbreviations:* NPFAD, nonpolar fatty acid dimers; NPFAM, nonpolar fatty acid monomers; FAP, fatty acid polymers; OxFAD, oxidized fatty acid dimers; OxFAM, oxidized fatty acid monomers.

esters. Methyl esters are separated by silica column chromatography, using 150 mL hexane/diethyl ether (88:12) to elute a nonpolar fraction and 150 mL diethyl ether to obtain the polar fraction. Analyses of the nonpolar and polar fractions are performed by HPSEC, using the chromatographic conditions described above. Quantitation of nonpolar fatty acid dimers (NPFAD) and nonpolar fatty acid monomers (NPFAM) is based on the gravimetric determination of the nonpolar fraction. Oxidized fatty acid monomers (OxFAM), oxidized fatty acid dimers (OxFAD), and fatty acid polymers (FAP) are likewise quantitated in the polar fraction. The methodology was described in detail, including calibration and reproducibility data, in an earlier publication (39).

The combined chromatographic analysis used permits quantitation of five groups of compounds differing in polarity or molecular weight. Fatty acid monomers are separated into nonpolar and polar monomers (the latter originated via oxidation) in the first and second fractions, respectively. Likewise, nonpolar dimers (which represent thermal alteration, since no oxygen is involved in their structure) are quantitated in the first fraction, and oxidative dimers are determined independently in the polar fraction. Finally, polymeric compounds are included and analyzed in this latter fraction. Therefore, global quantitation of the compounds eluted in the second fraction provides a measurement of the total oxidized fatty acyl groups included in triglyceride molecules.

Applications

Given that quantitation is based on gravimetric determinations, and considering the high contribution of unchanged fatty acyls in oxidized triglyceride molecules, it is important to note that applications of this methodology will present certain limitations in sensitivity for samples of low oxidation level until a good internal standard can be found. Nevertheless, two applications of special interest are worthy of remark; they are illustrated by the following examples:

Evolution of oxidation in model fatty acid methyl esters. A particularly useful application of the HPSEC analysis is to follow oxidation at low temperature using methyl ester standards, because a preliminary separation of the nonpolar fraction by adsorption chromatography is not necessary—only a direct HPSEC injection of the entire sample. Figure 5 shows HPSEC chromatograms of a starting sample (A) of methyl linoleate (ML) and samples, withdrawn after oxidation at room temperature, at low (B) and high (C) oxidation levels. As can be observed, unoxidized and oxidized monomers eluted at distinct retention times by virtue of the clear differentiation in molecular weight between the intact compounds and the oxidized monomeric molecules (mostly hydroperoxides under these conditions). Therefore, through a simple analysis by HPSEC, run in just over 15 min, a complete picture of the oxidative state can be attained (35).

Figure 6 shows the overall results obtained for ML at room temperature as shown by direct analysis using HPSEC. As with LLL at 60°C, the oxidized monomers continuously increased from the initial stages of oxidation; this measurement is thus shown to have great utility for early oxidation under ambient conditions as well as at higher temperatures. A parallel rise in peroxide value was monitored. Under these conditions, dimers did not appear until high amounts of oxidized monomers were accumulated (i.e., 70 mg/g specifically of oxidized methyl esters) and high peroxide

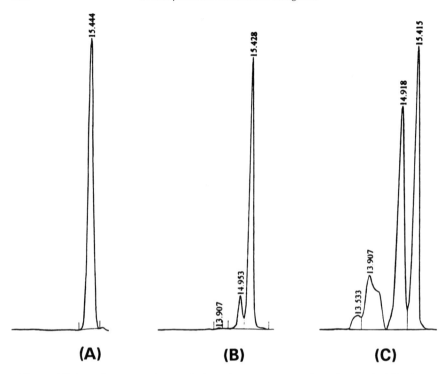

Fig. 5. High-performance size exclusion chromatograms of starting methyl linoleate (A) and methyl linoleate samples withdrawn at different points during oxidation at room temperature (B and C). Retention times (min): 13.5, ML polymers; 13.9, ML dimers; 14.9, ML oxidized monomers; 15.4, unoxidized ML.

values (close to 500 meq/kg) had been reached. This suggests that temperature may have an important influence on the oxidation kinetics.

Quantitation of oxidized fatty acids in used frying oils. Table 3 shows the results obtained through application of the complete procedure on fatty acid methyl esters for evaluation of used frying oils, also including quantitation of total level and distribution of polar glyceridic compounds (24). Samples were supplied by Food Inspection Services (Andalucía, Spain); those selected for this table contained polar-compound levels close to the limit for discarding the oil (25%). Although the type of oil used, the frying conditions, and the fried products were very variable, the samples included here all had similar polar-compound contents, ranging from 23.1 to 27.6%. Samples with the same percentages of total polar compounds showed different patterns of compound distribution. Thus, D and E had 27.5 and 27.6% polar compounds, respectively, but E showed more oxTGM and was relatively poor in TGP, whereas D had a higher level TGD and was relatively poor in DG. Such differences are not strange, considering that these samples are of unknown history and may have undergone very different treat-

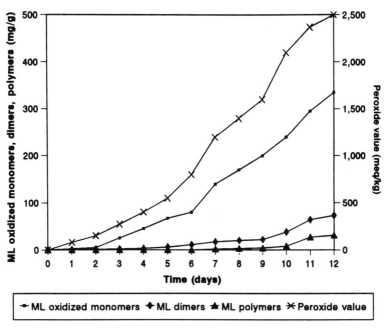

Fig. 6. Evolution of oxidized monomers, dimers, polymers, and peroxide value in samples of methyl linoleate (ML) throughout oxidation at room temperature.

ments in terms of frying temperature, frying periods, and total period of use. These variables are known to exert an important effect on the distribution of polar compounds; oxTGM are expected to be major products at low temperature, whereas TGD and TGP would increase greatly at high temperature. On the other hand, analyses of fatty acid composition, for which the data are not shown here, clearly indicated that sample D corresponded to conventional sunflower oil, with a higher tendency to form polymeric compounds, whereas sample E presented the fatty acid profile typical of an olive oil. Differences between samples were better reflected in the values of total altered fatty acids, which were lower for E (8.7%) than for D (10.7%). Clearly, the content of total altered fatty acids is an excellent indicator of the thermoxidative alteration level, as it measures exclusively the amount of fatty acyls affected.

Apart from the presence of substantial amounts of oxFAD and FAP, particularly relevant was the high occurrence of oxFAM (approximately 30 mg/g), given the nutritional significance of such compounds, which can be readily absorbed, presenting digestibility coefficients around 80% (40). Detailed quantitation of the wide array of oxygenated compounds included in this group is a difficult task, because of the complexity of the structures involved. In this respect, further studies have been initiated in our laboratory, as described in the next paragraph.

TABLE 3
Total Polar Compounds (wt% on oil), Polar Compound Distribution (mg/g oil), Total Altered Fatty Acids (wt% on oil), and Altered Fatty Acid Distribution (mg/g oil) in Used Frying Oils from Restaurants and Fried-Food Outlets

Sample	Polar compounds						Altered fatty acids				
	Total	Distribution					Total	Distribution			
		TGP	TGD	OxTGM	DG	FA		FAP	NPFAD	OxFAD	OxFAM
A	23.1	40.0	92.2	67.9	25.1	5.8	8.1	7.6	19.9	20.9	32.6
B	25.5	46.8	89.8	85.7	25.8	6.9	10.4	10.8	30.9	28.4	33.9
C	25.7	62.5	94.3	58.8	35.2	6.2	10.5	12.8	29.6	27.3	35.3
D	27.5	62.2	98.1	76.5	33.8	4.4	10.7	10.9	33.8	27.9	34.4
E	27.6	37.3	72.6	93.8	61.8	10.5	8.7	4.6	16.5	28.2	37.7

Abbreviations: TGP, triglyceride polymers; TGD, triglyceride dimers; OxTGM, oxidized triglyceride monomers; DG, diglycerides; FA, fatty acids; FAP, fatty acid polymers; NPFAD, nonpolar fatty acid dimers; OxFAD, oxidized fatty acid dimers; OxFAM, oxidized fatty acid monomers.

Analysis of the Major Structures of Oxidized Fatty Acid Monomers by Gas-Liquid Chromatography–Mass Spectrometry (GLC-MS)

Recently, we have introduced a further analytical step in the methodology detailed above, with the aim of analyzing the complex fraction of oxidized fatty acid monomers including epoxides, ketones, and hydroxyacids as well as polyoxygenated monomeric compounds (1). This further step consists of the isolation of the peak of oxidized fatty acid monomers by HPSEC and subsequent analysis by GLC-MS.

Methodology

Fractions of oxidized fatty acid monomers isolated by HPSEC were collected and further separated by GLC-MS using an HP 5890 gas chromatograph coupled to an AEI-MS 30/70 mass spectrometer with electron ionization at 4 kV and 70 eV electron energy. Column: DB-WAX, 25 m × 0.25 mm i.d., 0.25 micron film. Column program: 90°C (2 min); 4°C/min to 240°C (25 min).

Applications

A representative GLC-MS chromatogram showing the main structures included in the oxidized fatty acid monomer group from an used frying oil is presented in Fig. 7. Peaks eluting at retention times shorter than 33 min (prior to scan number 350) correspond to oxidized compounds of molecular weight lower than that of the original fatty acids. Among the most relevant compounds were the short-chain aldehydic acids (originally triglyceride-bound aldehydes resulting from hydroperoxide breakdown), namely, methyl 8-oxooctanoate (A), 9-oxononanoate (B), 10-oxo-8-decenoate (C), 11-oxo-9-undecenoate (D), 12-oxo-9-dodecenoate (E) and 13-oxo-9,11-tridecadienoate (F), at scan numbers of 193, 225, 293, 321, 346, and 394, respectively (41).

As to those compounds of molecular weight higher than that of the starting fatty acids, the main groups present corresponded to epoxyacids, ketoacids, and hydroxyacids. All these groups of compounds eluted at retention times longer than 40 min (scan number over 420) and were composed of many different individual compounds, depending on the position and number of functional groups and unsaturation degree. We are now attempting to elucidate their structure with the help of model systems and derivatization techniques (hydrogenation and silylation). Additionally, we have begun to evaluate oxidized fractions by HPLC using methyl linoleate and methyl oleate oxidized at room temperature, in order to control artifact formation, due to both FAME preparation and the high temperature required for GLC analysis, and to explore the possibilities of quantitation for samples oxidized at low temperature.

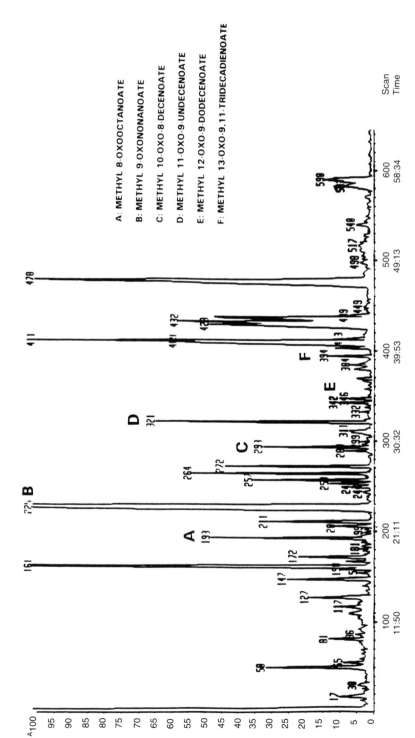

Fig. 7. GLC-MS chromatogram of a representative fraction of oxidized fatty acid methyl ester monomers from a used frying oil, plotted for total ion abundance vs. scan number (upper line) and retention time (lower line).

Acknowledgements

This study was supported by EC (Project AIR1-CT92-0687) and CICYT (Project ALI 95-0736). The authors would like to thank Ms. Mercedes Giménez for assistance.

References

1. Frankel, E.N. (1985) Flavor Chemistry of Fats and Oils, in *Flavor Chemistry of Fats and Oils*, Min, D.M., and Smouse, T.H., American Oil Chemists' Society, Champaign, pp. 1–37.
2. Addis, P.B. (1986) Ocurrence of Lipid Oxidation Products in Foods, *Food Chem. Toxicol. 24*, 1021–1030.
3. Grosch, W. (1987) Autoxidation of Unsaturated Lipids, in *Autoxidation of Unsaturated Lipids*, Chan, H.W.-S., Academic Press Inc., London, pp. 95–139.
4. Eriksson, C.E., (1987) Autoxidation of Unsaturated Lipids, in *Autoxidation of Unsaturated Lipids*, Chan, H.W.-S., Academic Press Inc., London, pp. 207–231.
5. Kubow, S. (1992) Routes of Formation and Toxic Consequences of Lipid Oxidation Products in Foods, *Free Radical Biol. Med. 12*, 63–81.
6. Esterbauer, H. (1993) Cytotoxicity and Genotoxicity of Lipid-Oxidation Products, *Am. J. Clin. Nutr. 57*, 779S–786S.
7. Frankel, E.N. (1985) Chemistry of Free Radical and Singlet Oxidation of Lipids, *Prog. Lipid Res. 23*, 197–221.
8. Foote, C.S. (1985) Chemical Changes in Food During Processing, in *Chemical Changes in Food during Processing*, Richardson, T., and Finley, J.W., The AVI Publishing Company Inc., Connecticut, pp. 17–31.
9. Frankel, E.N. (1991) Recent Advances in Lipid Oxidation, *J. Sci. Food Agric. 54*, 495–511.
10. Frankel, E.N. (1979) Autoxidation in Food and Biological Systems, in *Autoxidation in Food and Biological Systems*, Simic, M.G., and Karel, M., Plenum Press, New York, pp. 141–170.
11. Rossell, J.B. (1989) Rancidity in Foods, in *Rancidity in Foods*, Allen, J.C., and Hamilton, R.J., Elsevier Publisher, USA, pp. 21–45.
12. Marini, D. (1992) Food Analysis By HPLC, in *Food Analysis by HPLC*, Nollet, L.M.L., Marcel Dekker, New York, pp. 169–240.
13. Frankel, E.N. (1982) Volatile Lipid Oxidation Products, *Prog. Lipid Res. 22*, 1–33.
14. Frankel, E.N. (1984) Lipid Oxidation: Mechanisms, Products and Biological Significance, *J. Am. Oil Chem. Soc. 61*, 1908–1917.
15. Frankel, E.N. (1993) In Search of Better Methods to Evaluate Natural Antioxidants and Oxidative Stability in Food Lipids, *Trends Food Sci. Technol. 4*, 220–225.
16. Gray, J.L. (1978) Measurements of Lipid Oxidation: A Review, *J. Am. Oil Chem. Soc. 55*, 539–546.
17. Löliger, J. (1989) Méthodes instrumentales Pour L'analyse De L'etat D'oxydation De Produits Alimentaires, *Rev. Franç. Corps Gras 36*, 301–308.
18. Holmer, G. (1991) Proceedings of the 16th Scandinavian Symposium on Lipids, in *Proceedings of the 16th Scandinavian Symposium on Lipids*, Holmer, G., Lipidforum, Hardanger, pp. 114–137.
19. Kanner, J., and Rosenthal, I. (1992) An Assessment of Lipid Oxidation in Foods, *Pure Appl. Chem. 64*, 1959–1964.

20. Halliwell, B., and Chirico, S., (1993) Lipid Peroxidation: Its Mechanism, Measurement and Significance, *Am. J. Clin. Nutr. 57,* 715S–725S.
21. Perrin, J.-L., Perfetti, P., and Naudet, M. (1985) Etude Analytique Approfondie D'huiles Chauffées II-Etude Comparative De Corps Gras Différents Amenés À Des États D'altération Comparables, *Rev. Franç. Corps Gras 32,* 205–214.
22. Christopoulou, C.N., and Perkins, E.G. (1989) Isolation and Characterization of Dimers Formed in Used Soybean Oil, *J. Am. Oil Chem. Soc. 66,* 1360–1370.
23. Gardner, D.R., Sanders, R.A., Henry, D.E., Tallmadge, D.H., and Wharton, H.W. (1992) Characterization of Used Frying Oils. Part 1: Isolation and Identification of Compound Classes, *J. Am. Oil Chem. Soc. 69,* 499–508.
24. Márquez-Ruiz, G., Tasioula-Margari, M., and Dobarganes, M.C. (1995) Quantitation and Distribution of Altered Fatty Acids in Frying Fats, *J. Am. Oil Chem. Soc. 72,* 1171–1176.
25. Dobarganes, M.C., Pérez-Camino, M.C., and Márquez-Ruiz, G. (1988) High Performance Size Exclusion Chromatography of Polar Compounds in Heated and Non-Heated Fats, *Fat. Sci. Technol. 90,* 308–311.
26. Dobarganes, M.C., Pérez-Camino, M.C., Márquez-Ruiz, G., and Ruiz-Méndez, M.V. (1989) Edible Fats and Oils Processing: Basic Principles and Modern Practices, in *Edible Fats and Oils Processing: Basic Principles and Modern Practices,* edited by D.R. Erikson, American Oil Chemists' Society, Champaign, Illinois, pp. 427–429.
27. Hopia, A. (1993) Analysis of High Molecular Weight Autoxidation Products Using High Performance Size Exclusion Chromatography. II. Changes During Processing, *Food Sci. Technol. 26,* 568–571.
28. Pérez-Camino, M.C., Márquez-Ruiz, G., Ruiz-Méndez, M.V., and Dobarganes, M.C. (1990) Determinación Cuantitativa De Triglicéridos Oxidados Para La Evaluación Global Del Grado De Oxidación En Aceites Y Grasas Comestibles, *Grasas y Aceites 41,* 366–370.
29. Pérez-Camino, M.C., Márquez-Ruiz, G., Ruiz-Méndez, M.V., and Dobarganes, M.C. (1991) Proceedings of Euro Food Chem VI, in *Proceedings of Euro Food Chem VI, Vol. 2,* Baltes, W., Eklund, T., Fenwick, R., Pfannhauser, W., Ruiter, A., and Thier, H.P., Lebensmittelchemische Gesellschaft, Frankfurt, pp. 569–574.
30. Hopia, A. (1993) Analysis of High Molecular Weight Autoxidation Products Using High Performance Size Exclusion Chromatography. I. Changes During Autoxidation, *Food Sci. Technol. 26,* 563–567.
31. Hopia, A., Lampi, A.-M., Piirönen, V.I., Hyvönen, L.E.T., and Koivistoinen, P.E. (1993) Application of High-Performance Size-Exclusion Chromatography to Study the Autoxidation of Unsaturated Triacylglycerols, *J. Am. Oil Chem. Soc. 70,* 779–784.
32. Márquez-Ruiz, G., Jorge, N., Martín-Polvillo, M., and Dobarganes, M.C. (1996) Rapid, Quantitative Determination of Polar Compounds in Fats and Oils by Solid-Phase Extraction and Exclusion Chromatography Using Monostearin as Internal Standard, *J. Chromatogr. 749,* 55–60.
33. IUPAC (1987) Iupac Standard Method 2.507 in Standard Methods for the Analysis of Oils, Fats and Derivatives, 7th Edition, in *Standard Methods for the Analysis of Oils, Fats, and Derivatives,* 7th edition, International Union of Pure and Applied Chemistry, Blackwell, Oxford, Standard Method 2.507.
34. Ragnarsson, J.O., and Labuza, T.P. (1977) Accelerated Shelf Life Testing for Oxidative Rancidity in Foods, *Food Chem. 2,* 291–308.

35. Márquez-Ruiz, G., Martín-Polvillo, M., and Dobarganes, M.C. (1996) Quantitation of Oxidized Triglyceride Monomers and Dimers as An Useful Measurement for Early and Advanced Stages of Oxidation, *Grasas y Aceites 47,* 48–53.
36. Niemelä, J.R.K., Wester, I., and Lahtinen, R.M. (1996) Industrial Frying Trials with High Oleic Sunflower Oil, *Grasas y Aceites 47,* 1–4.
37. Martín-Polvillo, M., Márquez-Ruiz, G., Jorge, N., Ruiz-Méndez, M.V., and Dobarganes, M.C. (1996) Evolution of Oxidation During Storage of Crisps and French Fries Prepared with Sunflower Oil and High Oleic Sunflower Oil, *Grasas y Aceites 47,* 54–58.
38. Raoux, R., Morin, O., and Mordret, F. (1996) Sensory Assessment of Stored French Fries and Crisps Fried in Sunflower and High Oleic Sunflower Oils, *Grasas y Aceites 47,* 63–74.
39. Márquez-Ruiz, G., Pérez-Camino, M.C., and Dobarganes, M.C. (1990) Combination of Absorption and Size-Exclusion Chromatography for the Determination of Fatty Acid Monomers, Dimers and Polymers, *J. Chromatogr. 514,* 37–44.
40. Márquez-Ruiz, G., Pérez-Camino, M.C., and Dobarganes, M.C. (1992) Digestibility of Fatty Acid Monomers, Dimers and Polymers in the Rat, *J. Am. Oil Chem. Soc. 69,* 930–934.
41. Márquez-Ruiz, G., Ríos, J.J., Tasioula-Margari, M., and Dobarganes, M.C. (1995) Proceedings of Euro Food Chem VIII, in *Proceedings of Euro Food Chem VIII,* Austrian Chemical Society, Vienna, pp. 430–433.

Chapter 11
Analysis of *trans* Fatty Acids

J. Fritsche and H. Steinhart

Universität Hamburg, Institut für Biochemie und Lebensmittelchemie, Grindelallee 117, 20146 Hamburg, Federal Republic of Germany

Introduction

In the diet of most industrial nations, approximately 20–30% of total daily calories are provided by fatty acids that contain at least one double bond. The usual configuration of these double bonds is the *cis* configuration, and they are typically positioned at the third, sixth, or ninth carbon atom from the terminal methyl group, as in oleic acid (18:1n-9) or linoleic acid (18:2n-6). However, some fatty acids have one or more double bonds in the *trans* configuration, the so-called *trans* fatty acids (Fig. 1).

Trans fatty acids (TFA) are present in a variety of foodstuffs, but they contribute only 4–12% of total dietary fat intake (2–4% of total energy intake) in the United States (1–3). In Germany, as an example for European nutrition, the *trans* fatty acid intake is approximately 3–4 g/d (4). Some *trans* fatty acid contents in major food product categories are shown in Table 1 (5–7).

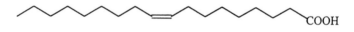

18:1*c*-9 (oleic acid)

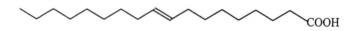

18:1*t*-9 (elaidic acid)

FIG. 1. Structure of oleic acid (18:1*c*9) and elaidic acid (18:1*t*9).

TABLE 1
Trans Fatty Acid Content of Selected Major Food Product Categories (5–7)

Product category	trans 18:1 (% of total fat)	t,c/c,t 18:2 (% of total fat)	Total trans (% of total fat)
Margarine, U.S.			
Stick, hard, soy (n = 52)[a]	24.1 (19–41)[b]	2.1 (0–7)	27.0 (19–49)
Tub, soft, soy (n = 44)	14.4 (9–21)	1.9 (0–9)	17.3 (11–28)
Margarine, Germany			
Diet margarine (n = 6)	0.6 (0–1.7)	0.3 (0.1–0.8)	1.3 (0.4–2.4)
Plant margarine (n = 14)	7.0 (1.8–20.1)	0.3 (0.1–0.8)	7.5 (0.7–21.0)
Shortening			
Commercial, soy (n = 2)	33.6 (30–38)	3.8 (3–4)	37.4 (34–42)
Household, soy (n = 28)	14.5 (9–27)	4.1 (1–7)	19.6 (3–30)
Oils			
Cooking, soy (n = 12)	8.0 (5–11)	2.8 (1–6)	11.9 (1–13)
Salad, soy (n = 3)	0.9 (0–3)	0.7 (0–2)	1.6 (0–5)
Plant oils (n = 6)	0.2 (0–1.5)	0.03 (0–0.1)	0.3 (0–1.5)
Bakery			
Cookies, sugar (n = 43)	15.0 (3–32)	1.8 (0–5)	18.4 (4–36)
Cake, coffee (n = 3)	9.6 (9–11)	1.0 (0.8–1.11)	11.0 (10–13)
Fast food			
Milk shake (n = 12)	2.0 (2–3)	0.3 (0–1)	2.6 (2–4)
Hamburger (n = 11)	3.6 (3–5)	0.3 (0–0.5)	3.9 (3–5)
Snacks			
Chips, potato (n = 24)	10.0 (0–34)	1.7 (0–7)	12.7 (0–40)
French fries (n = 11)	18.7 (3–32)	1.4 (0–3)	20.9 (3–34)
Dairy products			
Butter (n = 10)	2.9 (2–6)	0.3 (0–1)	3.5 (2–7)
Milk, whole (n = 4)	2.1 (2–3)	0.8 (0.7–1)	3.0 (2.7–3.4)
Meats			
Beef, lean, raw (n = 9)	3.2 (2–5)	0.2 (0–0.3)	4.0 (2–5)
Pork, lean, raw (n = 7)	0.2 (0–1)	0 (0)	0.2 (0.1–0.3)
Chicken, lean, raw (n = 3)	0.9 (1–11)	0.3 (0.2–0.4)	1.3 (0.7–1.4)

[a]n = Number of individual means from different data sets used to calculate a mean of means.
[b]Mean = range in parentheses.

Trans fatty acids arise in the first stomach of ruminant animals as intermediates in the hydrogenation (saturation) of dietary unsaturated fatty acids by the hydrogen produced during bacterial fermentation. As a result, the fat in butter, cheese, milk, beef, and mutton contains approximately 2 to 8% *trans* fatty acids by weight (7). *Trans* fatty acids are also formed during the industrial hydrogenation of vegetable or fish oils.

The development of the partial hydrogenation process by W. Normann in the 1930s had a great influence on the edible oil industry. The conversion of polyunsaturated fatty acids to saturated fatty acids increased human ability to alter the functional properties of vegetable oil. The first hydrogenation products were a blend of totally hydrogenated cottonseed oil and refined cottonseed oil and were not a significant source of *trans* fatty acids. Partial hydrogenation of vegetable oil results in the formation of products with different *trans* fatty acid contents; the extreme case is a fully hydrogenated oil, which contains zero percentage *trans* fatty acids. Note that the *trans* fatty acid content of a partially hydrogenated vegetable oil is theoretically limited by the thermodynamics of the *cis-trans* equilibrium to approximately 75% of the total number of double bonds (8,9). In practice, this theoretical limit is never achieved during partial hydrogenation except with a specially designed, sulfur-poisoned nickel catalyst.

The distributions of *cis* and *trans* positional 18:1 isomers in partially hydrogenated vegetable oil and butter are compared in Fig. 2 (10). The *trans* 18:1 positional isomers in partially hydrogenated vegetable oils form a Gaussian distribution that centers around 18:1*t*10 and 18:1*t*11. This pattern is distinctly different from the pattern of butter, which contains 18:1*t*11 (*trans*-vaccenic acid) as the predominant *trans* isomer. In both partially hydrogenated vegetable oil and butter, 18:1*c*9 is the predominant *cis* isomer. Partially hydrogenated vegetable oil contains *cis* positional isomers with double bonds located between the 7 and 16 positions. For butter, 18:1*c*11 is the only *cis* monoene isomer of significance other than 18:1*c*9.

In the 1940s, two-thirds of the per capita consumption of visible fat were of animal origin and about one-third was of vegetable origin; in the mid-1960s one-third was of animal origin and two-thirds were of vegetable origin. This change in primary fat sources resulted in a significant increase in the intake of *trans* fatty acids and polyunsaturated fatty acids and a decrease in the intake of saturated fat. This shift away from animal fats included the increasing use of salad and cooking oil, margarine, and vegetable shortenings (3). From the 1960s until the mid-1980s, most household salad and cooking oils were prepared from partially hydrogenated soybean oil.

However, during the mid-1980s most manufacturers of household salad and cooking oils switched to unhydrogenated vegetable oils, probably because of increasing public health concern about indications that *trans* fatty acids may have unfavorable effects on plasma lipoproteins. Recent interest in *trans* fatty acids was sparked by epidemiological evidence linking *trans* fatty acids with higher plasma total cholesterol and low-density lipoprotein (LDL) cholesterol concentrations and increased incidence of coronary heart disease (CHD) (1,2,11–14). Collectively, these studies raised many questions about the effects of *trans* fatty acids on plasma lipids and lipoproteins and the biological mechanisms that explain these interactions.

The *trans* fatty acid content of food products varies widely both within the same product category and between categories (cf. Table 1). The main reason for this variability is that the *trans* fatty acid content of partially hydrogenated vegetable oil depends on the conditions of hydrogenation (e.g., hydrogenation catalyst, temperature, hydrogen pressure, and time). Another reason for this variability is that products often contain a blend of partially hydrogenated vegetable oils of different sources (soybean, canola, corn, or sunflower oil). In addition, the proportions of hydrogenated and unhydrogenated oil in these blends are varied to obtain the desired physical properties.

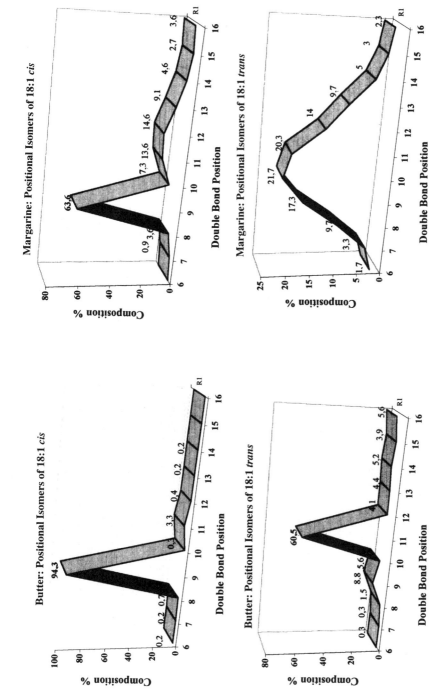

FIG. 2. Distribution of *cis* and *trans* positional isomers in the *cis* and *trans* monoene fractions isolated from partially hydrogenated vegetable oil and butter (10).

The TFA content of dairy fat is more constant. For example, butter contains approximately 3.5% (range 2–7%) total TFA. Dairy fat from grass-fed animals typically has a higher TFA content than that from grain-fed animals, which accounts for the difference in the TFA content of winter and summer butters (15).

In contrast to the unfavorable effects of *trans* fatty acids on the plasma lipoproteins, conjugated linoleic acid isomers (CLA)—above all 18:2c 9t11—seem to have anticarcinogenic and antioxidative properties in animal studies (16–20). Dairy fats are the main source of conjugated linoleic acids, which are also formed during hydrogenation by rumen bacteria (21–23).

To investigate the different effects of individual *cis/trans* fatty acid isomers on lipid metabolism and incorporation in cell membranes, it is necessary to provide chromatographic techniques coupled with sophisticated detection methods to separate and quantify the geometrical and positional isomers.

Infrared Spectroscopy

Principle

Isolated *trans* double bonds show absorption at 967 cm^{-1} (10.3 µm; cf. Fig. 3), deriving from the C-H out-of-plane deformation band for *trans* R_1 HC=CH-R_2 groups, accompanied by the CH_3 in-plane rocking band (1121 cm^{-1}) for saturated FAMEs. The content of isolated *trans* double bonds is determined by measurement of absorption or transmission intensity.

An infrared (IR) spectrum (transmission or absorption) is obtained by scanning from 1050 to 900 cm^{-1}. The infrared spectrophotometers used are double-beam IR (equipped with reference and sample cell) or Fourier transform IR (FTIR). In FTIR the initial scan from carbon disulfide (background spectrum) is stored in the memory of the instrument data-handling system. To obtain transmission and absorption values, the background spectrum is ratioed against the sample spectrum.

Before IR analysis, excessive levels of impurities (e.g., polymers) should be removed from the sample by using suitable cleanup procedures [e.g., thin-layer or column chromatography, especially solid-phase extraction (SPE) columns]. Generally, IR measurements are carried out using fatty acid methyl esters.

A major problem is that samples analyzed by IR as methyl esters produce *trans* levels that are 1.5–3.0% lower for *trans* values from 1 to 15% (24). Therefore, correction factors to compensate for the lower absorption of methyl esters have been proposed by the AOAC (Official Method 965.34) (25). Another problem is that conjugated *trans* double bonds absorb very close to the isolated *trans* bond (conjugated *trans/trans* near 990 cm^{-1} or conjugated *cis/trans* near 990 and 950 cm^{-1}) and can interfere with the isolated *trans* measurement (26). If fatty acids are investigated, a band near 935 cm^{-1}, arising from the O-H out-of-plane deformation in carboxylic acids, also interferes with the isolated *trans* band, particularly at low *trans* levels (less than 15% by weight). The overlap of the *trans* absorption with other bands in

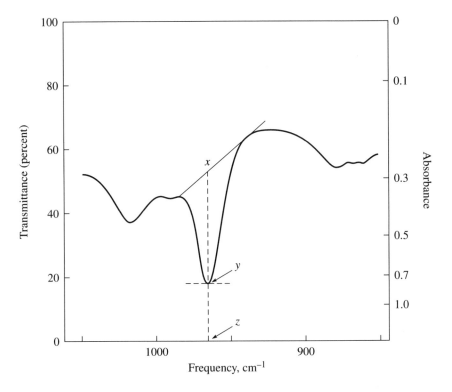

FIG. 3. Infrared (IR) spectrum of partially hydrogenated canola oil methyl esters (2% solution in CS_2) (42). Reprinted from *The Journal of AOAC INTERNATIONAL*, 1995, Volume 78, Number 3, page 786. Copyright 1995, by AOAC INTERNATIONAL, Inc.

the spectrum produces a strongly sloping background that converts the *trans* band into a shoulder at levels below 2% and reduces the accuracy of determination. Because of these interferences, methods for determination of total *trans* unsaturation of fatty acid methyl esters [e.g., AOCS Official Method Cd 14-61 (27)] have several drawbacks despite the use of the baseline technique to correct for any background absorption. The method is applicable only to samples containing less than 5% conjugated fatty acids.

IR spectroscopic determination of *trans* unsaturation yields a high bias for triacylglycerols and a low bias for fatty acid methyl esters, as has been known since 1965 (24). Many procedures in the literature have proposed changes, ranging from minor refinements to major modifications. These include applying arithmetical compensation to eliminate biases; using double-beam differential spectrophotometry to eliminate background interferences (28); applying the internal absorption band ratioing procedure (29,30); eliminating the volatile toxic carbon disulfide solvent by using attenuated total reflection (ATR) (30,31) or 0.1-mm transmission (32,33) cells; using the advantages offered by FTIR spectroscopy instrumentation (34–36); using regression analysis; using the band height or area (35–38) as the independent variable; modifying the calibration procedure for two-component or multiple-component (30–32) mixtures that contain a *trans* reference material instead of calculating the absorptivity of a standard (24,28); and applying partial least-squares chemometric procedures (33) or post-measurement spectral subtraction manipulations (35).

A two-component calibration procedure to overcome some of the drawbacks of the AOCS official IR method, especially the low bias for methyl esters and the need for calculation of correction factors, has been proposed by Madison et al. (37). *Trans* content was calculated using a calibration curve of absorption versus percentage isolated *trans* unsaturation. The calibration curve was developed using a series of carbon disulfide solutions containing different ratios of methyl elaidate and methyl linoleate. Calibration and test solutions are scanned from 900 to 1500 cm^{-1} against a carbon disulfide blank. A baseline is drawn between peak minima at about 935 and 1020 cm^{-1}, and the baseline-corrected absorbance of the *trans* peak (967 cm^{-1}) is obtained. The baseline for the test sample spectrum is drawn exactly as the baseline in the standard spectrum, by overlaying the two spectra. This procedure allows analysis of *trans* contents in the 0.5–36% range with increasing accuracy.

ATR

To eliminate the strongly sloping background of the 966 cm^{-1} *trans* band, the single-beam (SB) spectrum of the *trans*-containing fat is "ratioed" against that of an unhydrogenated oil or a reference fat or oil containing only *cis* double bonds. Thus, a symmetric absorption band on a horizontal background is obtained. The area under the *trans* band can then be accurately integrated between the same limits, 990 and 945 cm^{-1}, for all *trans* levels. To speed up the analysis, an attenuated total reflection (ATR) liquid cell is used, into which oils, melted fats, or their methyl esters are poured without weighing or quantitative dilution with carbon disulfide.

The *trans* levels determined by ATR are closer to those found by GC when the hydrogenated fat is measured against an unhydrogenated oil than when it is measured against a *cis* reference material. Small differences were observable between *trans* levels in hydrogenated fat test samples and the corresponding methyl ester derivatives (9.3 and 2.2% at about 2 and 41% *trans*, respectively). The lower limits of identification and quantification are 0.2 and 1%, respectively (39).

Optothermal Window

The complications of quantitative studies by traditional transmission IR measurements using a cell with short and difficult-to-reproduce path length (<10 m) can be circumvented by using the novel concept of optothermal window (OW) spectroscopy.

The operational principle of the OW technique, actually a variant of conventional photoacoustic spectroscopy, is as follows: Modulated laser radiation passes through the OW cell before impinging on the sample. The OW cell is actually an optically transparent disk that has a large thermal expansion coefficient. Its rear side is equipped with an annular piezoelectric transducer. Due to the absorption of radiation, the sample's temperature rises, and the generated heat diffuses into the disk, which expands. The induced stress is then detected at the modulation frequency by the piezoelectric transducer in conjunction with the lock-in amplifier.

The OW signal remains unaffected by thermal expansion of the sample and is also less susceptible to the effect of other thermal parameters of the sample. Finally, as long as it exceeds the sample's thermal diffusion length, the thickness of the sample is not relevant, making the OW technique practical for quantitative infrared analysis of strongly absorbing fluids and semifluids (40).

Combined FTIR and GC

The advantage of combining infrared spectroscopic determination of *trans*-monounsaturated fatty acid methyl esters with capillary gas chromatography is that any interferences, particularly those due to partial overlap of adjacent *trans* and *cis* octadecenoic isomer GC peaks, would not disturb the accuracy of the measurements (41). Thus, it is possible to determine the general fatty acid composition of partially hydrogenated vegetable oil by GC/IR.

Ratnayake et al. (42) proposed the combination of GC and infrared spectrophotometry for accurate determination of total 18:1*t* and 18:1*c* isomers in partially hydrogenated vegetable oils (PHVO).

Total *trans* unsaturation determined by IR (Σ *trans*) was related to the capillary GC weight percentages of the component *trans* fatty acid methyl esters by the mathematical formula

$$\Sigma trans = 18:1t + 0.84 \times 18:2t + 1.74 \times 18:2tt + 0.84 \times 18:3t \qquad (1)$$

where 0.84 and 1.74 are correction factors relating GC weight percentages to the IR *trans* equivalents for 18:2*t*, 18:3*t*, and 18:2*tt*, respectively. This formula is the basis for determining total 18:1*t* and 18:1*c* isomers, and total 18:1*t* is calculated from the mathematical formula

$$18:1t = \Sigma trans - 0.84 \times (18:2t + 18:3t) - 1.74 \times 18:2tt \qquad (2)$$

The content of 18:1*c* is then calculated as the difference between total 18:1 fatty acid methyl esters, determined by GC, and 18:1*t*.

Since in the foregoing combined GC and IR method 18:1*t* and IR *trans* unsaturation are linearly related, the accuracy of 18:1*t* determination is dependent solely on the accuracy of total *trans* unsaturation determination by IR spectroscopy.

Hyphenated GC–FTIR

Three fundamentally different types of interfaces between a gas chromatograph and a Fourier transform infrared spectrometer can be distinguished. The most common interface, the light pipe (LP), incorporates a flow-through light pipe gas cell that leads to vapor-phase spectra, which are measured in real time at intervals of approximately 1 s (43). For systems of this type, the minimum identifiable quantity (MIQ) is rarely less than 5 ng, with the MIQ typically increasing by an order of magnitude for compounds of low absorptivity eluting from the GC column as broad peaks. The relatively low sensitivity of the light pipe–based GC–FTIR interface (43) is usually insufficient for *trans* FAMEs, which are usually quantitated in the nanogram range (44).

Better sensitivity (MIQ in the subnanogram range) is achieved by the use of matrix isolation (MI) (45) or direct deposition (DD) (46) techniques. The MI method involves adding argon (1.5% by volume) to the GC carrier gas (helium) and trapping the effluent on the outer rim of a slowly rotating (about 3 mm/min) gold disk held at cryogenic temperatures (about 12 K). During a gas chromatographic run, helium is removed by vacuum pumps, and the analyte molecules, surrounded by an excess of argon atoms, are frozen into a solid matrix. The analytes isolated in the IR-transparent argon matrix are subsequently analyzed by IR spectroscopy. The reflection-absorption spectrum is then measured. The position of each analyte peak on the Cyrolect® collection disk is indexed by its observed GC retention time. By optimizing the performance of the system, including optical alignment, and reproducibly locating the peak maximum on the collection disk, the extent of postcolumn peak broadening can be minimized (45).

In the DD interface, GC eluates are condensed on a cooled moving window without dilution in a matrix of any type. The position and movement (e.g., 100 μm during the full-width at half-height (fwhh) of a typical GC peak) of the window are controlled and recorded during a GC run by the spectrometer's software (47). The width of the sample trace can be reduced to about 100 μm, leading to an approximately 10-fold increase in the absorbance per unit weight over a GC/MI–FTIR interface with a 300-μm-wide trace (48).

A further advantage that has been claimed for MI and DD interfaces in comparison to the corresponding measurements made with a light pipe is the increase in peak absorptivity, since the fwhh of bands in MI spectra (typically 0.8 to 8 cm^{-1}) and in DD spectra (4 to 8 cm^{-1}) is significantly less than the fwhh of corresponding bands in vapor-phase spectra (≥ 8 cm^{-1}) and the integrated area is approximately constant (49).

Although the baseline noise level is always lowest for spectra measured at low resolution in a given time, the increased absorbance of narrow bands measured at high resolution partially offsets the increased noise level (50).

For real-time measurements the acquisition time per spectrum must be kept below 1 s, so in practice it is rare to find any real-time GC–FTIR measurements made at a higher resolution than 4 cm^{-1}. Subnanogram MIQs have never been reported for measurements made in real time, and all reported spectra of subnanogram sample quantities have been measured with postrun signal averaging (44).

Ag^+ Chromatography

Since its introduction by Morris in 1963, silver ion (Ag^+) or argentation chromatography has been one of the most important tools available to lipid analysts for the separation of molecular species of lipids (51). The topic has been the subject of many reviews (52–54), most recently by Nikolova-Damyanova (55).

Fatty acids can be separated according to both the configuration and the number of their double bonds, and, with Ag–HPLC, also according to the position of the double bonds. The analysis can be carried out either with triacylglycerols or with fatty acids after suitable derivatization, such as FAMEs or phenacylesters (56–64).

In most of the published work, the Ag^+ chromatographic technique has been used in conjunction with thin–layer chromatography (TLC), with silver nitrate being incorporated into a silica gel layer by various means (66). Common TLC solvent systems are, for example, hexane/diethyl ether (90:10, v/v), benzene for *cis/trans* methyl linolenate, and chloroform/acetone/acetic acid (96:4:0.5 v/v/v) for *cis/trans*-methyl arachidonate fractionation. TLC plates are sprayed with 2',7'-dichlorofluorescein solution and dried, and the spots are visualized by UV at 254 nm. Using Ag–TLC, FAMEs can be separated only by the total number of *cis* and *trans* double bonds, not by the specific location, unless extreme conditions are applied (30% silver nitrate in the sorbent layer and −25°C (67).

R_f values of different components are highly variable, being dependent on such factors as the proportion of silver nitrate in the layer, its degree of hydration, atmospheric humidity and temperature, and the nature of the mobile phase—factors that are not easily controlled (55).

Recently, a stable silver-ion column for high-performance liquid chromatography (HPLC) has been developed, in which the silver ions are linked via ionic bonds to phenylsulfonic acid moieties, which are bound to a silica matrix (68). This column has been used extensively for the separation of fatty acid derivatives and triacylglycerols. In particular, excellent separations of positional and geometrical isomers of unsaturated fatty acid phenacylesters have been achieved (69). In contrast to Ag–TLC, it is possible to control many of the chromatographic parameters, especially mobile phase composition, flow rate, and column temperature, with a high degree of accuracy. Other advantages of Ag–HPLC are the reusability of the column, simple and rapid sample separation, short analysis time, and recoveries usually above 95% (Ag–TLC about 60%) (70).

If the samples are analyzed by GC before and after the separation, the amount of the *trans* fatty acids in complex mixtures can be determined. The separation of polyunsaturated *cis*- and *trans*-FAMEs is possible with a UV-compatible solvent system (e.g., mobile phase consisting of hexane with 1% acetonitrile) (70). Baseline separation of the 9-*trans* and 9-*cis* 18:1 isomers was achieved. The elution order when using this mobile phase for the methyl linoleate isomers was reported to be (65)

$$18:2t9,t12 > 18:2t9,c12 > 18:2c9,t12 \tag{3}$$

and differs from that obtained with capillary GC (SP 2330, SP 2340 or SP 2560), in which $18:2c9,t12$ reportedly elutes before $18:2t9,c12$ (71). The separation of all eight *cis/trans* isomers of methyl 18:3 was similar to the separation obtained on a 50 m CP Sil™ 88 capillary GC column (72). The resolution of 15 of the 16 possible *cis/trans* isomers of methyl 20:4 far exceeds the capabilities of HPLC (65).

Perhaps surprisingly for such a well-established technique, the mechanism of the interactions between unsaturated centers and silver ions in chromatographic systems is poorly understood. It is known that a single silver ion can interact with two double bonds simultaneously, because crystalline complexes with two moles of ethylene and other olefins have been isolated (73,74). In favorable circumstances, a silver ion may interact with two double bonds in the same molecule (75). It would be expected that the degree of retention would depend mainly on the strength of the interaction between the silver ions and the π-electrons of the double bonds. *Trans* bonds have a lower π-electron density than do *cis* double bonds; therefore, metal ions form weaker complexes with *trans* fatty acids than with *cis* fatty acids.

However, in silver-ion TLC, interactions between parts of the molecule and silanol groups on the surface of the silica must also be considered. In an HPLC system some interactions with residual silanol moieties are also possible, as is some form of bonding with the oxygen atoms of the sulfonic acid group or hydrophobic interactions with the phenyl moieties. The retention of fatty acids depends on the number, position, and configuration of double bonds and is additionally influenced by other functional groups of the molecule (76).

Morris et al. (67) suggested that the greater mobility of the isomers that have a double bond near the ester group in silver-ion TLC was due to the partial withdrawal or delocalization of the π electrons of the double bond caused by the inductive effect of the ester group. However, as such an effect decreases rapidly with increasing distance between the double bond and the ester group, it can only explain the elution order of the monoenoic fatty acids.

An interaction between one silver ion and two double bonds simultaneously may also explain the chromatographic behavior of dienoic derivatives with silver-ion HPLC. When the distance between the double bonds is optimum (e.g., with a 1,5-*cis*,*cis*-diene system), fatty acids are very strongly retained (75). The k'' values remained higher than those of methylene-interrupted dienes, however, and the effect of double bond position was diminished.

The theory of complexation between silver ions and *bis*-double bond systems could potentially be applied to polyenes. It would predict that a triene would be held twice as strongly as a diene, a tetraene three times as strongly, and so forth. Such a simple relationship was not found in experience, possibly because interactions with the ester moiety have to be taken into consideration and because the conformations of polyenes may permit some interactions between silver ions and double bonds that are remote from each other. With bulky molecules such as triacylglycerols, an ester effect might be less important (77). Not to be overlooked is the influence of temperature on the interactions and thus on the separation characteristics.

Gas Chromatography

The most important method to investigate fatty acid compositions is gas chromatography. This separation technique requires volatile analytes. Esterified lipids (e.g., triacylglycerols, phospholipids, cholesteryl esters) and free fatty acids have to be converted to suitable derivatives. Most common is the transesterification to methyl esters, which is the principal chemical reaction performed by lipid analysts (78–80). The fatty acid methyl esters (FAMEs) are generally detected by a flame ionization detector (FID).

The current American Oil Chemists' Society (AOCS) Official Method Ce 1c-89 (revised 1990) has been designed to evaluate the general fatty acid composition, including the levels of 18:1c and 18:1t isomers, and the total *trans* unsaturated fatty acid content in hydrogenated and unhydrogenated vegetable oils (81). For this measurement a direct, one-step capillary gas chromatography (GC) procedure was developed. The Official AOCS Method uses a 60 m × 0.25 mm i.d. fused-silica capillary column coated with SP-2340 (*bis*-cyanopropylphenyl polysiloxane). This direct GC method is based on the assumption that *cis* and *trans* octadecenoic acid isomers are completely separable on this high polar column.

In capillary columns coated with cyanoalkylsiloxane stationary phases (e.g., SP 2560 or CP™ Sil 88), 18:2t, 18:2tt, and 18:3t are separated as distinct groups without any serious interferences or overlaps (cf. Fig. 4), and levels of these *trans* polyunsaturates can be obtained directly by GC analysis. However, because of the multiplicity of positional and geometrical isomers present in partially hydrogenated vegetable oils, a satisfactory separation of 18:1t isomers as a group from that of *cis* isomers is not feasible on cyanoalkylsiloxane phases or any other currently available GC stationary phases (41,82).

Nevertheless, the separations obtained with the 60-m column were encouraging, and it seemed possible to improve the separations by increasing the length of the column. Recently 100-m capillary columns (SP 2560; CP Sil™ 88) have been made commercially available, and such columns can improve the resolution of *trans*-18:1 isomers (83,84).

18:1 Isomers

The study of *trans*-18:1 isomer distribution in partially hydrogenated oils, ruminant fats, or human tissue lipids generally requires, as a first step, the isolation of monoenoic acids, usually by argentation chromatography.

To quantify the individual *trans*-18:1 isomers, the fatty acid mixture can alternatively be purified by preparative GC and then submitted to ozonolysis, followed by reduction of the ozonides (85,86). The resulting fragments (generally aldehydes and aldehyde esters) are then analyzed by GC. However, in addition to the length and tediousness of the procedure, another drawback is that the shorter fragments are more or less volatile and may be partly lost during the procedure (87). Moreover, the accuracy of the method depends on the use of suitable correction factors for the unequal response of the FID.

Although several attempts have been made to analyze the individual *trans*-18:1 isomers directly by GC on capillary columns, only a few authors have obtained

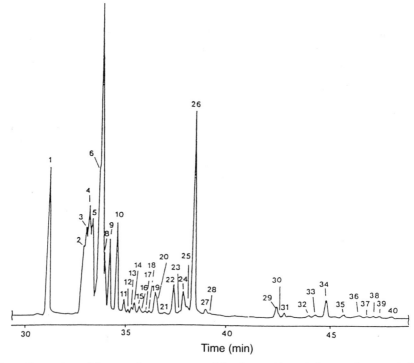

FIG. 4. C_{18} region of the gas chromatogram of the fatty acid methyl esters from partially hydrogenated soybean oil, using 100 m × 0.25 mm fused-silica capillary column coated with SP 2560 (42). Peak identification: 1. 18:0 2. 18:1Δ6-8t 3. 18:1Δ9t 4. 18:1Δt10 5. 18:1Δt11 6. 18:1Δ12t 7. 18:1Δ6–8c + 18:1Δ9c (major component) + 18:1Δ13–14t 8. 18:1Δ10c + 18:1Δ15t 9. 18:1Δ11c 10. 18:1Δ12c 11. 18:1Δ13c 12. 18:2tt 13. 18:1Δ14c + 18:1Δ16t 14. 18:2tt 15. 18:1Δ15c 16. 18:2tt 17. 18:2tt 18. 18:2Δ9t,12t 19. 18:2Δ9c,13t + 18:2Δ8t,12c 20. 18:2Δ9c,13t + 18:2Δ8t,12c 21. 18:2ct/tc 22. 18:2Δ9c,12t 23. 18:2Δ8c,13c 24. 18:2Δ9t,12c 25. 18:2Δ9t,15c + 18:2Δ10t,15c + 18:2Δ9c,13c 26. 18:2Δ9c,12c (linoleate) 27. 18:2Δ9c,15c 28. Not identified 29. 20:0 30. 18:3? 31. 18:3Δ9c,12c,15t 32. 18:3Δ9c,12t,15c 33. 18:3Δ9t,12c,15c 34. 18:3Δ9c,12c,15c 35. 20:1c 36–40. 18:2 conjugates. Reprinted from *The Journal of AOAC INTERNATIONAL*, 1995, Volume 78, Number 3, page 790. Copyright 1995, by AOAC INTERNATIONAL, Inc.

promising results (83,84). In recent studies on the content and distribution of *trans*-18:1 acids in ruminant fats and human milk lipids, rather good resolution of several, but not all isomers was obtained using a 50-m capillary column coated with 100% cyanopropyl polysiloxane (88). For example, the *trans*-10 18:1 isomer appeared as a simple shoulder on the leading edge of the main *trans*-11 18:1 acid, and it was seldom possible individually to quantitate these two isomers. Moreover, the *trans*-12 18:1 isomer was not particularly well resolved from the adjoining *trans*-11 and *trans*-13 plus *trans*-14 18:1 acids. The *trans* 18:1 isomers with a double bond in positions 12, 13 or higher coeluted with the *cis* isomers (*cis*-18:1>9). This coelution was observed for several oil types. The incomplete resolution of *trans*-12 18:1 and *trans*-

13 18:1 isomers resulted in an underestimation of the *trans* level by as much as approximately 32% of total *trans* content (89).

To minimize the underestimation of *trans* fatty acid levels, Duchateau et al. developed an optimized method suitable for hydrogenated, as well as for refined, processed oils (90). The developed method was to comply with the following requirements:

1. Maximum resolution was required between the *trans* and *cis* 18:1 isomers; especially the *trans*-13 18:1 isomer had to be separated from the *cis*-9 18:1 isomer.
2. 20:1 had to elute with sufficient resolution from linolenic acid isomers; if this were not possible, elution of 20:1 had to elute between the last eluting mono-*trans* 18:3 isomer (18:3*t*9,*c*12,*c*15) and 18:3*n*-6.

The optimized temperature conditions for a CP Sil® 88, SP 2340 and BPX 70 GC columns were simulated by computer software. Input to the simulation program consisted of the retention times of the specific isomers, together with general column and temperature conditions (e.g., column dimensions, column head pressure, linear velocity carrier gas, and injector and detector temperature).

The accuracy of the developed method was checked with *cis* and *trans* fatty acid fractions isolated by Ag–HPLC. The *trans* values obtained with the optimized method were in good agreement with the results obtained for the isolated fractions (deviation from isolated fractions: 0.8 to 6.5% (optimized method), in contrast to 21.3–26.4% (AOCS method) for *trans* amounts between 12.7 and 43.2%).

18:2 Isomers

Baseline separation of the geometric isomers of linoleate (18:2*t*9,*t*12, 18:2*c*9,*t*12, and 18:2*t*9*c*12) is possible with cyanopropyl-coated capillary columns (91). For this kind of capillary column the elution order of geometric isomers is (92)

$$18:2t9,t12 > 18:2c9,t12 > 18:2t9,c12 \tag{4}$$

The presence of conjugated double bond systems in the alkyl chain increases the retention time of FAMEs considerably over that of a similar compound with methylene-interrupted double bonds. For example, 18:2*c*9*t*11 has appreciably greater equivalent chain length (ECL) values than the corresponding values for methyl linoleate (93).

18:3 Isomers

Six out of the eight linolenic acid (18:3*n*-3) geometrical isomers (LAGI) are almost base-line resolved when the cyanopropylpolysiloxane-coated column (CP™Sil 88) is operated at relatively low temperature (150°C) and pressure (0.8 kg/cm^2 He) (94). The elution order of LAGI (either as FAMEs or as fatty acid propyl esters, FAPEs) on the CP™Sil 88 is

$$t,t,t < c,t,t < t,c,t < c,c,t < t,t,c < c,t,c < t,c,c < c,c,c \tag{5}$$

When FAPEs of LAGI are analyzed on the same column operated at 165°C with a carrier gas pressure of 1.05 kg/cm^{-2}, all eight isomers are partially resolved. The time of analysis (20 vs. 45 min) is significantly shortened. However, peak resolution of FAPEs is not as good as that obtained with FAMEs (94).

Reference LAGI were prepared by isolation from linseed oil (using urea inclusion and preparative HPLC) following nitrous acid isomerization or by isolation from heated linseed oil and were identified by combining R_f values determined after Ag$^+$ chromatography and literature data (95).

GC–MS

To elucidate the structure of fatty acids the use of a mass spectrometer connected to a GC is customary. Commonly, fatty acids are derivatized to the methyl or trimethylsilyl esters for GC analysis. Long-chain saturated methyl esters are easily identified, but fragmentation patterns of unsaturated FAMEs are not indicative of the position of double bonds (96). Double bond positions can be determined if the unsaturated fatty acids are converted to suitable derivatives. For instance, formation of pyrrolidide (97), picolinyl, and 4,4-dimethyloxazolinyl (DMOX) (98,99) derivatives were proposed for the GC–MS identification of fatty acids with different functional groups.

The derivatization to DMOX products proved to be a powerful method in GC–MS fatty acid analysis with a wide range of applicability (98,99). Natural and synthetic fatty acids with acetylenic, hydroxy, and keto groups, as well as methyl branched fatty acids, have been analyzed and identified by their low-resolution EI mass spectra (100). Mass spectra of DMOX derivatives of fatty acids give very intense ions at m/z 113 and 126, which are characteristic of DMOX derivatives of fatty acids (101,102). For the DMOX derivative of 20:4c5,c8,c11,t15 the Δ5 double bond can be localized by the appearance of the prominent ion at m/z 153. This ion is a diagnostic ion in the spectra of DMOX derivatives with their first double bond at the Δ5 position (101).

The geometry of double bonds in 20:4c5,c8,c11,t15 could be deduced by comparison of the experimental and calculated ECL values. This approach has been successfully applied to the identification of pre-separation C$_{20}$ *cis*-polyunsaturated fatty acids and geometrical isomers of 20:5n-3, as well as to some *cis* and *trans* C$_{18}$ isomers (103,104). Combining preseparation of 18:1 *cis/trans* isomers by Ag$^+$ chromatography and GC–MS of DMOX derivatives, the *trans*-18:1 positional isomers can be completely identified from Δ8–16 (105).

To confirm the identity of the minor Δ5, Δ6, and Δ7 *trans*-18:1 FAMEs, positional isomers found in the Ag–HPLC fraction containing the *trans* monounsaturated FAME mixture, ^{13}C-NMR can be applied (106).

Conclusion

Applying the sophisticated methods presented above, the determination of all geometrical and positional isomers of the common *trans* fatty acids is possible. However, the usual method for determination of *trans* fatty acids is the combination of Ag^+ chromatography and gas chromatography. Analysis of *trans* fatty acid compositions is performed mainly in foodstuffs and in human adipose tissues and plasma.

Trans fatty acids are formed by biohydrogenation in ruminants or by catalytic hardening of vegetable and fish oils. Thus, they occur in numerous foodstuffs that contain partially hydrogenated vegetable oil, dairy products, or meat from ruminants, in considerably fluctuating amounts. Based on current evidence, catalytic hydrogenation does not produce any fatty acid isomer that is not also produced by microbial hydrogenation in the rumen. For instance, there is no isomer present in margarine that is not also present in butter.

The content of *trans* fatty acids in vegetable oils may be reduced by changes in technology (e.g., interesterification to adjust functional properties) and changes in raw materials. These approaches were successfully applied in Germany and led to a visible decrease of *trans* fatty acids in margarine (6,7).

In human metabolism *trans* fatty acids are not biochemically equivalent to either their corresponding saturated or *cis* unsaturated fatty acids. Results from many biochemical studies suggest that *trans* fatty acids are best viewed as a separate class of fatty acids that are recognized, metabolized (e.g., by oxidation, acylation, elongation, and desaturation) and regulated by the same mechanisms that control the metabolism of other common dietary fatty acids (107).

Further investigations will have to be carried out to evaluate the effects of the individual *trans* fatty acid isomers on the human lipid metabolism, above all on the incidence of coronary heart disease, and to evaluate the role of conjugated linoleic acid isomers in cancer protection.

Acknowledgments

The authors acknowledge S. Hartmann for critical comments in the manuscript and encouragement.

References

1. Willett, C.W., Stampfer, M.J., Manson, J.E., Colditz, G.A., Speizer, F.E., Rosner, B.A., Sampson, L.A., and Hennekens, C.H. (1993) Intake of *Trans* Fatty Acids and Risk of Coronary Heart Disease Among Women, *Lancet 341*, 581–585.
2. Troisi, R., Willett, W.C., and Weiss, S.T. (1992) *Trans*-fatty Intake in Relation to Serum Lipid Concentrations in Adult Men, *Am. J. Clin. Nutr. 56*, 1019–1024.
3. Hunter, J.E., and Applewhite, (1991) T.H. Reassessment of *Trans* Fatty-acid Availability in the US Diet, *Am. J. Clin. Nutr. 44*, 707–717.

4. Steinhart, H., and Pfalzgraf, A. (1991) Aufnahme *Trans*-Isomerer Fettsäuren—Eine Abschätzung auf Basis der nationalen Verzehrsstudie *Z. Ernährungswiss. 31*, 196–204.
5. Dickey, L.E.(1995) *Trans* Fatty Isomer Content of Food Sources, *INFORM 6*, 484A.
6. Pfalzgraf, A., and Steinhart, H. (1995) Gehalte von *trans*-Fettsäuren in Margarinen, *Dtsch. Lebensm. Rundsch. 91*, 113–114.
7. Pfalzgraf, A., Timm, M., and Steinhart, H. (1994) Gehalte von *trans*-Fettsäuren in Lebensmitteln *Z. Ernährungswiss. 33*, 24–43.
8. Johnson, R.W., and Pryde, E.H. (1979) *Isomerization, Conjugation, and Cyclization*, 1st edn., American Oil Chemists' Society, Champaign, Illinois, pp. 342–352.
9. Okonek, D.V. (1986) in *Hydrogenation: Proceedings of an AOCS Colloquium* (Hastert, R., ed.), American Oil Chemists' Society, Champaign, Illinois, pp. 65–88.
10. Craig-Schmidt, M.C. (1992) in *Fatty Acids in Foods and their Health Implications* (Chow, C.K., ed.), Marcel Dekker Inc., New York, pp. 365–398.
11. Siguel, E.N., and Lerman, R.H. (1993) *Trans*-Fatty Acid Patterns in Patients with Angiographically Documented Coronary Artery Disease, *Am. J. Cardiol. 71*, 916–920.
12. Ascherio, A., Hennekens, C.H., Buring, J.E., Master, C., Stampfer, M.J., and Willett, W.C. (1994) *Trans*-Fatty Acid Intake and Risk of Myocardial Infarction, *Circulation 89*, 94–101.
13. Roberts, T.L., Wood, D.A., Reimersma, P.J., Gallagher, P.J., and Lampe, F.C. (1995) *Trans* Isomers of Oleic and Linoleic Acids in Adipose Tissues and Sudden Cardiac Death, *Lancet 345*, 278–282.
14. Tzonou, A., Kalandidi, A., and Trichopoulou, A. (1993) Diet and Coronary Heart Disease: A Case-Control Study in Athens, Greece, *Epidemiology 4*, 511–516.
15. Emken, E.A. (1984) Nutrition and Biochemistry of *Trans* and Positional Fatty Acid Isomers in Hydrogenated Oils, *Annu. Rev. Nutr. 4*, 339–376.
16. Ip, C., Singh, M., Thompson, H.J., and Scimeca, J.A. (1994) Conjugated Linoleic Acid Suppresses Mammary Carcinogenesis and Proliferative Activity of the Mammary Gland in the Rat, *Canc. Res. 54*, 1212–1215.
17. Belury, M.A. (1995) Conjugated Dienoic Linoleate: A Polyunsaturated Fatty Acid with Unique Chemoprotective Properties, *Nutr. Rev. 53*, 83–89.
18. Van den Berg, J.J.M., Cook, N.E., and Tribble, D.L. (1995) Reinvestigation of the Antioxidant Properties of Conjugated Linoleic Acid *Lipids 30*, 599–605.
19. Schonberg, S., and Krokan, H. (1995) The Inhibitory Effect of Conjugated Dienoic Derivatives (CLA) of Linoleic Acid on the Growth of Human Tumor Cell Lines is in Part Due to Increased Lipid Peroxidation, *Anticanc. Res. 15*, 1241–1246.
20. Fischer, S.M., Leyton, J., Lee, M.L., Locmiskar, M., Belury, M.A., Maldave, R.E., and Slaga, T.J. (1992) Differential Effects of Dietary Linoleic Acid on Mouse Skin-Tumor Promotion and Mammary Carcinogenesis, *Canc. Res. 52*, 2049–2054.
21. Chin, S.F., Liu, H., Storkson, J.M., Ha, Y.L., and Pariza, M.W. (1992) Dietary Sources of Conjugated Dienoic Isomers of Linoleic Acid, a Newly Recognized Class of Anticarcinogens, *J. Food Comp. Anal. 5*, 185–197.
22. Kepler, C.R., Hirons, K.P., McNeill, J.J., and Tove, S.B. (1966) Intermediates and Products of the Biohydrogenation of Linoleic Acid by *Butyrivibrio fribrisolvens*, *J. Biol. Chem. 241*, 1350–1354.
23. Hughes, P.E., Hunter, H.J., and Tove, S.B. (1982) Biohydrogenation of Unsaturated Fatty Acids, *J. Biol. Chem. 257*, 3643–3649.

24. Firestone, D., and LaBouliere, P. (1965) Determination of Isolated *Trans* Isomers by Infrared Spectrophotometry, *J. Assoc. Off. Anal. Chem. 48,* 437–443.
25. AOAC Method 965.34, in *Official Methods of Analysis of the Association of Official Analytical Chemists* (Helrich, K., ed.), 1994, Arlington, Virginia.
26. O'Connor, R.T. (1956) Application of Infrared Spectrophotometry To Fatty Acid Derivatisation, *J. Am. Oil Chem. Soc. 33,* 1–15.
27. Official Method Cd 14–61, in *Official Methods and Recommended Practices of the American Oil Chemists' Society* (Firestone, D., ed.), 1993, AOCS Press, Champaign, Illinois.
28. Huang, A., and Firestone, D. (1971) Determination of Low Level Isolated *Trans* Isomers in Vegetable Oils and Derived Methyl Esters by Differential Infrared Spectrophotometry, *J. Assoc. Off. Anal. Chem. 54,* 47–51.
29. Allen, R.R. (1969) The Determination of *Trans* Isomers in GLC Fractions of Unsaturated Esters *Lipids* 4, 627–628.
30. Belton, P.S., Wilson, R.H., Sadeghi-Jorabchi, H., and Peers, K.E. (1988) A Rapid Method for the Estimation of Isolated *Trans* Double Bonds in Oils and Fats Using FTIR Combined with ATR, *Lebensm. Wiss. Technol. 21,* 153–157.
31. Dutton, H.J. (1974) Analysis and Monitoring of *Trans*-Isomerization by IR ATR Spectrophotometry, *J. Am. Oil Chem. Soc. 51,* 406–409.
32. Sleeter, R.T., and Matlock, M.G. (1989) Automated Quantitative Analysis of Isolated (Nonconjugated) *Trans* Isomers Using FTIR Spectroscopy Incorporating Improvements in the Procedure, *J. Am. Oil Chem. Soc. 66,* 121–127.
33. Ulberth, F., and Haider, H.J. (1992) Determination of Low Level *Trans* Unsaturation in Fats by FTIR Spectroscopy, *J. Food Sci. 57,* 1444–1447.
34. Lanser, A.C., and Emken, E.A. (1988) Comparison of FTIR and Capillary GC Methods for Quantification of *Trans* Unsaturation in Fatty Acid Methyl Esters, *J. Am. Oil Chem. Soc. 65,* 1483–1487.
35. Toschi, T.G., Capella, P., Holt, C., and Christie, W.W. (1993) A Comparison of Silver Ion HPLC Plus GC with FTIR Spectroscopy for the Determination of *Trans* Double Bonds in Unsaturated Fatty Acids, *J. Sci. Food Agric. 61,* 261–266.
36. Official Method Cd 14-95, *Official Methods and Recommended Practices of the American Oil Chemists' Society* (Firestone, D., ed.), 1995, AOCS Press, Champaign, Illinois.
37. Madison, B.L., DePalma, R.A., and D'Alonzo, R.P. (1982) Accurate Determination of *Trans* Isomers in Shortenings and Edible Oils by Infrared Spectrophotometry, *J. Am. Oil Chem. Soc. 59,* 178–181.
38. IUPAC Method 2.207, *International Union of Pure and Applied Chemistry, Standard Methods for the Analysis of Oils, Fats and Derivatives* (Paquot, C., and Hautfenne, A., eds.), 1987, Blackwell Scientific Publications, Oxford.
39. Mossoba, M.M., Yurawecz, M.P., and McDonald, E. (1996) Rapid Determination of the Total *Trans* Content of Neat Hydrogenated Oils by Attenuated Total Reflection Spectroscopy, *J. Am. Oil Chem. Soc. 73,* 1003–1009.
40. Favier, J.P., Bicanic, D., van de Bovenkamp, P., Chirtoc, M., and Helander, P. (1996) Detection of Total *Trans* Fatty Acids Content in Margarine: An Intercomparison Study of GLC, GLC+TLC, FT–IR, and Optothermal Window (open photoacoustic cell), *Anal. Chem. 68,* 729–733.

41. Ratnayake, W.M.N., and Beare-Rogers, J.L. (1990) Problems of Analyzing C_{18} *Cis* and *Trans*-Fatty Acids of Margarine on the SP-2340 Capillary Column, *J. Chromatogr. Sci. 28*, 633–639.
42. Ratnayake, W.M.N., Alfieri, J., Bacler, S., Ballch, B., Debets, B., Dionne, D., Gould, S., Henry, D.E., Lapointe, M.R., Lahtinen, B., McDonald, R.E., Pelletier, G., and Wolf, D. (1995) Determination of *Trans* Unsaturation by Infrared Spectrophotometry and Determination of Fatty Acid Composition of Partially Hydrogenated Vegetable Oils and Animal Fats by Gas Chromatography/Infrared Spectrophotometry: Collaborative Study, *J. AOAC Int. 78*, 783–802.
43. Griffiths, P.R., de Haseth, J.A., and Azarraga, L.V. (1983) Capillary GC/FT-IR, *Anal. Chem. 55*, 1361A–1371A.
44. Mossoba, M.M., McDonald, R.E., and Prosser, A.R. (1993) Gas Chromatographic/Matrix Isolation/Fourier Transform Infrared Spectroscopic Determination of *Trans*-Monounsaturated and Saturated Fatty Acid Methyl Esters in Partially Hydrogenated Menhaden Oil, *J.Agric. Food Chem. 41*, 1998–2002.
45. Mossoba, M.M., Niemann, R.A., and Chen, J.-Y. T. (1989) Picogram Level Quantification of 2,3,7,8-Tetrachlorodibenzo-*p*-dioxin in Fish Extracts by Capillary Gas Chromatography/Matrix Isolation/Fourier Transform Infrared Spectrometry, *Anal. Chem. 61*, 1678–1685.
46. Bourne, S., Hefner, A.M., Norton, K.L., and Griffiths, P.R. (1990) Performance Characteristics of Real-Time Direct Deposition Gas Chromatography/Fourier Transform Infrared Spectrometry, *Anal.Chem. 62*, 2448–2452.
47. Fuoco, R., Shafer, K.H., and Griffiths, P.R. (1986) Capillary Gas Chromatography/Fourier Transform Infrared Microspectrometry at Subambient Temperature, *Anal. Chem. 58*, 3249–3254.
48. Griffiths, P.R., and Henry, D.E. (1986) Fourier Transform Infrared Spectroscopy, *Prog. Anal. Spectrosc. 9*, 455–482.
49. Reedy, G.T., Bourne, S., and Cunningham, P.T. (1979) Gas Chromatography/Infrared Matrix Isolation Spectrometry, *Anal. Chem. 51*, 1535–1540.
50. Griffiths, P.R., and de Haseth, J.A. (1986) *Fourier transform infrared spectrometry*, Wiley-Interscience, New York, pp. 248–259.
51. Morris, L.J. (1963) Separation of Isomeric Long-Chain Polyhydroxy Acids By Thin–Layer Chromatography, *J. Chromatogr. 12*, 321–328.
52. Morris, L.J., and Nichols, B. (1972) *Progress in Thin Layer Chromatography, Related Methods* (Niederwieser, A., ed.), Ann Arbor-Humphey Scientific Publishers, Ann Arbor, Michigan, pp. 74–79.
53. Christie, W.W. (1982) *Lipid Analysis*, 2nd edn., Pergamon Press, Oxford, pp. 74–75.
54. Nikolova-Damyanova, B. (1992) in *Advances in Lipid Methodology I*, (Christie, W.W., ed.), Oily Press, Ayr, UK, pp. 181–237.
55. Nikolova-Damyanova, B., Herslof, B.G., and Christie, W.W. (1992) Silver Ion High-Performance Liquid Chromatography of Isomeric Fatty Acids, *J. Chromatogr. 609*, 133–140.
56. Durst, H.D., Milano, M., Kikta, E.J., Conelly, S.A., and Grushka, E. (1975) Phenacyl Esters of Fatty Acids via Crown Ether Catalysts for Enhanced Ultraviolet Detection in Liquid Chromatography, *Anal. Chem. 47*, 1797–1801.
57. Borch, R.F. (1975) Separation of Long Chain Fatty Acids as Phenacyl Esters by Pressure Liquid Chromatography, *Anal. Chem. 47*, 2437–2439.

58. Pei, P.T.S., Kossa, W.C., Ramachandran, S., and Henly, R.S. (1976) High Pressure Reversed Phase Liquid Chromatography of Fatty Acid *p*-Bromophenacyl Esters, *Lipids 11*, 814–816.
59. Roggero, J.P., and Coen, S.V. (1976) Isocratic Separation of Fatty Acid Derivatives by Reversed Phase Liquid Chromatography. Influence of the Solvent on Selectivity and Rules for Elution Order, *J. Liq. Chromatogr. 4*, 1817–1829.
60. Engelhardt, H., and Elgass, H. (1978) Optimization of Gradient Eluent Separation of Fatty Acid Phenacyl Esters, *J. Chromatogr. 158*, 249–259.
61. Vioque, E., Maza, M.P., and Millan, F. (1985) High Performance Liquid Chromatography of Fatty Acids as Their *p*-Phenylazophenacyl Esters, *J. Chromatogr. 331*, 187–192.
62. Engelmann, G.J., Esmans, E.L., Alderweireldt, F.C., and Rillaerts, R. (1988) Rapid Method for the Analysis of Red Blood Cell Fatty Acids by Reversed Phase High Performance Liquid Chromatography, *J. Chromatogr. 432*, 29–36.
63. Hanis, T., Smrz, M., Klir, P., Macek, J., Klima, J., Base, J., and Deyl, Z. (1988) Determination of Fatty Acids as Phenacyl Esters in Rat Adipose Tissue and Blood Vessel Walls by High Performance Liquid Chromatography, *J. Chromatogr. 452*, 443–458.
64. Korte, K., Chein, K.R., and Casey, M.L. (1986) Separation and Quantitation of Fatty Acids by High Performance Liquid Chromatography, *J. Chromatogr. 375*, 225–231.
65. Adlof, R.O. (1994) Separation of *Cis* and *Trans* Unsaturated Fatty Acid Methyl Esters by Silver Ion High-Performance Liquid Chromatography, *J. Chromatogr. 659*, 95–99.
66. Christie, W.W., and Moore, J.H. (1971) Structure of Triglycerides Isolated from Various Sheep Tissues, *J. Sci. Food Agric. 22*, 120–124.
67. Morris, L.J., Wharry, D.M., and Hammond, E.W. (1967) Chromatographic Behaviour of Isomeric Long-Chain Aliphatic Compounds II. Argentation Thin–Layer Chromatography of Isomeric Octadecenoates, *J. Chromatogr. 31*, 69–76.
68. Christie, W.W. (1987) A stable silver-loaded column for the separation of lipids by high performance liquid chromatography *J. High Res. Chromatogr. Chromatogr. Commun. 10*, 148–150.
69. Christie, W.W., and Breckenridge, G.H.M. (1989) Separation of *Cis* and *Trans* Isomers of Unsaturated Fatty Acids by HPLC in the Silver Ion Mode, *J. Chromatogr. 469*, 261–269.
70. Aveldano, M.I., Van Rollins, M., and Horrocks, L.A. (1983) Separation and Quantitation of Free Fatty Acids and Fatty Methyl Esters by Reversed Phase High Pressure Liquid Chromatography, *J. Lipid Res. 24*, 83–93.
71. Chen, Z.Y., Pelletier, G., Hollywood, R., and Ratnayake, W.M.N. (1995) *Trans* Fatty Acid Isomers in Canadian Human Milk, *Lipids* 30, 15–21.
72. Wolff, R.L. (1994) Analysis of α-Linolenic Acid Geometrical Isomers in Deodorized Oils by GLC on Cyanoalkyl Polysiloxane Stationary Phase: A Note of Caution, *J. Am. Oil Chem. Soc. 71*, 907–909.
73. Kasai, P.H., McLeod, D., and Watanabe, T. (1980) Acetylene and Ethylene Complexes of Copper and Silver Atoms, *J. Am. Chem. Soc. 102*, 179–185.
74. Gains, P., and Dunitz, J.D. (1967) Die Struktur der mittleren Ringverbindung XIV. Struktur des Silbernitrat-*Trans*-Cyclodecen-Addukts, *Helv. Chim. Acta 50*, 2379–2386.
75. π-Komplexe mit ungesättigten organischen Liganden in *Gmelin's Handbuch der Anorganischen Chemie*, Vol. 61, B5, Springer Verlag, Berlin, 1975, p. 26.

76. Gunstone, F.D., Ismail, I.A., and Lie Ken Jie, M.S.F. (1967) Thin Layer and Gas–Liquid Chromatographic Properties of the *Cis* and *Trans* Methyl Octadecenoates and of Some Acetylenic Esters, *Chem. Phys. Lipids 1*, 376–388.
77. Christie, W.W. (1988) Separation of Molecular Species of Triacylglycerols by High-Performance Liquid Chromatography with a Silver Ion Column, *J. Chromatogr. 454*, 273–284.
78. Morrison, H.R., and Smith, L.M. (1964) Preparation of Fatty Acid Methyl Esters and Dimethylacetals from Lipids with Boron Fluoride-Methanol, *J. Lipid Res. 5*, 600–608.
79. Lepage, G., and Roys, C.C. (1986) Direct Transesterification of All Classes of Lipids in a One-Step Reaction, *J. Lipid Res. 27*, 114–120.
80. Park, P.H., and Goins, R.E. (1994) In Situ Preparation of Fatty Acid Methyl Esters for Analysis of Fatty Acid Composition in Foods, *J. Food Sci. 59*, 1262–1266.
81. Official Method Ce 1c-89, *Official Methods and Recommended Practices of the American Oil Chemists' Society*, 4th edn., 1990, American Oil Chemists' Society, Champaign, Illinois.
82. Ratnayake, W.M.N., Hollywood, R., O'Grady, E., and Beare-Rogers, J.L. (1990) Determination of *Cis* and *Trans*-Octadecenoic Acids in Margarines by Gas Liquid Chromatography–Infrared Spectrophotometry, *J. Am. Oil Chem. Soc. 67*, 804–810.
83. Wolff, R.L., and Bayard, C.C. (1995) Improvement in the Resolution of Individual *Trans*-18:1 Isomers by Capillary Gas–Liquid Chromatography: Use of a 100-m CP-Sil 88 column, *J. Am. Oil Chem. Soc. 72*, 1197–1201.
84. Adlof, R.O., Copes, L.C., and Emken, E.A. (1995) Analysis of the Monoenoic Fatty Acid Distribution in Hydrogenated Vegetable Oils by Ag–HPLC, *J. Am. Oil Chem. Soc. 72*, 571–574.
85. Parodi, P.W. (1976) Distribution of Isomeric Octadecenoic Fatty Acids in Milk Fat, *J. Dairy Sci. 59*, 1870–1873.
86. Grandgirard, A., Piconneaux, A., Sebedio, J.L., O'Keefe, S.F. Semon, E., and Quere, J.L. (1989) Occurrence of Geometrical Isomers of Eicosapentaenoic and Docosahexaenoic Acids in Liver Lipids of Rats Fed Heated Linseed Oil, *Lipids* 24, 799–804.
87. Heckers, H., Melcher, F.W., and Schloeder, U. (1977) SP 2340 in the Glass Capillary Chromatography of Fatty Acid Methyl Esters, *J. Chromatogr. 136*, 311–317.
88. Wolff, R.L. (1995) Content and Distribution of *Trans*-18:1 Acids in Ruminant Milk and Meat Fats. Their Importance in European Diets and their Effect on Human Milk, *J. Am. Oil Chem. Soc. 72*, 259–272.
89. Ratnayake, W.M.N. (1992) AOCS Method Ce 1c-89 Underestimates the *Trans*-Octadecenoate Content in Favor of the *Cis* Isomers on Partially Hydrogenated Vegetable Fats, *J. Am. Oil Chem. Soc. 69*, 192–199.
90. Duchateau, G.S.M.J.E., van Oosten, H.J., and Vasconcellos, M.A. (1996) Analysis of *Cis*- and *Trans*-Fatty Acid Isomers in Hydrogenated and Refined Vegetable Oils by Capillary Gas–Liquid Chromatography, *J. Am. Oil Chem. Soc. 73*, 275–282.
91. Lanza, E., and Slover, H.T. (1981) The Use of SP 2340 Glass Capillary Columns for the Estimation of the *Trans* Fatty Acid Content of Foods, *Lipids* 16, 260–267.
92. Sampugna, J., Pallansch, L.A., Enig, M.G., and Keeney, M. (1982) Rapid Analysis of *Trans* Fatty Acids on SP-2340 Glass Capillary Columns, *J. Chromatogr. 249*, 245–255.
93. Christie, W.W. (1973) The Structure of BILE Phosphatidylcholines, *Biochem. Biophys. Acta 316*, 204–211.

94. Wolff, R.L. (1992) Resolution of Linolenic Acid Geometrical Isomers by Gas-Liquid Chromatography on a Capillary Column Coated with a 100% Cyanopropyl Polysiloxane Film (CP™Sil 88), *J. Chromatogr. 30,* 17–22.
95. Grandgirard, A., Julliard, F., Prevost, J., and Sebedio, J.L. (1987) Preparation of Geometrical Isomers of Linolenic Acid, *J. Am. Oil Chem. Soc. 64,* 1434–1440.
96. Marx, F., and Claßen, E. (1994) Analysis of Epoxy Fatty Acids by GC–MS of Their 4,4-Dimethyloxazoline Derivatives, *Fat Sci. Technol. 96,* 207–211.
97. Anderson, B.A., and Holmann, R.T. (1974) Pyrrolidides for Mass Spectrometric Determination of the Position of the Double Bond in Monounsaturated Fatty Acids, *Lipids 9,* 185–195.
98. Zhang, J.Y., Yu, Q.T., Liu, B.N., and Huang, Z.H. (1988) Chemical Modification in Mass Spectrometry IV—2-Alkenyl-4,4-Dimethyloxazolines as Derivatives for the Double Bond Location of Long-chain Olefinic Acids, *Biomed. Environ. Mass Spectrom. 15,* 33–44.
99. Yu, Q.T., Liu, B.N., Zhang, J.Y., and Huang, Z.H. (1988) Location of Methyl Branchings in Fatty Acids: Fatty Acids in Uropygial Secretion of Shanghai Duck by GC–MS of 4,4-Dimethyloxazoline Derivatives, *Lipids 23,* 804–810.
100. Spitzer, V., Marx, F., Maia, J.G.S., and Pfeilsticker, K. (1990) *Currupira tefeensis* (Olacacae)—A Rich Source of Very Long Chain Fatty Acids, *Fett Wiss. Technol. 92,* 165–168.
101. Fay, L., and Richli, U. (1991) Location of Double Bonds in Polyunsaturated Fatty Acids by Gas Chromatography–Mass Spectrometry after 4,4-Dimethyloxazoline Derivatization, *J. Chromatogr. 541,* 89–98.
102. Luthria, D.L., and Sprecher, H. (1993) 2-Alkenyl-4,4,-dimethyloxazolines as Derivatives for the Structural Elucidation of Isomeric Unsaturated Fatty Acids, *Lipids 28,* 561–564.
103. Ackman. R.G., and Hooper, S.N. (1973) Additivity of Retention Data for Ethylenic Functions in Aliphatic Fatty Acids, *J. Chromatogr. 86,* 83–88.
104. Sebedio, J.L., and Ackman, R.G. (1982) Calculation of GLC Retention Data for Some Accessible C_{20} Isomeric *Cis*-Unsaturated Fatty Acids, *J. Chromatogr. Sci. 20,* 231–234.
105. Mossoba, M.M., McDonald, R.E., Roach, J.A.G., Fingerhut, D.D., Yurawecz, M.P., and Sehat, N. (1997) Spectral Confirmation of *Trans* Monounsaturated C_{18} Fatty Acid Positional Isomers, *J. Am.Oil Chem. Soc.*, in press.
106. McDonald, R.E., Armstrong, D.J., and Kreishmann, G.P. (1989) Identification of *Trans*-Diene Isomers in Hydrogenated Soybean Oil by Gas Chromatography, Silver Nitrate Thin-Layer Chromatography, and ^{13}C NMR Spectroscopy *J. Agr. Food Chem. 37,* 637–642.
107. Emken, E.A. (1995) Physiochemical Properties, Intake, and Metabolism of *Trans* Fatty Acids, *Am. J. Clin. Nutr.* (Suppl.) *62,* 659S–669S.

Chapter 12

Separation of Fatty Acid Methyl Esters and Triacylglycerols by Ag-HPLC: Silver-Ion and Normal-Phase Contributions to Retention

R.O. Adlof

National Center for Agricultural Utilization Research, Agricultural Research Service, U.S. Department of Agriculture, 1815 N. University St. Peoria, IL 61604

Introduction

Silver-ion chromatography, a technology dating back more than 50 years (1), has been widely applied to the analysis of lipids and other unsaturated compounds. Thin-layer chromatography (TLC) and high-performance liquid chromatography (HPLC) account for the majority of applications, and improved methodologies and new applications continue to be published. Several excellent reviews of silver-ion TLC and HPLC applications are available (2–6). Whether one is separating unsaturated molecules using silver ions with TLC, resin chromatography, high-performance liquid chromatography (Ag-HPLC), reversed-phase chromatography, supercritical fluid chromatography (SFC), or solid-phase extraction (SPE), the underlying concept remains the same: retention is directly related to the stability of the complex formed between the silver (Ag) ions and the π electrons of the carbon-carbon double bond(s) of the molecule. Retention characteristics of unsaturated molecules (7,8) are considered to depend on:

1. The number of double bonds in the molecule (the more double bonds, the greater the retention)
2. The geometry of the double bonds (*trans* isomers elute before *cis* isomers)
3. The position of the double bond(s)
4. The atom(s) or group(s) attached to the carbon atoms of the double bond(s)
5. The interaction of silver ions with unpaired electrons of the fatty acid (FA) carbonyl oxygen(s).

We have, over the years, synthesized a large number of deuterium-labeled FA isomers (geometric and positional) to study their metabolism and interaction in humans (9). Silver-resin chromatography is a valuable technology (10) for the preparative purification of a wide variety of labeled fatty acid methyl esters (FAME) and triacylglycerols (TAG). The resin utilized is Rohm and Haas XN1010, a macroreticular, polystyrene-divinylbenzene copolymer containing sulfonic acid groups. The coarse (300 to 1,000 µm) resin is ground and fractionated into appropriate mesh sizes by wet-sieving (water). A glass column is slurry-packed (water), the sulfonic acid protons of the resin are exchanged with silver ions, and the resin is flushed with the solvent of choice (acetone or methanol). Unlike silver nitrate–impregnated silica TLC or LC systems, the silver ions are bound to the resin and do not dissolve in the stronger solvents (i.e., methanol) required to elute polyunsaturated compounds. The high porosity of the resin yields silver-ion loading percentages of 35 to 38%. A 5 × 25 cm silver-resin column is thus capable of fractionating 20 to 30 g samples of *cis/trans* FAME (10). The large (75 to 150 μm) and irregularly shaped resin particles, however, result in peak broadening and loss of resolution. This problem limits silver resin's application as an analytical tool.

Ag-HPLC, as defined today, utilizes a column packed with 5 to 10 μm Nucleosil SA^{TM} (phenylsulfonic acid groups bonded to a silica substrate) or similar substrate in which the sulfonic acid protons have been exchanged with Ag ions (11,12). This technology provides a tremendously powerful technique for the analysis of unsaturated compounds. With the advent of commercially available HPLC columns, Ag-HPLC has, over the last decade, been applied to the separation of *cis* and *trans* fatty acid methyl esters (12–15), FAME positional isomers from partially hydrogenated vegetable oils (15), conjugated FAMEs (7), TAG isomers (16,17), and FAMEs labeled with deuterium atoms on the double bond carbons (18). The power of this methodology is demonstrated (14) by its ability to separate 15 of the 16 *cis/trans* isomers of isomerized methyl arachidonate (20:4). This feat cannot be duplicated by gas–liquid capillary chromatography or by any other existing technology.

Ag–HPLC has also been applied to the analysis of deuterium-labeled monoacylglycerol (MAG) and diacylglycerol (DAG) positional isomers synthesized in our laboratory (19). While Ag–HPLC could be used to separate underivatized DAG isomers, MAG isomers were difficult to elute and resulted in poor peak shapes and resolution with our solvent system (acetonitrile in hexane). Conversion of the MAG and DAG isomers to their di- or monoacetate derivatives (acetic anhydride/pyridine/80°C) resulted in improved peak shape and resolution. The retention of MAG and DAG positional isomer pairs on our silver-ion HPLC column was directly related to the number of acetate groups (or inversely related to the number of carbon atoms in the FA on the TAG), with the more acetate-containing TAGs being retained longest. Previous work (16) with mixed TAGs such as 16:0/18:1/16:0 and 16:0/16:0/18:1 showed 16:0/18:1/16:0 to elute first (the FA on the 1(3) position of the TAG having a greater influence on retention than the FA at the 2-position).

Utilizing a dual-column Ag–HPLC system at a solvent flow of 1 mL of 1.2% acetonitrile (ACN) in hexane per minute, tripalmitoylglycerol (3 × 16:0) was found to elute first (*ca.* 11 min), while triacetin (3 × 2:0) eluted at *ca.* 70 min (19). The retention of unsaturated FAMEs and TAGs on silver-ion chromatographic systems has generally been related to the stability of the complex formed between the silver ions and the π electrons of the carbon-carbon double bond(s), a condition absent in the 16:0 series. Under these conditions, the effect of chain length of saturated fatty acids on the retention characteristics of FAMEs and TAGs was found to be influenced more by "normal-phase" than by "silver-ion" effects (20).

Methodology

Materials and Reagents

Hexane (Allied Fisher Scientific, Orangeburg, NY), ACN, benzene, acetic anhydride and pyridine (all E. Merck, Darmstadt, Germany) were used as received. MAG and DAG positional isomer samples (trioleoylglycerol, 1-monooleoylglycerol, 2-monooleoylglycerol, 1,2-dioleoylglycerol, 1,3-dioleoylglycerol, and the analogous 16:0 glycerols) were obtained (99%+ pure) from Sigma Chemical Co., St. Louis, MO. Homogeneous TAG (3 × 18:0, 3 × 16:0, 3 × 14:0, 3 × 12:0, 3 × 10:0, 3 × 8:0, 3 × 6:0, and 3 × 2:0), mixed TAG (16:0/18:1/16:0 and 16:0/16:0/18:1) and FAME (16:0, 10:0, 6:0, 4:0, and 2:0) were also obtained from Sigma. The acetate derivatives were prepared as in Ref. 19.

High-Performance Liquid Chromatography

A Spectra-Physics P2000 solvent delivery system (Thermo Electron Corp., Waltham, MA, USA), and a Rheodyne 7125 injector (Rheodyne, Inc., Cotati, CA) with a 20-µL injection loop were used. The ChromSpher lipids columns (Cat. No. 28313; 4.6 mm i.d. × 250 mm stainless steel; 5 µm particle size; silver ion–impregnated) were purchased from Chrompack International, Middelburg, The Netherlands, and used as received. To improve peak resolutions and to allow the injection of larger sample sizes, two lipids columns were connected in series. Solvent flow was standardized at 1.0 mL/min; isocratic conditions (0.05% to 1.2% ACN in hexane; 23°C) were used to minimize variations in TAG and FAME retention(s) and resolution(s). The void volume of this pair of columns was 4.2 mL. Either a flame ionization detector (Model 945, Tremetrics Corp., Austin, TX), or a UV detector (ISCO V4 Absorbance Detector (ISCO, Inc., Lincoln, Nebraska) set at 206 nm, was utilized as HPLC detector. Detector signal output was monitored by computer (Grams/386 for Chromatography, Galactic Industries Corporation, Salem, NH).

Results And Discussion

The resolving power of Ag–HPLC is demonstrated (Fig. 1) by the fractionation of 15 of the 16 possible *cis/trans* isomers of 20:4.

Figure 2 depicts the pattern obtained from separation of an 18:1 TAG mixture composed of trioleoylglycerol, mono- and dioleoylglycerols (as their di- and monoacetate derivatives, respectively), and triacetin. If one assumes that retention is due to Ag/FA double bond interactions and that the FA(s) located in the 1(3)-positions of the TAG have the greatest influence on retention characteristics (Fig. 3), then 18:1/18:1/2:0 should elute before 18:1/2:0/18:1 and 2:0/18:1/2:0 before

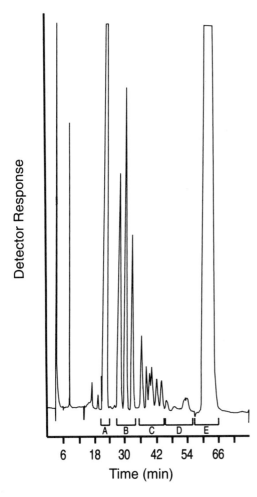

FIG. 1. Separation of isomerized methyl arachidonate. Sample size, 100 µg. Flow rate: 1.0 mL/min 0.125% ACN in hexane. UV detection at 210 nm. Group A = (4-*trans*); B = (3-*trans*, 1-*cis*); C = (2-*trans*, 2-*cis*); D = (1-*trans*, 3-*cis*); E = (4-*cis*).

18:1/2:0/2:0. This was found not to be the case. A similar elution pattern was noted when a 16:0 TAG standard mixture was analyzed (Fig. 4). In the 16:0 example no carbon-carbon double bonds are present, yet the pattern persists. Thus, the 2:0 exerts a greater influence on retention than either 16:0 or 18:1.

The separation pattern of 18:0/2:0/18:0, 18:0/18:0/2:0, 16:0/2:0/16:0, and 16:0/16:0/2:0 using this solvent system is shown in Fig. 5. The pattern is similar to that found in the 16:0 and 18:1 TAG samples, and the stearate and palmitate TAGs are also well separated. Given this data and the large differences in retention observed with saturated TAGs and FAMEs composed of FAs of differing chain lengths (solvent system of ACN in hexane), previous assumptions (14) that saturated FAs (as TAGs or FAMEs) are unretained under these solvent conditions are certainly incorrect.

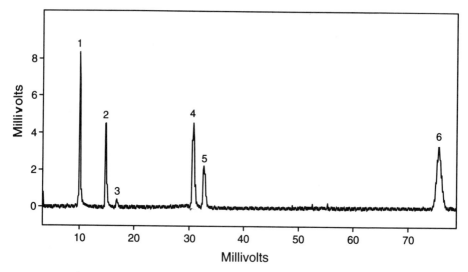

FIG. 2. Analysis of oleate/acetate isomer mix containing 3 × 18:1 (No. 1), 18:1/2:0/18:1 (2), 18:1/18:1/2:0 (3), 18:1/2:0/2:0 (4), 2:0/18:1/2:0 (5), 3 × 2:0 (6) by dual-column, silver-ion high-performance liquid chromatography. Sample size: 17 µg. Flow rate: 1.0 mL/min 1.2% ACN in hexane. Flame ionization detection.

Figure 6 illustrates the separation achieved with a mixture of homogeneous TAGs (3 × 18:0, 3 × 16:0, 3 × 14:0, 3 × 12:0, 3 × 10:0, 3 × 8:0, and 3 × 2:0). Insert A shows the optimized separation of the same TAG mixture (without 3 × 2:0). With

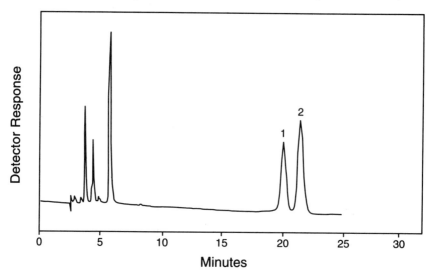

FIG. 3. Separation of 16:0/18:1/16:0 (No. 1) and 16:0/16:0/18:1 (2) mixture by dual-column Ag-HPLC. Sample size: 120 µg. Flow rate: 1.0 mL/min 0.5% ACN in hexane. UV detection at 206 nm.

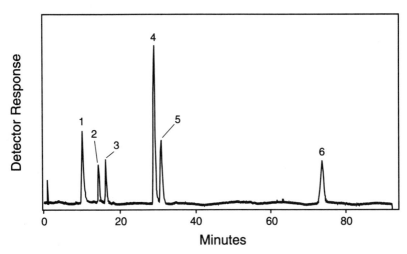

FIG. 4. Analysis of isomer mix containing 3 × 16:0 (No. 1), 16:0/2:0/16:0 (2), 16:0/16:0/2:0 (3), 16:0/2:0/2:0 (4), 2:0/16:0/2:0 (5) and 3 × 2:0 (6) by dual-column, silver-ion high-performance liquid chromatography. Sample size: 36 µg. Flow rate: 1.0 mL/min 1.2% ACN in hexane. Flame ionization detection.

this system, triacetin elutes at *ca.* 70 min. Thus, the effects of FA chain length are not linear but increase rapidly at very short FA chain lengths. The three chromatograms in Fig. 7 demonstrate the effect of increasing the percentage of ACN in the ACN/ hexane solvent system on the separation of a mixture of saturated FAMEs (16:0, 10:0, 6:0, 4:0, and 2:0) of differing chain lengths. At a solvent strength of

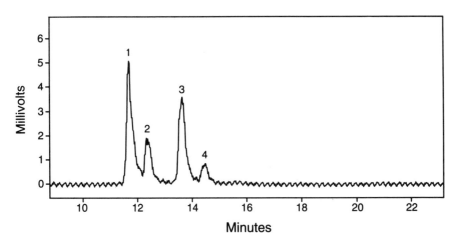

FIG. 5. Analysis of isomer mix containing 18:0/2:0/18:0 (1), 18:0/18:0/2:0 (2), 16:0/2:0/16:0 (3) and 16:0/16:0/2:0 (4) by dual-column, silver-ion high-performance liquid chromatography. Sample size: 15 µg. Flow rate: 1.0 mL/min 1.2% ACN in hexane. Flame ionization detection.

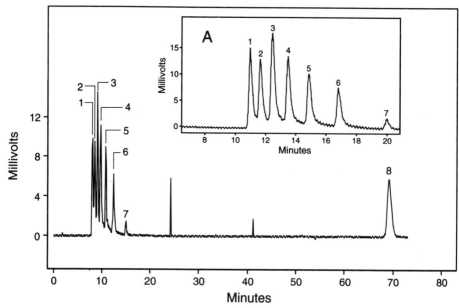

FIG. 6. Analysis of homogeneous, saturated TAG mix [3 × 18:0 (1), 3 × 16:0 (2), 3 × 14:0 (3), 3 × 12:0 (4), 3 × 10:0 (5), 3 × 8:0 (6), 3 × 6:0 (7), 3 × 2:0 (8)] by dual-column Ag-HPLC. Sample size: 44 μg. Flow rate: 1.0 mL/min 1.2% ACN in hexane. Flame ionization detection. Insert A: Optimized separation of TAG mix (no 3 × 2:0). Sample size: 38 μg. Flow rate: 1.0 mL/min 0.7% ACN in hexane. Flame ionization detection.

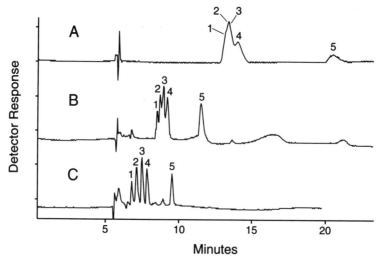

FIG. 7. Analysis of FAME mixture [16:0 (1), 10:0 (2), 6:0 (3), 4:0 (4), 2:0 (5)] by dual-column, Ag–HPLC at differing contents of ACN in hexane. Sample size: 20 μg. Flow rate: 1.0 mL/min. Flame ionization detection. A: 0.05% ACN in hexane. B: 0.2% ACN in hexane. C: 1.0% ACN in hexane.

0.05% ACN in hexane, only 2:0 was completely resolved, but resolution improved as the percentage of ACN was increased.

In silica-based TLC and HPLC, silanol group ("normal phase") effects (2–4,21,22) have been used to explain the elution patterns of unsaturated FAME and TAG mixtures containing FAs of differing chain lengths. Such effects have also been applied to the separation of unsaturated TAGs by Ag–TLC (7) and Ag-HPLC (7,19) and to TAGs separated by supercritical fluid chromatography (using a silica-based cation exchanger in the silver ion form) (23). The improved FAME resolution we observed with increasing percentages of ACN is also a characteristic of normal-phase systems (2, 21). In the absence of carbon-carbon double or triple bonds, substrate interactions with the more polar silanol groups of the Ag-HPLC column packing (some 30% of the original silanol groups of the silica remain unreacted) predominate, resulting in normal-phase elution patterns. While silver-ion effects might still be present (a recent paper by Nikolova-Damyanova (24) documents the contribution of the unpaired electrons of the carbonyl oxygen to FA retention), they can be considered minor when compared to the "normal-phase" characteristics of the silanol groups.

Whether such "normal-phase" phenomena are observed in the chlorinated hydrocarbon/ACN solvent systems utilized by Christie, Nikolova-Damyanova, and others (2–5,7,8,12,13,24) is unclear. Netting and Duffield (25), however, observed limited separation and elution orders similar to those obtained by us on a silica gel column for a series of straight-chain fatty acids (as their pentafluorobenzyl esters) using a solvent system of dichloromethane-hexane half-saturated with water. The silanol group ("normal-phase") effect and the hydrophobic effect from the alkylphenyl part of the alkylphenylsulfonic acid moieties bound to the silica backbone may be only minor contributors to retention with chlorinated hydrocarbon–based solvent systems, the latter effect increasing with increasing ACN content of the mobile phase. With hydrocarbon-based mobile phases, the adsorption ("normal-phase") effect is probably more important than the hydrophobic one (W.W. Christie, unpublished correspondence).

In the case of unsaturated FAs and TAGs, the contributions by the silanol groups (increased %ACN required to improve resolution) would seem to conflict with, rather than augment, the resolving power of the silver ions. Due to the nonlinear relationship demonstrated between FA chain length and retention, utilization of a longer-chain FA rather than triacetin (i.e., C_4 or C_5) in MAG and DAG analyses would decrease retention and still permit resolution of mixed-FA MAGs and DAGs.

Applications of the dual functionality of the ChromSpher Ag–HPLC column will remain limited until a more thorough understanding of the contributions and interactions of solvent, silver, silanol, and substrate. Normal-phase effects must certainly be included in any attempts to predict the elution orders of FAMEs and TAGs containing FAs of widely differing chain lengths. More work is thus required to characterize the varied and complex interactions that contribute to the concept of "Ag-HPLC."

Acknowledgments

The author wishes to thank T. Lamm for technical assistance.

References

1. Lucas, H.J., Moore, R.S., and Pressman, D. (1943) The Coordination Of Silver Ions with Unsaturated Compounds II, *Cis-* and *Trans*-2-Pentene, *J. Am. Chem. Soc. 65*, 227–229.
2. Christie, W.W. (1995)in *New Trends in Lipid and Lipoprotein Analyses* (Sebedio, J.-L., and Perkins, E.G., eds.), AOCS Press, Champaign, IL, Chs. 6, 7.
3. Dobson, G., Christie, W.W., and Nikolova-Damyanova, B. (1995) Silver-Ion Chromatography of Lipids and Fatty Acids, *J. Chromatogr. B. 671*, 197–222.
4. Christie, W.W. (1987) *High Performance Liquid Chromatography and Lipids*, Pergamon Press, Oxford, England, pp. 156, 202–206.
5. Nikolova-Damyanova, B. (1992) in *Advances in Lipid Methodology—One* (Christie, W.W., ed.), The Oily Press, Ayr, Scotland, UK, pp. 181–237.
6. Touchstone, J.C. (1995) Thin-Layer Chromatographic Procedures for Lipid Separations, *J. Chromatogr. B. 671*, 169–195.
7. Nikolova-Damyanova, B., Herslof, B.G., and Christie, W.W. (1992) Silver-Ion High-Performance Liquid Chromatography of Derivatives of Isomeric Fatty Acids, *J. Chromatogr. 609*, 133–140.
8. Nikolova-Damyanova, B., Christie, W.W., and Herslof, B.G. (1995) Retention Properties of Triacylglycerols on Silver-Ion High-Performance Liquid Chromatography, *J. Chromatogr. A. 694*, 375–380.
9. Emken, E.A., Adlof, R.O., Rohwedder, W.K., and Gulley, R.M. (1993) in *Proceedings of the Third International Conference on Essential Fatty Acids and Eicosanoids* (Sinclair, A., and Gibson, R., eds.), American Oil Chemists' Society Monograph, Champaign, Illinois, pp. 23–25.
10. Adlof, R.O. (1994) in *Separation Technology* (Vansant, V.F., ed.), Elsevier Science, B.V., Antwerp, Belgium, pp. 777–781.
11. Powell, W.S. (1981) Separation of Eicosenoic Acids, Monohydroxyicosenoic Acids and Prostaglandins by High-Pressure Liquid Chromatography on a Silver-Ion-Loaded Cation-Exchange Column, *Anal. Biochem. 115*, 267–277.
12. Christie, W.W. (1987) A Stable Silver-Loaded Column for the Separation of Lipids by High Performance Liquid Chromatography, *J. High Resol. Chromatogr. Chromatogr. Commun. 10*, 148–149.
13. Juaneda, P., Sebedio, J.L., and Christie, W.W. (1994) Complete Separation of the Geometrical Isomers of Linolenic Acid by High Performance Liquid Chromatography with a Silver-Ion Column, *J. High. Resol. Chromatogr. Chromatogr. Commun. 17*, 321–324.
14. Adlof, R.O. (1994) Separation of *Cis* and *Trans* Unsaturated Fatty Acid Methyl Esters by Silver-Ion High Performance Liquid Chromatography, *J. Chromatogr. A 659*, 95–99.
15. Adlof, R.O., Copes, L.C., and Emken, E.A. (1995) Analysis of the Monoenoic Fatty Acid Distribution in Hydrogenated Vegetable Oils by Silver-Ion High-Performance Liquid Chromatography, *J. Am. Oil Chem. Soc. 72*, 571–574.

16. Adlof, R.O. (1995) Analysis of Triacylglycerol Positional Isomers by Silver-Ion High-Performance Liquid Chromatography, *J. High. Res. Chromatogr. 18*, 105–107.
17. Blomberg, L.G., Demirbüker, M., and Andersson, P.E. (1993) Argentation Supercritical Fluid Chromatography for Quantitative Analysis of Triacylglycerols, *J. Am. Oil Chem. Soc. 70*, 939–946.
18. Adlof, R.O. (1994) Silver-Ion HPLC Separation of Fatty Acid Methyl Esters with Deuterium Atoms on the Double Bonds, *J. Chromatogr. A 685*, 178–181.
19. Adlof, R.O. (1996) Analysis of Fatty Acid Mono- and Diacylglycerol Positional Isomers by Silver-Ion HPLC, *J. Chromatogr. A 741*, 135–138.
20. Adlof, R.O. (1996) Normal-Phase Separation Effects with Lipids on a Silver-Ion High-Performance Liquid Chromatography Column, *J. Chromatogr. A*, in press.
21. Scott, R.P.W., and Kucera, P. (1975) Solute Interactions with the Mobile and Stationary Phases in Liquid-Solid Chromatography, *J. Chromatogr. 112*, 425–442.
22. Kramer, J.K.G., Fouchard, R.C., and Farnworth, E.R. (1985) The Effect of Fatty Acid Chain Length and Unsaturation on the Chromatic Behavior of Triglycerides on Iatroscan Chromarods, *Lipids 20*, 617–619.
23. Demirbüker, M., and Blomberg, L.G. (1991) Separation of Triacylglycerols by Supercritical-Fluid Argentation Chromatography, *J. Chromatogr. 550*, 765–774.
24. Nikolova-Damyanova, B., Herslof, B.G., and Christie, W.W. (1996) Mechanistic Aspects of Fatty Acid Retention in Silver-Ion Chromatography, *J. Chromatogr. A. 749*, 47–54.
25. Netting, A.G., and Duffield, A.M. (1983) Pentafluorobenzyl Esters As Derivatives for the Semi-Preparative High-Performance Liquid Chromatography of Fatty Acids, *J. Chromatogr. 257*, 174–179.

Chapter 13

Near Infrared Analysis of Oilseeds: Current Status and Future Directions

J.K. Daun and P. Williams

    Canadian Grain Commission, Grain Research Laboratory, 1404-303 Main Street, Winnipeg, Manitoba R3C 3G8 Canada,

Introduction

Improvements in instrumentation over the past 7 to 10 years have made near infrared spectroscopy an attractive option for oilseeds analysis. Early instrumentation required grinding of the sample; although that was a satisfactory option for cereal grains and perhaps for soybeans, grinding of soft seeds such as canola presented problems. Over the past few years, several instruments have been developed with mechanisms for scanning large samples of whole seed. In addition, instruments with the ability to scan a wide spectral range have become available at prices suitable for many laboratories carrying out routine testing.

    This chapter summarizes the current state of instrumentation, components measured, and systems utilizing near infrared (NI) analyzers to test oilseeds. The Canadian Grain Commission's Grain Research Laboratory (GRL) has used near infrared reflectance (NIR) analysis to test Canadian oilseeds, including soybeans, canola, flax, and mustard seed, as well as cereal grains and legumes, for many years. The experience gained provides examples to illustrate the application of NIR analysis to different situations, including research, crop monirtoring and quality control.

Theory of Near Infrared Spectroscopy

Spectroscopy depends on the absorption of light energy by molecules. In the case of near infrared radiation, the absorptions are due to changes in the vibrational and rotational quantum numbers of the molecules (1). The wavelengths employed in near infrared analysis correspond to overtones (usually the first to the third) of the fundamental infrared absorptions, as well as combination bands between 2500 nm and 15000 nm. These effects result in extremely complex spectra that have limited qualitative use but can be used quantitatively by employing regression analysis.

    In most food-related bio-organic materials, useful absorption (or reflectance) extends from 2500 nm down to below 1000 nm. Below 1000 nm, modern instruments also allow the analyst to utilize the visible portion of the spectrum. Examination of the near infrared and visible reflectance spectra of different oilseeds (Fig. 1) shows obvious differences between the seeds. In the visible range (400 to

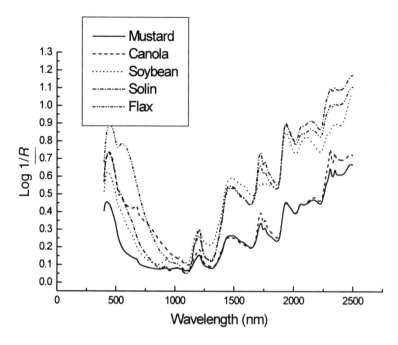

Fig. 1. Near infrared and visible spectra of several Canadian oilseeds. Spectra were prepared from samples of whole seed scanned on a NIRS6500 spectrometer.

700 nm) an absorption at 430 nm due to carotenoids is accompanied by one at 600 nm due to red seed coat pigments, which is especially strong in flax. In canola, a band near 700 nm is assigned to chlorophyll.

The NI spectral region (1000 nm to 2500 nm) is complex. Differences between the low-oil soybean and the high-oil flax, solin, canola, and mustard seed are obvious. The difference in intensity between the canola and mustard curves and the flax, solin, and soybean curves is probably due to seed size. Comparison of the spectrum of canola seed with the spectrum of commercial canola meal highlights the impact of oil in the spectrum (Fig. 2). The complex nature of the calibration is indicated by the overlap of bands from various components. Much of the fine structure in the bands related to oil content is due to contributions from different combinations of fatty acids. Changes in fatty acid composition can complicate the accurate determination of oil content (2).

Instrumentation

Instrumentation for near infrared spectrometry has been reviewed recently(3,4). Near infrared instruments operate on the same principle as any other spectrometer. Light energy is provided over the spectral region of interest (source). Wavelength selection is by filter, grating, or diode array. The light beam may be split (dual

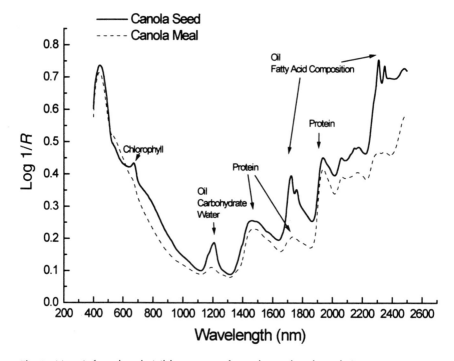

Fig. 2. Near infrared and visible spectra of canola seed and meal. Spectra were prepared from samples of whole seed scanned on an NIRS6500 spectrometer.

beam), with one part as the reference and one part as energy directed at the sample. Energy reflected from or transmitted through the sample is collected and compared to a reference energy. For reflectance, the reflected light may be concentrated by using a spherical mirror. Transmittance units require different cell path lengths for different sample types. Reflectance instruments employ ceramic or special plastic as a reference whereas transmittance devices use air as a reference.

Reflected or transmitted light is most commonly detected by lead sulfide detectors between 1000 and 2500 nm and by silicon detectors at lower wavelengths. The indium/gallium arsenide detector is also coming into common use. Units of measurement are reflectance (R) and transmittance (T), or more commonly log $1/R$ or log $1/T$, both of which correspond to absorbance from classical spectrometry. The most important feature that has allowed NI spectrometry to become successful is the ability, in a very short time, to electronically average a large number of measurements made over a wide wavelength range. This allows comparatively minor differences in spectra to be utilized quantitatively.

For purposes of comparison it is convenient to separate NI instruments into two categories: ground-seed filter instruments and whole-seed scanning instruments. The first NI instruments used for routinely measuring components of agriculture products employed narrow band-pass filters to discriminate spectral components;

because of the size of the optical scanning path, they required that the sample be ground in order to present a representative sample to the instrument. The DICKEY-john Instalab 600 (DICKEY-john Canada Inc., Cornwall, Ontario) is an example of such an instrument; it has been used in the Grain Research Laboratory. The first major success of these instruments in agricultural commodities was in the measurement of protein and moisture in wheat samples (5,6). Many instruments of this type are still in use for this purpose.

At the GRL, one of the first routine uses of NI instrumentation on oilseeds was in the determination of chlorophyll in canola seed. This method was originally developed using a classical spectrometer (7), but with the addition of two filters at the appropriate wavelengths it was possible to adapt a DICKEY-john Instalab 600 spectrometer to carry out this analysis (8). This method gives results with comparable accuracy and precision to the reference method and is still in routine use in our laboratory (9).

Ground-seed filter-type instruments have several advantages, including their relatively low cost, reliability, and ease of operation, which have allowed them to achieve common usage for determining protein and moisture at country delivery points. They also have several disadvantages, which may result in their eventual displacement by more sophisticated units.

The major disadvantage of the early instruments was the requirement of a ground sample. This was necessary to meet the sampling requirements presented by a scanning area of between 2 and 10 cm^2. Problems encountered in grinding hard seeds such as wheat (10) are compounded in oilseeds by the presence of oil, which not only acts as a pasting agent when present at contents greater than 45% but also acts as an agglomerating agent at lower concentrations, resulting in inconsistent particle size and characteristics (10). Agglomeration of particles is complicated if the ground oilseed is stored for any period prior to analysis. In addition, oil may be extruded onto the quartz or glass windows of sample cells, necessitating their cleaning and drying between samples. This problem has been alleviated by the use of open-cup sample cells (such as in the DICKEY-john 600), but this type of cell leads to the buildup of dust on the interior of the instrument and is also prone to changes in the surface due to moisture loss.

The use of filters for specific wavelengths resulted in limited flexibility and could lead to errors (2). For example, when chlorophyll filters were added to the DICKEY-john instrument, it was necessary to remove the filters used for moisture determination. Also, the instrument is fitted with lead sulfide detectors, which do not operate with optimum efficiency in the low-wavelength range. While newer models of filter instruments allow addition of more filters, this is an additional expense. Finally, in filter-type instruments, calibrations are not readily transferable from instrument to instrument (11) due to minor differences in the filters. The characteristics of the filters have a tendency to change over time, resulting in the need for continual checking and adjustment of slope and bias.

Whole-seed scanning instruments have several advantages over ground-seed filter instruments. The first is flexibility. Because of the use of a monochromator, it is possible to utilize any combination of wavelengths in the spectral range, enabling the instrument to be used to test a large number of components. Modern NI instruments have the capability of being monitored and calibrated from a remote site via a network system.

Adaptation of sample presentation mechanisms to allow analysis of whole seeds results in a more rapid analysis overall, with the advantage that the sample is not destroyed in the process. This allows the actual test sample to be stored for further analysis or, in the case of plant breeding, for growing. In addition to saving the sample, most instruments also allow the complete scan to be saved for future analysis of different components and for use in calibrations not yet developed. The larger sample used in these instruments (up to 100 g), along with optical improvements over the earlier filter instruments, result in improvements in accuracy and precision. Furthermore, use of whole-grain analyzers avoids the need for cell window cleanup between samples as well as eliminating errors due to grinding.

The major disadvantage of most whole-seed scanning instruments is their relatively high cost, although this has been decreasing, especially in units dedicated to a particular type of analysis. Use of these instruments also requires the availability of a high-speed (486 or better) personal computer. Modern NI software will not run on less sophisticated computers or on mainframes. There are also some restrictions on the use of scanning spectophotometer instruments in transmittance mode, including the necessity of variable cell path lengths.

The final disadvantage is the same for all NI instrumentation. The technique is secondary from an analytical point of view, and it is necessary to calibrate for each component and parameter.

The two categories used above are not cast in stone. Filter instruments have been used to test whole seeds (12), and some have been devised with sufficiently large sample cells (9). Scanning instruments can also be used to analyze ground samples (13,14). The categories reflect the past and present trends in the field. The use of filters and ground samples is characteristic of past generations of NI instruments, whereas the use of whole seed and scanning instruments is more representative of the present and future. The most recent introductions include a whole-seed filter instrument (12 filters) and a scanning diode-array instrument that can accommodate whole or ground seeds (15).

Calibration Tools

The complexity of NIR spectra calls for sophistication in calibration methods. Most NIR instruments have associated software to assist with development of calibrations. The software is used in selecting wavelength points, or ranges, and also for optimizing mathematical treatment of the $1/R$ optical data. First and second derivatives can be very useful in reducing baseline drift and in correction for aberrations caused by differences in particle characteristics. The second derivative is also useful in qualitative studies, since it can be used to reduce band overlap. Figures 3 and 4 illustrate how derivative

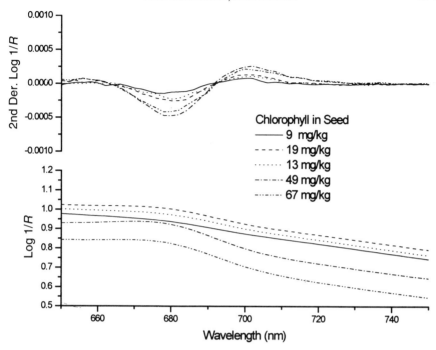

Fig. 3. Reflectance spectra of samples of canola seed with varying levels of chlorophyll. The second derivative suggests that a maximum in the region of 680 nm is appropriate for the red absorption of chlorophyll pigments. Spectra were prepared from samples of whole seed scanned on an NIRS6500 spectrometer.

spectroscopy can allow selection of wavelengths for analysis of components present in small amounts and causing only small perturbations in the overall spectrum.

Near infrared software is of three types: built-in, dedicated, and generic. Built-in software is included in the instrument design and usually includes provision for calibration, operation, and diagnostics. The older instruments had no provision for calibration, and the operator was obliged to record the optical signals in calibrate mode and to perform the multiple linear regression on any available computer. Modern instruments have provision for self-calibration, and the operator has to assemble sample sets for calibration and validation. Software for use with scanning spectrophotometers enable the operator to compute derivatives from the optical signals, to optimize the dimensions of the "segment" (smoothing) and "gap" (derivative size). The software includes provision for multiple linear regression, system diagnostics, scatter correction, and principal component analysis/partial least squares (PCA/PLS) regression. It also includes the capability of displaying and manipulating spectra, editing files, and other features.

The steps in calibration of a computerized spectrophotometer include assembling sample sets, scanning the samples (after sample preparation), optimizing the

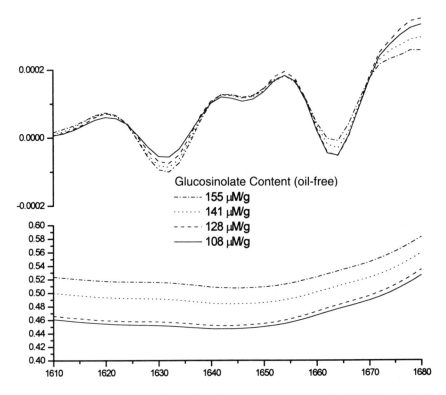

Fig. 4. Reflectance spectra of samples of oriental mustard seed with different levels of glucosinolates. The peaks at 1632 nm and 1664 nm are in the region suggested for glucosinolates (27). Spectra were prepared from samples of whole seed scanned on an NIRS6500 spectrometer.

mathematical treatment of the optical signal, optimizing the wavelength range, and verifying the efficiency of the equation. Modern software includes provision for all of these steps. Principal component analysis involves computing the degree to which different aspects of the spectral data contribute to the total variance of the system. Partial least squares regression utilizes the components to assemble equations for the prediction of the desired composition and functional parameters. The principal components can be displayed and used to study some of the functional aspects of the calibration to determine what factors are affecting the system (16,17).

The calibration process usually involves the selection of a set of 50 to 100 samples, covering the composition likely to be encountered. The sample set should contain a full range of values for all components for which calibrations are to be developed. Where multiple calibrations are required, it may be necessary to select a different set of calibration samples for each component. For many oilseeds the inverse relationship between oil and protein content makes it possible to select a single set of samples for these two components. Scanning instruments allow selection of a set of samples based on spectral differences. The original sample set is usually subdivided into a subset used to develop the calibration (calibration set). The remaining samples are used to test the calibration (prediction set).

Several statistical tools are in common use to evaluate NI calibration equations (10,18). Aside from the correlation coefficient (R), and R^2, the coefficient of determination, the most commonly used are

- Standard error prediction (SEP) (the standard deviation of differences between NI and reference data of the prediction set)
- Ratio of the standard deviation of the original reference data for the prediction set to the SEP (RPD) (values greater than 4 suggest that the calibration is capable of results suitable for quality control purposes)
- Ratio of the range of the prediction set (as determined by the reference method) to the SEP (RER or R/SE) (values in excess of 12 show that the equation may be used for quality control)
- Repeatability, a measure of the ability of the calibration to predict the same result when presented with the same sample
- Reproducibility, a measure of the ability of the calibration set to be transferred from instrument to instrument as well as from operator to operator, using methods that include sample preparation steps

One of the major recent advances in NI technology has been the improvement in software. Computation of principal component analysis (PCA) and partial least squares (PLS) regression, which took several hours at the time of their introduction to NIR technology, can now be completed in a minute or so. There are now at least five different software packages available to operators of computerized spectrophotometers. Some of these incorporate very comprehensive graphics, while others have excellent systems for scatter correction. All have capabilities for computing multiple linear regression (MLR), PCA, and PLS. Some of them, such as the GRAMS series, are generic, while others, such as NSAS, are dedicated to a particular series of instruments. The neural networks feature is the latest addition to NIR software and will likely find more applications in the future, as application probes more complicated spheres. Software for at-line applications has often been developed for specific situations, not applicable elsewhere (15).

Components Measured

The use of NIR spectrometry for determination of oil content, protein content, and moisture has been reported for many oilseeds including soybeans (19), canola, rapeseed (9,13,20,21), flaxseed (20,22), and crambe (23).

At the GRL, NIR has been used since 1992 to determine oil, protein, and moisture content of oilseeds in harvest surveys. In each of three years of study, the calibration for oil and protein was augmented by the addition of further samples from each harvest. In the first year (1992) a large prediction set was drawn from a random selection of samples as they were received. Since that time, prediction sets have been selected with a range of values from flaxseed and from both species of canola and across the entire growing area, resulting in a need for fewer samples than would be the case if a random selection procedure were used. The resulting prediction sets (illustrated for canola in Table 1) give a good estimation of the precision of the method as well as giving an indication of any needed bias adjustments (Fig. 5).

It is interesting that, although the precision of oil content determination is greater than that of protein content determination on an absolute basis, on a relative basis (i.e., about 45 g/100 g for oil content and 22 g/100 g for protein content) the standard deviation for oil (averaged for 1994 and1995) is ±1.3% relative, and that for protein is nearly the same at ±1.5% relative. One factor that has been shown to influence the accuracy of NIR results is the presence of other seed types in the sample (24). Increases in contamination with wild mustard was related to increases in differences between the predicted and actual values for oil content, protein content, and glucosinolate content.

Determination of chlorophyll in seeds is a problem that has particular interest to canola and rapeseed. Our earlier work (7,8,9) established the ability to use reflectance spectroscopy to determine chlorophyll in ground seed. For whole seed, the influence of seed color causes significant distortion in the visible region of the reflectance spectrum (Fig. 6). The use of derivative spectroscopy, however, allows much of the difference to be removed. Recently we have demonstrated that reflectance spectroscopy could be used determine chlorophyll in soybeans (Fig. 7).

Glucosinolate analysis in rapeseed has become a relatively common measurement (13,21,25,26) since the spectral area associated with glucosinolates was identified (27). Unfortunately only moderate success has been achieved in determining the glucosinolates in canola (9), possibly because the low level of glucosinolates in

TABLE 1
Summary of Prediction Data for Oil and Protein Content in Canola from GRL Harvest Surveys, 1992–1995

	Year	No. samples	R^2	Std. dev. g/100g	RPD
Oil content %,	1995	48	0.975	0.61	6.28
dry basis	1994	74	0.961	0.58	8.51
	1993	94	0.871	1.10	2.79
	1992	230	0.893	0.80	3.84
Protein content,	1995	48	0.985	0.37	8.02
%N × 6.25, dry basis	1994	74	0.985	0.49	8.12
	1993	94	0.905	0.91	3.26
	1992	230	0.945	0.60	4.91

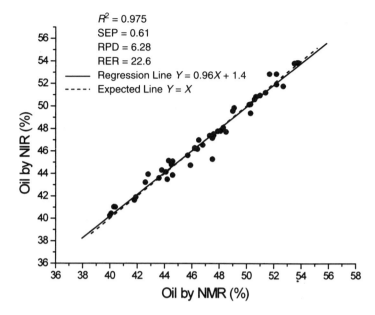

Fig. 5. Prediction data for determination of oil content in canola seed from the GRL's 1995 harvest survey. Samples were scanned on an NIRS6500 spectrometer, and oil contents were determined using a calibration developed from samples collected in 1992–1995 (9).

canola (less than 20 µm/g) are near the lower practical level of detection for NIR analysis. A comparative prediction plot for canola and mustard glucosinolates shows that although the canola had a lower SEP than the mustard, the correlation and RPD for mustard were much greater because of the wider range of glucosinolate values in the samples (Fig. 8).

In the past few years there has been an increasing interest in the determination of fatty acid composition both in oilseeds and in fats and oils by NIR spectroscopy. Erucic acid has been successfully determined in rapeseed (28) and other *Brassica* seeds (12). Our own previous work (9) demonstrated the importance of considering the fatty acids as components of the seed rather than of the variable oil content. In a study on the determination of the fatty acid composition of individual husked sunflower seeds, Sato and coworkers (29) also pointed out a major problem when dealing with NIR spectroscopy of sunflower seeds. The thick hull or husk on sunflower seed complicates direct analysis of the kernel without dehulling or grinding the sample.

Fatty acid composition seems to contribute to the success of NIR spectroscopy in discriminating between different vegetable oils (16,17). We have utilized the overall fatty acid composition data in developing calibrations for the determination of iodine value in flaxseed. The calibration developed is very robust and has

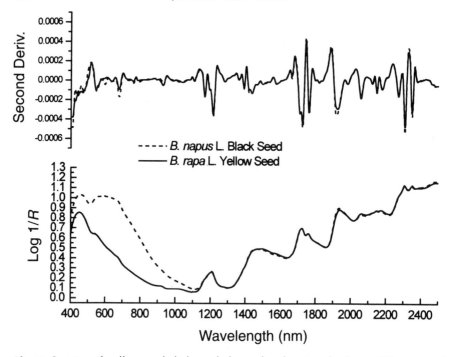

Fig. 6. Spectra of yellow and dark-seeded canola, showing the large differences of intensity in the visual region. Spectra were prepared from samples of whole seed scanned on an NIRS6500 spectrometer.

required little change over three years of usage. Fig. 9 shows prediction data from the 1995 harvest survey for flaxseed.

Other oilseed components determined by NIR spectroscopy include free fatty acids, soluble nitrogen, digestible nitrogen, germination capacity, and acid value of soybeans (30). NIR has also been used to determine fiber (14) and, more recently, soluble carbohydrates in soybeans (W. J. Mullin, Agriculture and Agri-Food Canada, Ottawa, ON, personal communication). NIR has also been shown to be an effective means for determining the extent of contamination of canola with wild mustard (24).

Utilization of NIR Analysis of Oilseeds

NIR is currently used by the USDA for testing of soybeans to establish market value (31). It has been established that, at a high-throughput country elevator, NIR testing of soybeans, coupled with a simple segregation into two protein categories, can be economically viable (31). In 1995, as a result of the low levels of oil in canola grown in Manitoba, several grain companies utilized NIR as a screening method to select samples with oil contents high enough to meet export contract specifications.

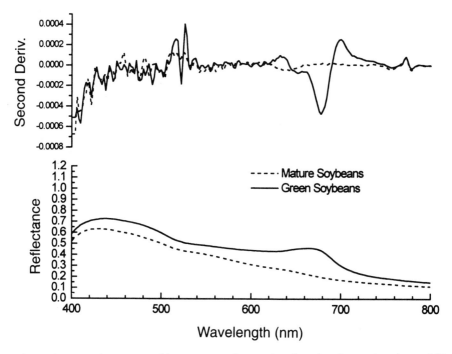

Fig. 7. Spectra of mature and immature soybeans showing absorbance band near 680 nm due to chlorophyll. Spectra were prepared from samples of whole seed scanned on an NIRS6500 spectrometer.

The GRL uses NIR as a tool for rapid analysis of several quality factors for harvest survey purposes (32,33,34). NIR has been used to test mustard seed and solin samples from harvest surveys as well as canola, flax, and soybeans. NIR has also been used in surveys of the quality of U.S. soybeans (35).

In the future it is likely that payment systems for seed delivery will be adopted in which NIR is used to determine factors such as oil and protein content, chlorophyll level, and possibly fatty acid composition. Such a system has been proposed for soybeans (31) and, by the Canadian Grain Commission, for chlorophyll in canola. The adoption of such a system will depend on the willingness of the grain trade and its customers to set premiums for quality factors (36,37). In Canada, where local processors are often in direct competition with exporters for a restricted seed supply, it is likely that the processors will initiate quality payment systems in order to attract higher-value seed to their facilities. Processors also have the technical staff needed to calibrate and maintain NIR instruments at their facilities; many of the NIR instruments used for testing protein content in wheat at country elevators where export seed is delivered are not suitable for testing oilseeds, since they do not operate over the appropriate wavelength range.

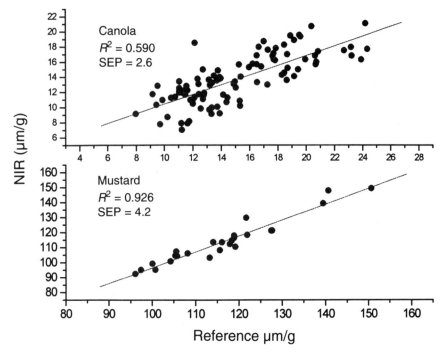

Fig. 8. Prediction data for determination of glucosinolates in canola and mustard seed. Samples were scanned on an NIRS6500 spectrometer, and glucosinolates were determined from calibrations developed for canola or mustard seed.

NIR has been used as a research tool in agronomic, plant breeding, and nutrition studies. NIR was used in evaluating the effect of wild mustard (*Brassica kaber*) competition on yield and quality of triazine-tolerant and triazine-susceptible canola (*B. napus* and *B. rapa*) (38). NIR has been used in screening samples for oil and protein in flaxseed (22) and rapeseed breeding programs (39) and in selection of low–erucic acid lines of *B. carinata* (40). An application of NIR that could be applied to oilseed meals is estimation of the true or apparent metabolizable energy of poultry rations (41).

The future of NI technology in the analysis and study of oilseeds, other grains, and derived products is assured. The speed, continuously improving accuracy, and flexibility place NI technology in a unique position. Other techniques, such as Fourier transform NI, have specific applications, but the main advantage of NI lies in its flexibility. The need for access to NI instrumentation for an extremely wide diversity of sample types has spawned myriad sample cells, to the extent that NI instruments are now capable of accepting material of types ranging from liquids through slurries to fresh fruits, vegetables, and forages; all types of grains; and derived products, including flour. Provision is made for bringing the instrument into direct contact with the sample, through interactance probes, and also for transmitting the signal through fiber-optic cables.

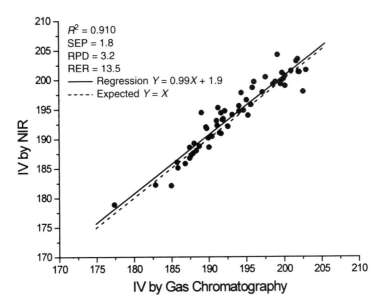

Fig. 9. Prediction data for determination of iodine value in flaxseed from the GRL's 1995 harvest survey. Samples were scanned on an NIRS6500 spectrometer, and oil contents were determined using a calibration developed from samples collected in 1992–1995 (9).

Software will continue to improve to the extent that the analysis ultimately will be fully automated. All the operator will have to do is to assemble calibration and prediction (validation) sample sets and scan the samples. The software will process all options of the mathematical treatment of the optical data, and report the best equation. Operators will have access to all other equations to test them if they wish.

The hardware of the future will include the monochromator but will also feature diode array devices. One such instrument, the Perten DA-7000, is already available. It scans an area from 400 to 1700 nm 600 times per second, and completes the test in a second. The instrument works in reflectance and transmittance mode and can accommodate both large (over 100 g) and small (10 g or less) samples. The DA-7000 can serve either as a "stand-alone" NI analyzer or as a computerized monchromator, in which mode it can be used as a research tool. Typical results for the DA-7000 on canola seed are included in Table 2.

The future will also see more use of networking among instruments and of analysis by remote control. The instrument will be set up with the automatic sampling system and will transmit the results to where they are needed—possibly to another automated section of the process, in situations where the NI instrument is being used in process control. Acousto-optical tunable and liquid optical tunable filter devices have been developed and will also add to the NI options available. Although these instruments are as fast as the DA-7000, it is unlikely that they will

TABLE 2
Preliminary Data for Prediction of Canola Parameters by the Perten DA-7000 Diode Array Visible–NIR Instrument

Statistic	Oil	Protein	Chlorophyll
R^2	0.96	0.96	0.92
SEP	0.56	0.39	0.92
RPD	8.1	9.0	4.7
n	74	74	74

supplant NI as the instrument of choice for most operations, because of sample presentation difficulties.

References

1. Murray, I., and Williams, P. (1987) in *Near-Infrared Technology in the Agricultural and Food Industries*, Williams, P., and Norris, K., American Association of Cereal Chemists, Inc., St. Paul, Minnesota, pp. 17–34.
2. Panford, J.A., and deMan, J.M. (1990) Determination of Oil Content of Seeds by NIR: Influence of Fatty Acid Composition on Wavelength Selection, *J. Am. Oil Chem. Soc. 67*, 473–482.
3. McClure, W.F. (1994) Near-Infrared Spectroscopy: The Giant is Running Strong, *Anal. Chem. 66*, 43a–53a.
4. Noble, D. (1995) Illuminating Near-I, *Anal. Chem. 67*, 735a–740a.
5. Williams, P. (1976) in *Advances in Automated Analysis*, Technicon International Congress, pp. 118–121.
6. Williams, P. (1976) in *Advances in Automated Analysis*, Technicon International Congress, pp. 131–136.
7. Daun, J.K. (1976) A Rapid Procedure for the Determination of Chlorophyll in Rapeseed by Reflectance Spectroscopy, *J. Am. Oil Chem Soc. 53*, 767–770.
8. Tkachuk, R., Mellish, V.J., Daun, J.K., and Macri, L.J. (1988) Determination of Chlorophyll in Ground Rapeseed Using a Modified Near-Infrared Reflectance Spectrophotometer, *J. Am. Oil Chem. Soc. 65*, 381–385.
9. Daun, J.K., Clear, K.M., and Williams, P. (1994) Comparison of Three Whole Seed Near-Infrared Analyzers for Measuring Quality Components of Canola Seed, *J. Am. Oil Chem. Soc. 71*, 1063–1068.
10. Williams, P. (1987) in *Near-Infrared Technology in the Agricultural and Food Industries*, Williams, P., and Norris, K., American Association of Cereal Chemists, Inc., St. Paul, Minnesota, pp. 143–167.
11. Bouveresse, E., Hartmann, C., Massart, D.L., Last, I.R., and Prebble, K.A. (1996) Standardization of Near-Infrared Spectrometric Instruments, *Anal. Chem. 67*, 682–690.
12. Velasco, L., Fernandez-Martinez, J., and DeHaro, A. (1996) Screening Ethiopian Mustard for Erucic Acid by Near-Infrared Reflectance Spectroscopy, *Crop Sci. 36*, 1068–1071.
13. McGregor, D.I. (1990) in *Canola and Rapeseed Production, Chemistry, Nutrition and Processing Technology*, Shahidi, F., Van Rostrand Reinhold, New York, pp. 221–231.

14. Panford, J.A., Williams, P.C., and deMan, J.M. (1988) Analysis of Oilseeds for Protein, Oil, Fiber and Moisture by Near-Infrared Reflectance Spectroscopy, *J. Am. Oil Chem. Soc. 65*, 1627–1634.
15. Williams, P. (1996) paper presented at the International Japanese Conference on Near-Infrared Reflectance, Tokyo, Nov. 20–21, 1996 (www.cgc.ca/Prodser/pubs/confpaper/nir-e.htm).
16. Sato, T. (1994) Application of Principal-Component Analysis on Near Infrared Spectroscopic Data of Vegetable Oils for Their Classification, *J. Am. Oil Chem. Soc. 71*, 293–298.
17. Cowe, I.A., Koester, S., Paul, C., McNicol, J., and Cuthbertson, D.C. (1988) Principal Component Analysis of Near Infrared Spectra of Whole and Ground Oilseed Rape *Brassica napus* Samples, *Chemom. Intell. Lab. Syst. 3*, 233–242.
18. Williams, P., and Sobering, D.C. (1993) Comparison of Commercial Near Infrared Transmittance and Reflectance Instruments for Analysis of Whole Grains and Seeds, *J. Near Infrared Spec. 1*, 25–32.
19. Hurburgh, C.R., and Brumm, T.J. (1990) Protein and Oil Content of Soybeans Received at Country Elevators, *Appl. Eng. Agric. 6*, 65–68.
20. Panford, J.A. (1989) in *Methods for Protein Analysis*, AOAC International, Washington, pp. 1–12.
21. Starr, C., Suttle, J., Morgan, A.G., and Smitan, D.B. (1985) A Comparison of Sample Preparation and Calibration Techniques for the Estimation of Nitrogen, Oil and Glucosinolate Content of Rapeseed by Near Infrared Spectroscopy, *J. Agric. Sci., Camb. 104*, 317–323.
22. Bhatty, R.S. (1991) Measurement of Oil in Whole Flaxseed by Near-Infrared Reflectance Spectroscopy, *J. Am. Oil Chem. Soc. 68*, 34–38.
23. Hartwig, R.A., and Hurburgh, C.R. (1990) Near-Infrared Reflectance Measurement of Moisture, Protein and Oil Content of Ground Crambe Seed, *J. Am. Oil Chem. Soc. 67*, 435–437.
24. Daun, J.K., DeClercq, D.R., Howard, H.K., Clear, K.M., and Thorsteinson, C.T. (1994) Wild Mustard (*Sinapis arvensis* L.) and the Analysis of Canola (*Brassica napus* L. and *Brassica rapa* L.) by NIR and NMR, *GCIRC Bulletin 10*, 178–182.
25. Biston, R., Dardenne, P., Cwikowski, M., Marlier, M., Severin, M., and Wathelet, J.P. (1988) Fast Analysis of Rapeseed Glucosinolates by Near Infrared Reflectance Spectroscopy, *J. Am. Oil Chem. Soc. 65*, 1599–1600.
26. Evans, E.J., Grant, A., Fenwick, G.R., Kwiatkowska, C.A., Spinks, E.A., Heaney, R.K., and Bilsborrow, P. (1989) A Comparison of Rapid-Release (X-ray-Fluorescence, Near-Infrared Reflectance) and Glucose-Release Methods for the Determination of the Glucosinolate Content of Oilseed Rape (*Brassica-Oleracea*), *J. Sci. Food Agric. 49*, 297–305.
27. Lila, M., and Furstoss, V. (1986) Détermination de Longueurs d'Onde Spécifiques pour la Mesure des Glucosinolates du Colza par Spectrophotométrie de Réflexion dans le Proche Infrarouge, *Agronomie 8*, 703–707.
28. Reinhardt, T.-C., and Röbellen, G. (1991) in *GCIRC Eighth International Rapeseed Congress*, McGregor, D.I., Saskatoon, Saskatchewan, pp. 1380–1384.
29. Sato, T., Takahata, Y., Noda, T., Yanagisawa, T., Morishita, T., and Sakai, S. (1995) Nondestructive Determination of Fatty Acid Composition of Husked Sunflower (*Helianthus auuna* L.) Seeds by Near Infrared Spectroscopy, *J. Am. Oil Chem. Soc. 72*, 1177–1183.

30. Sato, T., Abe, H., Kawano, S., Ueno, G., Suzuki, K., and Iwamoto, M. (1994) Near-Infrared Spectroscopic Analysis of Deterioration Indices of Soybeans for Process Control in Oil Milling Plant, *J. Am. Oil Chem. Soc. 71*, 1049–1055.
31. Hurburgh, C.R. (1994) Identification and Segregation of High-Values Soybeans at a Country Elevator, *J. Am. Oil Chem. Soc. 71*, 1073–1078.
32. DeClercq, D.R., Daun, J.K., and Tipples, K.H. (1996) *Quality of Canadian Soybeans, 1995*, Canadian Grain Commission, Winnipeg, Crop Bulletin No. 226.
33. DeClercq, D.R., Daun, J.K., and Tipples, K.H. (1995) *Quality of Western Canadian Canola, 1995*, Canadian Grain Commission, Winnipeg, Crop Bulletin No. 224.
34. DeClercq, D.R., Daun, J.K., and Tipples, K.H. (1996) *Quality of Western Canadian Flaxseed, 1995*, Canadian Grain Commission, Winnipeg, Crop Bulletin No. 225.
35. Hurburgh, C.R., Paynter, L.N., and Schmitt, S.G. (1987) Quality Characteristics of Midwestern Soybeans, *Appl. Eng. Agric. 3*, 159–165.
36. Feldmann, T. (1990) Soybean Protein and Oil Testing Operation, Shipment, and Marketing Logistics, *Grain Quality Newsletter* (May), 56–59.
37. Moeller, R. (1990) Oil and Protein Content in Soybeans—The Processor's Perspective, *Grain Quality Newsletter* (May), 41–44.
38. Mcmullan, P.M., Daun, J.K., and Declercq, D.R. (1994) Effect of Wild Mustard (*Brassica kaber*) Competition on Yield and Quality of Triazine-Tolerant and Triazine-Susceptible Canola (*Brassica napus* and *Brassica rapa*), *Can. J. Plant Sci. 74*, 369–374.
39. Bengtsson, L. (1985) Some Experiences of Using Different Analytical Methods in Screening for Oil and Protein Content in Rapeseed, *Fette Seifen Anstrichm. 87*, 262–265.
40. Velasco, L., Fernandez-Martinez, J., and de Haro, A. (1995) Isolation of Induced Mutants in Ethiopian Mustard (*Brassica carinata* L.) with Low Levels of Erucic Acid, *Plant Breeding 114*, 454–456.
41. Edney, M.J., Morgan, J.E., Williams, P.C., and Campbell, L.D. (1994) Analysis of Feed Barley by Near Infrared Reflectance Technology, *J. Near Infrared Spec. 2*, 33–41.
42. Buchmann, N.B. (1995) in *Near-Infrared Spectroscopy: The Future Waves*, Davies, A.M.C., and Williams, P., Montreal, pp. 479–483.

Chapter 14

Application of Fourier Transform Infrared Spectroscopy in Edible-Oil Analysis

J. Sedman, F.R. van de Voort, and A.A. Ismail

McGill IR Group, Department of Food Science and Agricultural Chemistry, Macdonald Campus, McGill University, 21,111 Lakeshore Road, Ste. Anne de Bellevue, Québec H9X 3V9, Canada

Introduction

The development of Fourier transform infrared (FTIR) spectroscopy during the past two decades has revitalized infrared (IR) quantitative analysis, and reductions in the cost of the instrumentation during the past five years, coupled with advances in software and chemometrics, have led to renewed interest in routine quality control (QC) applications of IR spectroscopy as a quantitative analytical tool. The McGill IR Group, formed in 1990, recognized the potential benefits of quantitative FTIR spectroscopy for food analysis in general (1), which include rapid and reproducible analyses, amenability to automation, and a reduction in chemical waste and analytical costs. Our research program has been directed toward the development of FTIR-based analytical methodologies suitable for implementation in QC laboratories. Most of this methodology development work has focused on fats and oils analysis. This chapter reviews a variety of FTIR applications that have been developed, including methods for the determination of *trans* and *cis* content, iodine value, saponification number, solid fat index/content, anisidine value, and peroxide value, bringing into the discussion basic principles of FTIR spectroscopy, sample handling techniques, chemometrics, automation, and calibration transfer concepts. The end result of much of our work is the production of preprogrammed analytical packages that are simple to implement, eliminating much of the development work that commonly holds back the use of new technologies. A number of processors are using these packages, and it is the contention of the McGill IR Group that FTIR spectroscopy will play an increasingly important role in the routine analysis of fats and oils and will in time replace a variety of conventional QC methods as it gains acceptance by the industry.

Fourier Transform Infrared (FTIR) Spectroscopy

An infrared spectrometer essentially consists of a source of continuous infrared radiation, a means for resolving the infrared radiation into its component wavelengths, and a detector. The first commercial scanning infrared spectrometers, which became available in the 1940s, employed a prism to resolve the infrared radiation into its component wavelengths. In instruments manufactured in the late 1950s and 1960s, the prism was replaced by a diffraction grating, leading to improved resolution. A more drastic change in instrument design, however, occurred in the late 1960s, with the development of FTIR instruments.

FTIR spectroscopy is based on interferometry and thus differs fundamentally from traditional dispersive infrared spectroscopy. The Michelson interferometer, employed in most FTIR spectrometers, uses a beamsplitter to divide the beam of radiation from the infrared source into two parts, one part being reflected to a stationary mirror and the other part being transmitted to a moving mirror (Fig. 1). When the beams are reflected back, they recombine at the beamsplitter, producing a constructive/destructive interference pattern due to the varying difference between the distances traveled by the two components of the beam. Part of the recombined beam then passes to the detector. After the infrared energy has been selectively absorbed by a sample placed between the beamsplitter and the detector, fluctuations in the intensity of the energy reaching the detector are digitized in real time, yielding an *interferogram* (Fig. 2A). The interferogram contains all the information that is required to produce the infrared spectrum of the sample, but this information is in the time domain. In order to obtain a conventional infrared spectrum, the interferogram is converted to the frequency domain by Fourier transformation. In FTIR spectroscopy, the interferogram is usually a plot of intensity as a function of the path difference between the stationary and the moving mirror, known as the retardation δ, which is proportional to time t because the moving mirror travels at constant velocity v; i.e.,

$$\delta = 2vt \text{ cm} \qquad (1)$$

Fourier transformation of the interferogram $I(\delta)$ then yields a spectrum with the x-axis in units of wavenumbers (cm^{-1}), $I(\bar{v})$, in accordance with the following relationship:

$$I(\delta) = 0.5H(\bar{v})I(\bar{v}) \cos 2\pi v \delta \qquad (2)$$

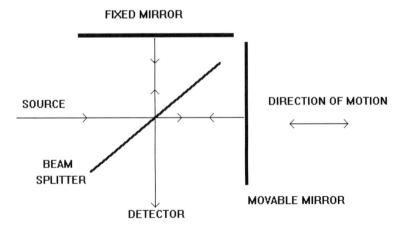

FIG. 1. Schematic diagram of a Michelson interferometer.

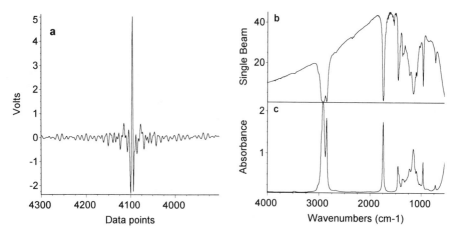

FIG. 2. (a) An interferogram produced by an FTIR spectrometer; (b) the corresponding single-beam spectrum; and (c) the absorption spectrum obtained by ratioing the spectrum in (b) against the single-beam background (i.e., open-beam) spectrum. The sample is trielaidin, an 18:1*t* triglyceride commonly present in fats and oils that have been hydrogenated.

where $H(\bar{\nu})$ is a single, wavenumber-dependent correction factor that accounts for instrumental characteristics. Only with the development of the Cooley-Tukey fast Fourier transform (FFT) algorithm and with the widespread availability of computers in the laboratory beginning in the 1970s did FTIR spectroscopy become practical. Since that time, FTIR spectroscopy has undergone very rapid development, and FTIR spectrometers have now replaced dispersive instruments almost completely.

Because FTIR spectrometers are single-beam instruments, $I(\bar{\nu})$ in equation 2 represents a *single-beam spectrum* (Fig. 2B), which consists of the emittance profile of the IR source on which are superimposed the absorption bands of the sample as well as the air background spectrum (i.e., absorption bands of carbon dioxide and water vapor present in the optical path). In order to eliminate the contributions of the source and the air background, the single-beam spectrum of the sample is digitally ratioed against a single-beam spectrum recorded with no sample in the beam [$I_0(\bar{\nu})$], and the absorption spectrum $A(\bar{\nu})$ is then computed (Fig. 2C):

$$A(\bar{\nu}) = -\log[I(\bar{\nu})/I_0(\bar{\nu})] \qquad (3)$$

Alternatively, the single-beam spectrum of the sample [$I(\bar{\nu})^S$] may be ratioed against the single-beam spectrum of a reference material [$I(\bar{\nu})^R$], thereby eliminating the spectral features common to the sample and the reference as well as the contributions from the source and air background. This operation is equivalent to a 1:1 subtraction of the reference spectrum ratioed against an open-beam spectrum from the sample spectrum ratioed against the same open-beam spectrum [$A(\bar{\nu})^S - A(\bar{\nu})^R$] but requires only a single mathematical manipulation:

$$A(\bar{v})^S - A(\bar{v})^R = -\log[I(\bar{v})^S/I_0(\bar{v})] - \{-\log[I(\bar{v})^R/I_0(\bar{v})]\} = -\log[I(\bar{v})^S/I(\bar{v})^R] \quad (4)$$

The major reasons for the present dominance of FTIR instrumentation are the significant advantages that interferometers have over dispersive instruments. In an interferometer all frequencies are measured simultaneously (the multiplexing advantage), and thus the entire FTIR spectrum of a sample can be collected in a single one-second scan. In addition, because the signal-to-noise ratio (S/N) of a spectrum increases as the square root of the number of scans, the multiplexing advantage also allows high S/N to be achieved in much shorter times than required with dispersive spectrometers. FTIR spectrometers also have higher energy throughput and are therefore more versatile in terms of the types of samples and sample handling techniques that can be employed. Another important advantage for data manipulation and quantitative applications is that wavelength precision is maintained by using an internal reference laser, so wavelength drifts over time are eliminated as a possible source of error. The substantial computing power of FTIR systems has also been an important factor in the success of FTIR spectroscopy. FTIR software packages provide a wide variety of data-handling routines, which facilitate spectral acquisition and interpretation, as well as powerful chemometric tools, which have substantially enhanced the utility of IR spectroscopy as a quantitative analysis tool.

Quantitative Analysis

IR spectroscopy is a secondary method of analysis, so the development of quantitative analysis methods requires calibration with a set of standards of known composition, prepared gravimetrically or analyzed by a primary chemical method, in order to establish the relationship between IR band intensities and the compositional variable(s) of interest. Various calibration approaches may be employed, ranging from a simple Beer's law plot, which is an adequate basis for calibration in the case of a simple system such as a single component dissolved in a noninteracting solvent, to the sophisticated multivariate analysis techniques that are required for more complex systems. The data-handling capabilities of FTIR systems have allowed these latter techniques to be implemented in the instrument software and applied directly to spectral data, bringing a resurgence to quantitative infrared spectroscopy during the past decade.

As with other types of absorption spectroscopy, the basis of quantitative analysis in IR spectroscopy is the Bouguer-Beer-Lambert law, or Beer's law:

$$A_{\bar{v}} = \varepsilon_{\bar{v}} bc \quad (5)$$

where $A_{\bar{v}}$ is the absorbance measured at wavenumber $\bar{v}$, $\varepsilon_{\bar{v}}$ is the molar absorption coefficient of the absorbing species at this wavenumber, b is the pathlength of the IR cell, and c is the concentration of the absorbing species. Application of Beer's law for the determination of the amount of a compound present in a sample requires that $\varepsilon_{\bar{v}}$ be determined by measuring the absorbance of a calibration standard of known concentration. Of course, in order to attain better accuracy, it is preferable to prepare a series of standards of different concentrations, spanning the concentration range of interest, and obtain $\varepsilon_{\bar{v}} b$ from a plot of absorbance vs. concentration by linear least-squares regression. This procedure averages out the errors due to instrumental noise, measurement errors, and other sources of random variation. In addition, it allows deviations from Beer's law in the concentration range of interest to be detected, such as may arise from hydrogen bonding, dimerization, and other intermolecular interactions. Such interactions may then be modeled by the introduction of higher-order terms into the equation that relates absorbance to concentration. An example taken from work by the McGill IR Group on the determination of free fatty acids in edible oils (2) is presented in Fig. 3.

When the concentration of more than one species will vary in the samples to be analyzed, a multivariate calibration approach may be required, because the above univariate approach cannot account for any contributions of additional components to $A_{\bar{v}}$, nor can it model interactions between components. Among the variety of multivariate calibration techniques that have been applied in the analysis of multi-

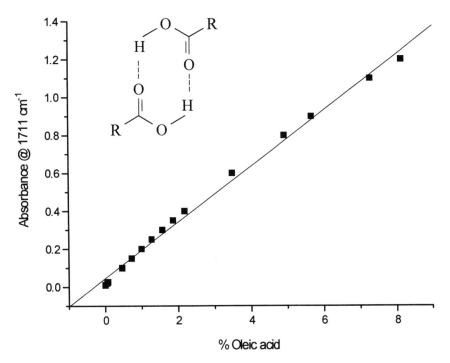

FIG. 3. Beer's law plot for oleic acid, obtained by spiking olive oil with oleic acid at concentrations ranging from 0.2 to 8.0%. The peak height of the band at 1711 cm^{-1}, which is characteristic of oleic acid dimers, is plotted vs. oleic acid concentration. The plot exhibits curvilinearity due to the monomer-dimer equilibrium, although the curvature is slight because the equilibrium is shifted far to the right. (Adapted from Ref. 2.)

component systems (3), the most powerful is partial-least-squares (PLS) regression, a form of factor analysis. A PLS calibration model is developed by compressing the spectral data for the calibration standards into a set of mathematical "spectra," known as the loading spectra or factors, which are linear combinations of the calibration spectra. The spectrum of each calibration standard is then decomposed into a weighted sum of the loading spectra, and the weights given to each loading spectrum, known as "scores," are regressed against the concentration data for the standards. When this calibration model is applied for the prediction of an unknown, the amounts of each loading spectrum employed in reconstructing the spectrum of the unknown, i.e., the "scores," are used to predict the concentration of the unknown. The mathematical details of the PLS algorithm are given in Ref. 4, and a discussion of its implementation is presented in Ref. 5.

Overfitting of the spectral data is always possible in developing a PLS calibration model, as the fit between the actual and the predicted values for the calibration standards will necessarily be improved as the number of loading spectra included in the model is increased, a perfect fit being achieved when the number of loading spectra equals the number of calibration standards. However, because the concentration data for the standards are used in generating the loading spectra, most of the spectral variability associated with the concentration variations is incorporated in the early loading spectra. Thus, the last loading spectra generated will mostly represent noise in the spectra of the calibration standards or spectral information that is unrelated to concentration, and their inclusion in the PLS calibration model will degrade its performance in the prediction of unknowns. Thus, validation of a PLS calibration model is always required in order to select the optimum number of loading spectra. This may be achieved by testing the performance of the model with standards not included in the calibration set. Alternatively, the "leave-one-out" cross-validation technique may be employed, whereby the calibration is performed m times with $m - 1$ standards and the mth standard is predicted as an unknown. The predicted residual error sum of squares (PRESS) is then computed from the errors in the predictions obtained for the m standards by cross-validation and plotted as a function of the number of factors employed in the calibration (Fig. 4). The optimum number of factors corresponds to the point at which the PRESS plot reaches a minimum or begins to level off.

A PLS calibration can, in principle, be based on the whole spectrum, although in practice the analysis is restricted to regions of the spectrum that exhibit variations with changes in the concentrations of the components of interest. As such, the use of PLS can provide significant improvements in precision relative to methods that use only a limited number of frequencies (6). Furthermore, PLS is able to compensate for spectral interferences such as overlapping bands and band shifts due to intermolecular interactions, provided all sources of such interferences that may be present in the samples to be analyzed are present in the calibration standards. The powerful data reduction capabilities of PLS can also be exploited to establish relationships between quality attributes or physicochemical properties and FTIR spectral data. In this context, the term "quality attribute" refers to a parameter that is not a direct measure

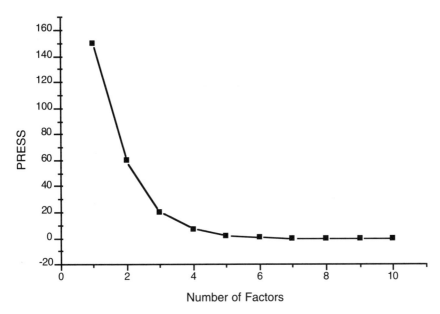

FIG. 4. Typical PRESS plot obtained from cross-validation of a PLS calibration, showing the predicted residual error sum of squares (PRESS) as a function of the number of factors included in the calibration model.

of the concentration of a particular component in the sample but rather is related in some manner to the concentrations of several components or to the overall composition of the sample. Although similar correlations can be achieved through the use of multiple linear regression by establishing relationships between the value of a quality attribute and absorbance values at several frequencies, such an approach requires that the compositional variables contributing to the quality attribute be identified. By contrast, PLS can be used to develop a calibration model for the prediction of a quality attribute without any prior knowledge of the underlying relationship between its value and compositional variables. These various advantages have made PLS the technique of choice in the FTIR analysis of complex multicomponent systems (6–10).

Sample Handling

Numerous books cover the topic of sample handling techniques in IR spectroscopy (11–13), and a detailed description of all the various alternatives is beyond the scope of this chapter. Instead, this section will focus on the two sample-handling techniques that are most commonly employed in edible-oil analysis applications: the use of transmission cells and attenuated total reflectance (ATR).

When IR measurements are made in the transmission mode, the IR beam passes *through* the sample, and only the IR radiation transmitted by the sample is measured by the detector. In the case of fluid samples, transmission measurements are typically made by placing the sample between two polished windows, made of an infrared-transmitting material (e.g., NaCl, KBr, or other salt crystals) and separated by a Teflon spacer, which fixes the pathlength (b in equation 5). The low energy output of traditional mid-IR radiation sources imposes fairly severe limitations on the pathlength that may be employed, because if all the available energy is absorbed by the sample, no signal reaches the detector. This is illustrated in Fig. 5, which shows the FTIR spectrum of a vegetable oil in a 0.010-mm transmission cell. From the percent transmittance values on the y-axis of this transmission spectrum, it can be seen that any increase in the pathlength beyond 0.010 mm will result in virtually no transmission of energy by the sample between ~3000 and ~2850 cm^{-1}, owing to the strong C-H absorptions in this region of the spectrum. If these exceptionally strong bands are excluded from consideration, the optimum pathlength for recording the spectra of neat oils is on the order of 0.025 mm. When the absorption bands to be measured are those of minor components (e.g., free fatty acids or oxidation products), the optimal pathlength depends on both the concentration range of the component of interest and the infrared absorption characteristics of the oil in the spectral region of interest, but it usually does not exceed 1 mm.

The attenuated total reflectance (ATR) sampling technique, which was developed in the 1960s, is based on the totally different optical principle of total internal reflection (14). In the ATR technique, the sample is placed in contact with a crystal

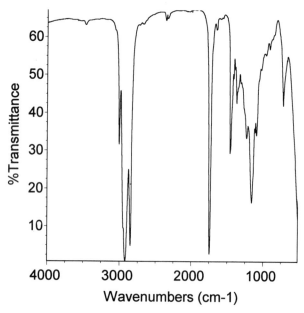

FIG. 5. Transmission spectrum of a sample of soybean oil in a 0.010-mm transmission cell, showing the percent transmittance of infrared radiation by the sample.

of a high–refractive index material, known as an internal reflection element (IRE). As illustrated in Fig. 6, light from the infrared source is launched into the IRE at an angle such that it undergoes multiple internal reflection as it travels down the crystal, giving rise to an evanescent wave at the surface of the IRE which decays exponentially as it propagates away from the surface of the IRE through the sample. The distance from the surface at which the intensity of the evanescent wave decays to $1/e$ of its value at the surface is defined as the depth of penetration, d_p. The evanescent wave is attenuated by the absorption of radiation by species on or near the surface of the IRE, and measurement of this attenuation as a function of wavelength yields the infrared spectrum of these species. The effective pathlength in an ATR measurement is given by the product of the number of internal reflections and the depth of penetration, d_p, given by

$$d_p = \frac{\lambda}{2\pi n_1 [\sin^2(\theta) - (n_2/n_1)]^{1/2}} \qquad (6)$$

where λ is the wavelength of the radiation in the IRE, n_1 is the refractive index of the IRE material, n_2 is the refractive index of the medium surrounding the IRE, and θ is the angle at which the incident light strikes the interface. The effective pathlength can thus be altered by varying the angle of incidence as well as by the choice of the IRE material. It should be noted that the depth of penetration, and hence the effective pathlength, increases on going from the high-frequency end of the spectrum to the low-frequency end. Therefore, the relative intensities of the peaks in the ATR spectrum of a sample will not be the same as those in the spectrum recorded for the same sample in a transmission cell. For example, the effective pathlength of a typical ZnSe ATR accessory with 8 reflections and $\theta = 45°$ can be calculated from equation 6 to be approximately 0.0024 mm for $\lambda = 3$ µm and 0.006 mm for $\lambda = 6$ µm. Because of their short effective pathlengths, ATR sampling accessories have proved particularly useful in work with aqueous solutions, owing to the strong absorption of water across large portions of the mid-IR spectrum.

From a sample-handing perspective, edible oils are, in principle, ideal candidates for FTIR spectroscopic analysis, because they can either be applied directly in their neat form onto an ATR crystal or be pumped through a transmission flow cell. Fats can be handled in an analogous manner if the ATR crystal or flow cell is heated and thermostated to a temperature above the melting point of the fat. Our experience

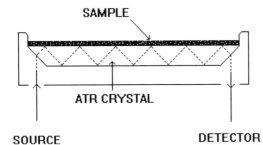

FIG. 6. Schematic drawing of a horizontal attenuated total reflectance (ATR) sampling accessory.

during the course of extensive work on edible-oil analysis by FTIR spectroscopy has led us to favor transmission cells for routine analyses because of a number of disadvantages of the ATR sampling technique. These include

1. The need for thorough cleaning of the ATR crystal with solvent after each sample
2. The general tendency of lipids to strongly adhere to the surface of the ATR crystal (the "memory effect"), which increases the risk of sample cross-contamination
3. The variation in the effective pathlength with slight changes in alignment of the crystal.

In addition, when operating at elevated temperatures, it is difficult to maintain good temperature control at the surface of the crystal due to evaporative cooling during cleaning with solvent. Unfortunately, ATR measurements are particularly sensitive to temperature variations because the effective pathlength is affected by changes in the refractive index of the sample or the ATR crystal in a wavelength-dependent manner; thus, temperature fluctuations of less than 1°C have been observed to cause significant curvature of the spectral baseline (15), leading to poor analytical reproducibility.

The McGill IR Group has designed a sample-handling system specifically for use in the analysis of fats and oils by transmission FTIR spectroscopy, because it became apparent during the course of method development work that the commercially available FTIR sample-handling accessories would not be particularly well suited for routine analysis of fats and oils in an industrial setting. In particular, most transmission flow cells are not designed for operation at the elevated temperatures employed for the analysis of fats, which requires that the IR cell be maintained at a temperature above the melting point of the samples to avoid the spectral complications arising from crystallization phenomena. In addition, it is also desirable that fats be premelted prior to sampling, not only in order to allow them to flow but also to achieve representative sampling; thus, the cell input and output lines must be maintained at an elevated temperature to avoid plugging due to crystallization. A schematic drawing of an FTIR spectrometer equipped with the sample-handling system that we have had constructed is shown in Fig. 7. Figure 8 depicts a side view of the sample-handling accessory within the optical compartment of the spectrometer, showing the cell and its housing, which is composed of a temperature control block and a removable cell insert, both made of stainless steel. The cell insert can be easily removed, in order to record an open-beam spectrum, and has a spring-loaded face plate to make it readily demountable so that the cell windows can be replaced whenever necessary. Figure 9 presents a schematic diagram of the flow path through the cell and its housing, including the inlet line, a three-way valve used to direct the oil flow through either the cell or a bypass line, and an outlet line emptying into a collection vessel, which in turn is connected to a vacuum line or pump. The purpose of the bypass line is to facilitate the process of flushing each sample out of the system with the next sample, as the flow through the bypass line is much less constricted

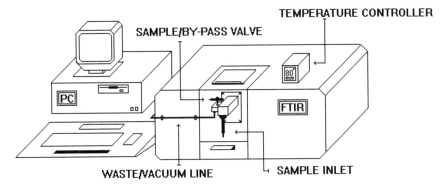

FIG. 7. Schematic of an FTIR spectrometer equipped with the heated oil analysis sample-handling accessory.

than that through the narrow-pathlength cell. This has made it possible to avoid the use of a solvent for rinsing out the IR cell between samples. All components of the accessory are heated to 80°C (±0.2°C) so that fats are in their liquid state and flow without crystallization in the lines or cell. The sample is prewarmed in a microwave oven or heated in an 80°C water bath and then presented at the input spigot of the accessory, where it is loaded by aspiration. The sample is then allowed to reach thermal equilibrium in the IR cell (~30 s) prior to spectral data collection.

This rugged sample-handling accessory has been designed to make it possible to analyze fats and oils routinely in their neat form under precisely controlled conditions. It allows for rapid and reproducible sample loading simply by turning a valve; a fully automatic version, driven by computer-controlled solenoid valves, is also available for high-volume applications (Dwight Analytical, Toronto, Canada). The McGill IR Group has tested this system extensively by analyzing >1000 samples and has found it to provide a practical and reliable means of sample handling that meets the requirements imposed by the physical characteristics of fats and oils.

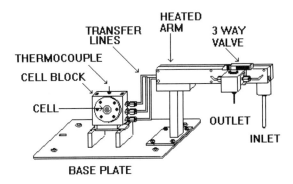

FIG. 8. Schematic of the heated oil analysis sample-handling accessory, showing input spigot, loading/bypass valve, cell block, and circular removable cell insert.

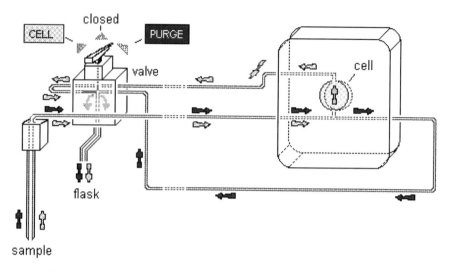

FIG. 9. Flow path through the oil analysis cell, illustrating cell bypass.

Applications

Determination of trans Content

At present, the only widespread application of IR spectroscopy in fats and oils analysis is the determination of isolated *trans* isomers, which is an official method of the American Oil Chemists' Society (AOCS) (16). Most unsaturated bonds in naturally occurring fats and oils are in the *cis* configuration, but *trans* double bonds may be present in processed fats and oils as a result of isomerization during processing or as the consequence of oxidation. Furthermore, when oils are hardened by hydrogenation for use in margarines and shortenings or partially hydrogenated to stabilize them to oxidation, *cis* double bonds are also converted to *trans*. The levels of *trans* fatty acids are of increasing concern to health professionals due to their association with heart disease (17), and the U.S. Food and Drug Administration has been petitioned to require the inclusion of *trans* fatty acids as part of saturated fat content in the labeling of fat-based products, such as margarines, spreads, and frying fats (18, 19). Although the traditional IR method for the determination of *trans* content has a number of well-known limitations, modifications of this method that take advantages of the capabilities of FTIR instrumentation have the potential to meet the present need for rapid, accurate *trans* analysis.

Trans analysis by IR spectroscopy dates back almost 50 years (20) and is based on the measurement of the characteristic absorption of isolated *trans* bonds at 10.3 µm (966 cm^{-1}), which is due to their C=C-H bending vibration (Fig. 10). An implicit assumption of the original work was that fats and oils containing no *trans* double bonds would have zero absorbance at the analytical wavelength, but it was soon estab-

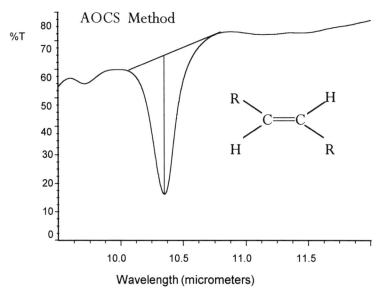

FIG. 10. Traditional IR method for the determination of isolated *trans* isomers in fats and oils.

lished that this is not the case, as all triglycerides exhibit a weak absorption band that overlaps with the *trans* absorption band (21). The intensity of this overlapping absorption band, which varies with the triglyceride composition of the sample, can contribute as much as 3 to 5 percentage points to the measured *trans* values. Because the underlying triglyceride absorptions will thus cause a substantial overestimation of the *trans* content of samples having low levels of *trans* isomers, the original AOCS method (Method Cd 14-61, recently replaced by Method Cd 14-95) required that samples containing less than 15% isolated *trans* isomers be converted to methyl esters prior to analysis. However, although this procedure eliminates the overlapping triglyceride absorptions, the *trans* band still sits on a sloping baseline, making accurate measurement of its intensity difficult. Furthermore, when a baseline is drawn between the limits indicated in the original AOCS method, the *trans* values obtained are low by 1.5 to 3 percentage points (21).

Over the years, a number of approaches aimed at increasing the accuracy of the IR determination of *trans* content have been investigated (22), and some of the improvements that have been suggested are incorporated in the revised AOCS method (Cd 14-95). The newer method requires the conversion of all samples to methyl esters, regardless of *trans* content. As in the original method, samples are dissolved in CS_2 and their spectra recorded in a fixed-pathlength (1 mm) transmission cell. Separate calibration equations are derived for the analysis of samples containing ≤10% *trans* isomers and >10% *trans* isomers. Instead of the single calibration standard employed in the original method (a CS_2 solution of trielaidin or methyl elaidate), a series of standards consisting of mixtures of methyl elaidate and methyl oleate in CS_2 are prepared, with the total concentration of methyl esters kept constant throughout the series. This two-component calibration approach allows for better modeling of the spectral baseline, and hence more accurate *trans* analyses, because the total concentration of methyl esters in each standard and in the samples to be analyzed is the same (0.20 g/10 mL). Other changes in the AOCS method concern the selection of baseline points in the measurement of the *trans* peak height; whereas fixed baseline points were formerly specified, the position at which the baseline is drawn in the modified method depends on the size of the *trans* peak.

Beyond updating the experimental protocol to reflect the data-handling capabilities of modern IR spectrometers (both FTIR and dispersive instruments), the recent modifications to the AOCS method are largely directed toward improving the accuracy of IR *trans* analysis. A number of investigators have also made efforts to simplify the experimental procedure, particularly by analyzing samples in their neat form and thereby eliminating the use of the volatile and noxious/toxic CS_2. For example, Sleeter and Matlock (23) developed an FTIR procedure for measuring the *trans* content of oils, analyzed as neat methyl esters using a 0.1-mm KBr transmission cell. This FTIR method was shown to provide better precision and a significant reduction in total analysis time (2.5 min/sample) in comparison with the traditional AOCS method, as well as having the advantage of being amenable to automation (23). The method was calibrated with standards prepared by dissolving varying amounts of methyl elaidate in methyl linoleate; the best calibration equation was obtained by a quadratic fit of the *trans* peak area vs. concentration data, owing to a slight curvature in the calibration plot over the concentration range investigated (0–50% *trans*). Sleeter and Matlock reported that more than 700 samples were run in the same cell over a period of nine months without any need for recalibration, demonstrating the stability of the FTIR spectrometer (23). Although this FTIR method greatly simplified the *trans* analysis, particularly by eliminating the need for CS_2, the analysis was still performed on methyl esters rather than neat fats and oils.

A means of eliminating the requirement for conversion to methyl esters is to employ the spectral ratioing capability of FTIR spectrometers to remove the contributions of triglyceride absorptions to the *trans* peak by ratioing the single-beam FTIR spectrum of the fat or oil being analyzed against the single-beam spectrum of a similar reference oil that is free of *trans* groups [24–25, Sedman, J., van de Voort, F. R., and Ismail, A. A. (1996) Upgrading the AOCS IR *trans* Method for Analysis of Neat Fats and Oils by FTIR Spectroscopy, *J. Am. Oil Chem. Soc.*, submitted]. The effectiveness of this ratioing procedure is illustrated in Fig. 11. Figure 11A shows the *trans* absorption region (995–937 cm^{-1}) in the overlaid spectra of a set of standards prepared by adding varying amounts of trielaidin to a *trans*-free soybean oil. The corresponding spectra obtained by ratioing the single-beam spectra of these calibration standards against the single-beam spectrum of the base oil are presented in Fig. 11B, illustrating the horizontal baseline produced by the

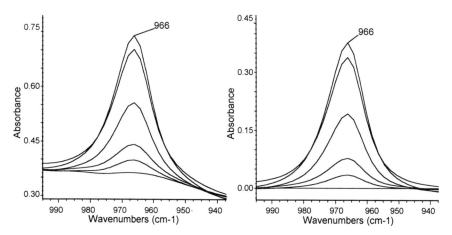

FIG. 11. (A) Overlaid spectra in the *trans* absorption region of calibration standards prepared by addition of various amounts of trielaidin to a *trans*-free base oil; (B) same spectra as in (A) after ratioing out the spectrum of the base oil.

ratioing procedure, which eliminates some of the uncertainty in the measurement of the *trans* peak height. The equations of the standard curves derived from the spectra shown in Figs. 11A and 11B are presented in equations 7 and 8, respectively

$$\% \ trans = -3.917 + 131.276 \ A_u(966) \quad R = 0.999 \quad SE = 0.430 \quad (7)$$

$$\% \ trans = -0.230 + 130.973 \ A_r(966) \quad R = 0.999 \quad SE = 0.398 \quad (8)$$

where

$\% \ trans$ = *trans* content expressed as % trielaidin

$A_u(966)$ = Absorbance @ 966 cm^{-1} relative to a baseline drawn between 995 and 937 cm^{-1} (raw spectrum)

$A_r(966)$ = Absorbance @ 966 cm^{-1} relative to a baseline drawn between 995 and 937 cm^{-1} (ratioed spectrum)

It may be noted from equation 7 that the underlying triglyceride absorption in the spectrum of the *trans*-free oil corresponds to an intercept value of 4% *trans,* whereas the intercept is eliminated from the calibration equation by the ratioing procedure. In early work with double-beam dispersive spectrometers, the similar approach of placing a CS_2 solution containing a *trans*-free oil in the reference beam to null the triglyceride absorptions in the spectrum of the sample was employed (26). However, the FTIR ratioing procedure provides a much simpler and more accurate means of removing these bands from the sample spectrum, and it has been proposed for adoption by the AOCS as Recommended Practice Cd 14b-95 for the quantitation of isolated *trans* isomers at levels equal to or greater than 1% (27).

In their investigation of the ratioing method, Mossoba et al. (24) employed the ATR sampling technique, which requires only that a neat sample be poured onto the surface of the ATR crystal and the spectrum scanned. For the analysis of fats, the ATR crystal was preheated in an oven. Calibration was performed using standards prepared by spiking triolein with varying amounts of trielaidin (0.4 to 44%). The single-beam spectra of these standards were ratioed against the single-beam spectrum of neat triolein, and a calibration plot was obtained by plotting the integrated area of the *trans* band (990–945 cm^{-1}) in these ratioed spectra against concentration. An analogous procedure was followed with methyl elaidate/methyl oleate mixtures to develop a calibration for use in the analysis of methyl esters. The two calibrations derived were employed in the analysis of partially hydrogenated soybean oils and the corresponding methyl esters, respectively, and yielded similar *trans* values, indicating that the ratioing method adequately compensates for the triglyceride absorptions that overlap with the *trans* band in the spectra of oils (24).

Although the ATR sampling technique employed by Mossoba et al. (24) in the work just described is convenient, our experience with heated ATR cells has led us to recommend the use of a flow-through transmission cell for a number of reasons, as outlined in the preceding section on sample handling. We have therefore developed a calibration for the FTIR determination of *trans* content by the ratioing approach for use with the custom-designed sample-handling accessory described in the preceding section on sample handling, operated at 80°C and equipped with a 0.025-mm KCl flow cell [Sedman, J., van de Voort, F. R., and Ismail, A. A. (1996) Upgrading the AOCS IR *trans* Method for Analysis of Neat Fats and Oils by FTIR Spectroscopy, *J. Am. Oil Chem. Soc.*, submitted]. The performance of this calibration was assessed with the seven AOCS Smalley Check Samples from the 1995 "*trans* by IR" series. The spectra of all samples analyzed were ratioed against the spectrum of an unhydrogenated soybean oil (having a *trans* content of <0.1%, as determined by GC), which had been used as the base oil in the preparation of the calibration standards. In assessing the accuracy of the *trans* predictions obtained, the means of the values reported by the participants in the Smalley Check Sample program were used as the reference values. Figure 12 presents a plot of the *trans* values determined by the FTIR ratioing method versus the Smalley means; the values corresponding to ±1 standard deviation (SD) around the Smalley means are also plotted on the *y*-axis to illustrate the spreads around the Smalley means graphically. This plot illustrates that the FTIR ratioing method produces results that fall within 1 SD of the mean values obtained by other laboratories (15 to 18 laboratories, depending on the sample in question) analyzing the same samples using the conventional AOCS IR method.

The main drawback of the ratioing method is the need to select a *trans*-free oil that is similar in composition to the samples to be analyzed. When analyzing partially hydrogenated soybean oils, Mossoba et al. (24) found that the *trans* values obtained by using triolein as the reference oil were 2.6 percentage points higher than those obtained when a refined, bleached, and deodorized soybean oil served as the reference material. They attributed this difference primarily to the greater similarity in composition between the soybean reference oil and the samples being analyzed,

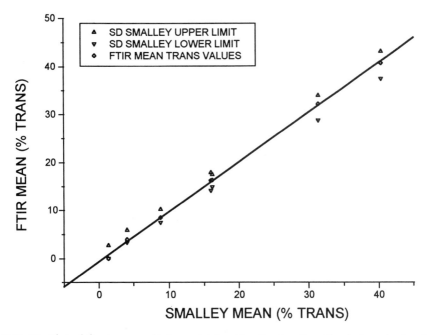

FIG. 12. Plot of the *trans* predictions obtained for the Smalley Check Samples by the FTIR ratioing method vs. the Smalley means. The triangles above and below the regression line represent ±1 SD around the Smalley means.

making the ratioing-out of the overlapping triglyceride absorptions more accurate. Thus, the limitations imposed on the ratioing method by the variability of the triglyceride absorptions among oils of different composition require more thorough examination, and the criteria for selection of an appropriate reference oil for a particular set of samples need to be defined.

An alternative approach to the FTIR prediction of *trans* content that has been investigated by the McGill IR Group (28) entails the development of a PLS calibration model that accounts for the overlapping triglyceride absorptions that contribute to the intensity of the *trans* absorption band. As discussed previously in the "Quantitative Analysis" section, PLS is often the technique of choice for the development of quantitative analysis methods for complex multicomponent systems whose analysis is complicated by overlapping bands and the effects of interactions. Thus, the PLS *trans* method is based on accounting within the calibration model for the triglyceride absorption bands that overlap the *trans* absorption, rather than ratioing out the interfering absorptions; accordingly, this approach eliminates the need for a *trans*-free reference oil. In order to model the variability in the triglyceride absorptions fully, the PLS calibration was based on a set of pure triglyceride standards, including saturated triglycerides (8:0,10:0,...,22:0), *cis*-unsaturated triglycerides (tripalmitolein, triolein, trilinolein, trilinolenin, trieicosenoin, and trierucin), and *trans*-unsaturated triglycerides (tripalmitelaidin, trielaidin, and tribrassidin). The original calibration set also included trilinolelaidin (18:2t,t), but this standard was subsequently removed, because validation studies indicated that it caused the calibration model to overestimate the *trans* content of samples [Sedman, J., van de Voort, F.R., Ismail, A.A., and Maes, P. (1996) Validation of FTIR *trans* and Iodine Value Analyses, *J. Am. Oil Chem. Soc.*, submitted]. In these validation studies, the PLS-predicted values were also found to be affected by baseline fluctuations; sensitivity to baseline variation can be minimized by basing the analysis on second-derivative spectra instead of raw absorption spectra, so the PLS method was recalibrated using the second derivatives of the spectra of the calibration standards. This optimized PLS *trans* method was extensively validated against the ratioing method just described, using more than 100 samples of hydrogenated rapeseed and soybean oil. An excellent concurrence between the two FTIR methods was obtained, as illustrated in Fig. 13, with a mean difference of 0.61% between the paired sets of *trans* predictions. Since these two methods compensate for the spectral interference due to triglyceride absorptions in distinctly different ways, the agreement between them indicates that both approaches successfully do so, allowing for the accurate determination of *trans* content from neat fats and oils without the need for their conversion to methyl esters.

In the foregoing validation study, the *trans* predictions obtained from both FTIR methods were also compared to the *trans* values derived from capillary GC analysis, which was performed for approximately one-third of the 100 validation samples. The GC *trans* data were linearly related to the FTIR *trans* data, whether obtained by the PLS or the ratioing method, by a factor of 0.88, and accordingly there was an increasing discrepancy between the GC and FTIR data as the *trans* content of the samples increased. Such discrepancies between GC and IR *trans* determinations are attributable to the unsatisfactory separation of 18:1t and 18:1c isomers on the currently available GC stationary phases (29). To evaluate the accuracy of IR *trans* data, Ratnayake and Pelletier (30) employed the combined procedure of silver nitrate thin-layer chromatography (which affords complete separation of 18:1t and 18:1c isomers) and capillary gas chromatography as the reference method. These authors found that the IR *trans* predictions for margarine samples, analyzed as methyl esters by the AOCS method, were about 16% lower than those obtained by the TLC/GC procedure. However, when the IR method was calibrated using a fatty acid methyl ester mixture derived from a partially hydrogenated canola oil, instead of methyl elaidate, the average difference between the IR and TLC/GC methods was only about 3%. Ratnayake and Pelletier (30) attributed this finding to the fact that the traditional calibration approach using a single *trans* isomer (i.e., 9t-18:1) does not account for differences in IR absorptivity among the various *trans* isomers present in partially hydrogenated vegetable oils (PHVO). These authors thus proposed that for IR *trans* analysis of PHVO a more suitable approach would be to base the calibration on standards prepared from a well-characterized PHVO.

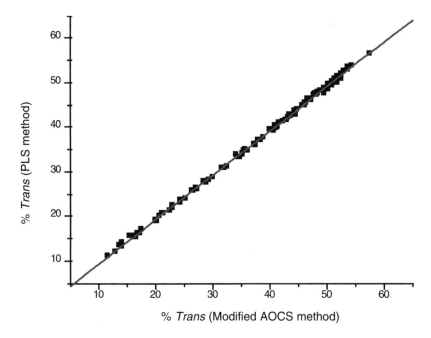

FIG. 13. Plot of % *trans* values predicted for hydrogenated rapeseed and soybean oils from a PLS calibration vs. the % *trans* values obtained by the FTIR ratioing method.

In summary, IR *trans* analysis has been dramatically simplified by recent developments in sample handling and analytical methodology, made possible by exploiting the advantages of FTIR spectrometers. With an FTIR spectrometer, high-quality spectra can be recorded from neat fats and oils, and the triglyceride absorptions overlapping the *trans* peak can be either eliminated by a spectral ratioing procedure or compensated for through a PLS-based calibration. Thus, the sample preparation steps of the AOCS IR *trans* method (i.e., conversion to methyl esters, dissolution in a solvent) are not necessary, making it possible to carry out the complete analysis in under 2 min/sample. The *trans* peak height or area can be accurately measured from the FTIR software, and the analysis can be standardized and automated by programming of the spectrometer (23). All these improvements have been encompassed in a *trans* analysis protocol developed by the McGill IR Group, with the objective of upgrading the IR method that is currently widely employed in the fats and oils industry.

Bulk Characterization of Fats and Oils: cis *and* trans *Content, Iodine Value, and Saponification Number*

In the previous section a calibration set was described that consisted of 18 pure triglyceride standards, employed in the development of a PLS calibration model for the prediction of the *trans* content of fats and oils. The McGill IR Group has employed the same calibration set to develop calibration models for the prediction of *cis* content and total unsaturation [iodine value (IV)], as well as saponification number (SN), a measure of weight-average molecular weight (15,28). The spectra of three of these calibration standards are presented in Fig. 14, illustrating the major absorption bands that are related to *cis* and *trans* content, IV, and SN. The use of these pure triglycerides as calibration standards has several advantages. First, the calibrations devised are "universal," as they are applicable to all triglyceride-based oils and fats. Second, this calibration approach has the benefit of eliminating the need for chemical analyses of the calibration standards, because the reference values for the pure triglycerides are known from their molecular structure; therefore, the accuracy of the IR method is not limited by the precision of a reference chemical method. Another key feature of our calibration approach is the use of PLS to establish correlations between the spectral data for these calibration standards and the corresponding reference values, as opposed to the traditional approach of attempting to identify a single absorption band whose height or area can be related to the measure of interest. For example, other IR methods that have been described in the literature as measuring total unsaturation are based on a simple calibration equation relating the extent of unsaturation to the height of the olefinic C-H stretching absorption (31) or the ratio of the height of this band to that of a CH_2 stretching absorption

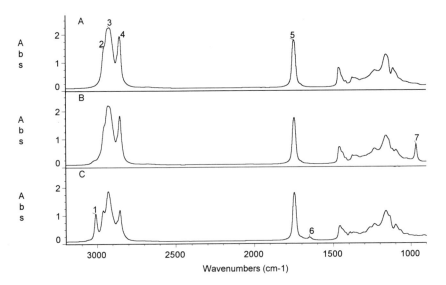

FIG. 14. FTIR spectra of tripalmitin (A), trielaidin (B), and trilinolenin (C), recorded in a 0.025-mm transmission cell. The labeled peaks are characteristic functional-group absorption bands: (1) CH stretch in *cis*-HC=CH; (2) CH_3 stretch; (3) CH_2 stretch; (4) CH_2/CH_3 stretch; (5) ester linkage C=O stretch; (6) *cis*-C=C stretch; (7) *trans*-C=CH bend.

(32,33). However, owing to the different absorptivities and slight differences in the position of the olefinic C-H stretching absorption in the spectra of *cis* and *trans* isomers, these methods do not accurately measure total unsaturation in hydrogenated oils or other samples containing significant amounts of *trans* isomers. These types of calibration approaches could, in principle, be extended to account for the contribution of *trans* isomers to total unsaturation via the application of multiple linear regression techniques. However, the McGill IR Group has adopted the alternative approach of exploiting the powerful capabilities of PLS regression to correlate spectral features to quality attributes (as discussed earlier in this chapter), given that total unsaturation or iodine value (IV), which is a chemical measure of total unsaturation (defined as the number of centigrams of iodine absorbed per gram of sample), is a quality attribute in the sense that the term is used here.

Our initial PLS calibration models for the determination of SN and IV were developed using the ATR sampling technique and validated against the official AOCS method with a variety of oil types (15). Although a horizontal ATR sampling accessory provides a rapid and convenient means of analyzing fats and oils, as samples are simply poured onto the surface of a heated horizontal ATR crystal (Fig. 6), it did not lend itself readily to automation; in addition, the analytical reproducibility and accuracy were highly dependent on precise temperature control and adequate cleaning of the crystal between samples. For this reason, new calibrations were subsequently developed for use with the transmission cell–based sample-handling accessory already described. PLS calibration models were simultaneously developed (28) for the prediction of *cis* content, expressed as percent triolein, as well as *trans* content, as described in the preceding section. In validation studies with more than 100 hydrogenated rapeseed and soybean samples, excellent internal consistency was obtained between the IV and *cis* and *trans* data predicted from these calibrations [Sedman, J., van de Voort, F.R., Ismail, A.A., and Maes, P. (1996) Validation of FTIR *trans* and Iodine Value Analyses, *J. Am. Oil Chem. Soc.*, submitted]. In addition, for ~30 GC-analyzed samples, the PLS IV predictions matched the IV calculated from the GC data within 1 IV unit (Fig. 15). These results, taken together with the close agreement between the *trans* values predicted from the PLS calibration model and those obtained by the modified (ratioing) AOCS FTIR method, as described in the preceding section, provide strong experimental evidence that IV, *cis* content, and *trans* content are all predicted accurately from their respective PLS calibration models. The SN calibration was validated with a set of 37 oil samples of different types (15), covering a wide range (185–253) of SN values, as the rapeseed and soybean validation samples did not span a sufficient range of SN values to allow an assessment of the predictive accuracy of this calibration. The validation data for the accuracy of the FTIR SN method relative to the chemical method (mean difference = ~2.7 SN units, standard deviation of the differences = ~2.0 SN units) were considered satisfactory, in view of the poor reproducibility of the chemical analyses (the standard deviation of the differences for duplicate analyses being 2.050 vs. 0.333 for the FTIR method).

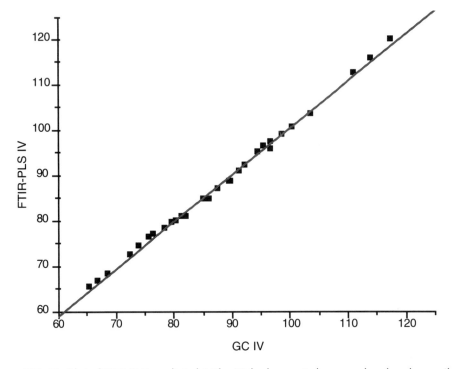

FIG. 15. Plot of FTIR-PLS predicted IV for 31 hydrogenated rapeseed and soybean oils vs. IV calculated from GC data.

With the availability of FTIR methods for the determination of SN, IV, and *cis* and *trans* content, FTIR spectroscopy has the potential to be employed routinely for the bulk chemical characterization of fats and oils. In order to demonstrate the routine implementation of FTIR oil analysis methodology, the McGill IR Group has converted an FTIR spectrometer into an automated and dedicated edible oil analyzer, capable of providing a total analysis (SN, IV, % *cis*, % *trans*) of a neat fat or oil sample in 1 to 2 min (34,35). A detailed description of this system will be presented in a later section of this chapter.

Solids Content of Fats

Another potentially important area of application of FTIR spectroscopy that has been investigated by the McGill IR Group is the determination of the solids content of fats (36). The solids content of fats as a function of temperature has an important bearing on the functional characteristics of margarines, shortenings, and other fat blends; therefore, solids content is an important quality control parameter in the edible fats and oils industry. In North America, solids content determinations have traditionally been performed using dilatometry to obtain the solid fat index (SFI), an empirical measure of the change in the specific volume of the fat as a function of temperature (37). In Europe, solid fat content (SFC), based on NMR spectroscopy, has been widely used as the preferred method of determining fat solids, and it has recently been approved as an official method by the AOCS (38). Both the SFC and SFI procedures involve measurements at a series of set temperatures and are fairly lengthy because tempering of the sample at each temperature is required in order to obtain reproducible values. In contrast, the FTIR method that has been proposed by the McGill IR Group as an alternative to the SFI method allows a four-temperature SFI profile of a fat to be predicted from a single FTIR measurement on the neat, melted sample (36).

The underlying principle of SFI determination by FTIR spectroscopy is that the SFI profile of a fat is defined by its fatty acid composition and distribution, which in turn is characterized by the mid-IR spectrum of the melted fat, as the spectrum represents the superposition of all the contributions of the individual triglycerides making up the fat. Calibration is thus based on the development of individual PLS calibration models relating the spectral features of melted fat samples to their known dilatometric SFI values obtained at each of the four common temperatures of measurement (50, 70, 80, and 92°F; or 10, 21, 27, and 33°C). By employing this approach, an FTIR SFI method was developed using 72 samples of partially hydrogenated soybean oil from 11 hydrogenation runs, obtained from a major U.S. vegetable oil processor and preanalyzed for SFI by the AOCS dilatometric method. Half of these samples were employed as calibration standards, with the remainder serving as validation samples. The calibration statistics for the PLS models derived to predict the dilatometric SFI data at each of the four temperatures are presented in Table 1 (36). Relatively few factors ($n = 6$–10) were required to account for the variation in the SFI data, and "leave-one-out" cross-validation of the four calibrations yielded an average root mean square error of ±0.71 SFI units, which is only slightly higher than the generally accepted value for the reproducibility of the dilatometric method (~±0.50). Note in Table 1 that the number of calibration spectra employed in deriving the SFI92 calibration model is lower than in the other three cases. The reason is that over one-third of the calibration standards had no measurable solids at the highest measurement temperature; these standards had to be excluded from the SFI92 calibration set, because it is not possible to model zero-

TABLE 1
Leave-One-Out Cross-Validation Data for the SFI Calibrations Derived for 50, 70, 80 and 92°F

Calibration	No. of spectra	R^2	RMSE[a]	Number of PLS factors
SFI50	34	0.984	0.905	8
SFI70	34	0.973	0.627	5
SFI80	34	0.944	0.580	10
SFI92[b]	22	0.910	0.730	6

[a]Root mean square error from the PRESS test.
[b]Calibration based on nonzero SFI data only.

SFI data by PLS—one would be forced to relate varying spectral information to a constant. Another consequence of this inability to model zero-SFI data is that the PLS calibration models developed predict small *negative* SFI values for samples having a dilatometric SFI value of zero. Based on duplicate analyses of the validation samples one week apart, the accuracy and reproducibility of the FTIR SFI method were determined to be ±0.60 and ±0.38 SFI units, respectively (36). Plots of the FTIR-predicted SFI vs. the dilatometric SFI data at the four temperatures are presented in Fig. 16. Similar results were subsequently obtained by calibrating against NMR SFC data (van de Voort, F.R., Sedman, J., Ismail, A.A., and Maes, P., manuscript in preparation). Thus, the FTIR method has the potential to serve as a viable substitute for either the traditional dilatometric or the NMR method for the determination of solids content, with the advantage of a reduction in the analysis time from hours to minutes by the elimination of the tempering steps required in these procedures.

As an alternative approach, the relationship between solids content and the SN, *cis,* and *trans* data obtained from the PLS calibration models described in the

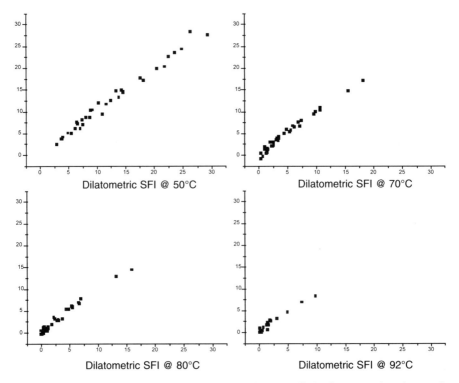

FIG. 16. Validation plots of FTIR-predicted SFI for partially hydrogenated soybean oils vs. dilatometric SFI measured at 50, 70, 80, and 92°F (or 10, 21, 27, and 33°C) (from Ref. 36).

preceding section was also investigated. It is well known that SN and the degree and form of unsaturation have an important bearing on the physical state of a lipid; for example, increasing unsaturation reduces the melting point, with the relative amounts of the *cis* and *trans* forms having opposing effects, while an increase in the weight-average molecular weight (lower SN) tends to increase the melting point if the *cis/trans* ratio is constant. Thus, multiple linear regression (MLR) equations relating SFI at each of four temperatures to the FTIR-predicted values of SN, *cis* content, and *trans* content were derived for the same sets of hydrogenated samples as employed in the PLS calibration. It was concluded that although the SN, *cis*, and *trans* data in combination could be related to SFI for samples from a single hydrogenation time course, no simple quantitative relationship could be formulated when samples of the same oil from different hydrogenation runs were combined (36). However, in subsequent investigations it was found that the use of additional terms in the MLR equations involving various transforms (square and natural logarithm) of SN, *cis*, and *trans* data provided a good fit to SFI for the combined sets of samples. The advantage of this MLR approach is that the calibration is more straightforward to perform than the development of PLS calibration models and also appears to be more robust (van de Voort, F.R., Sedman, J., Ismail, A.A., and Maes, P., manuscript in preparation).

One of the limitations of the FTIR approach to the determination of solids content is that both the PLS calibration models and the MLR equations derived are applicable only to samples with similar characteristics to those of the standards used to derive the calibration. Therefore, a separate calibration must be developed for each type of oil or blend. In this context, it is important to note that the calibrations that we have derived in both our SFI and SFC development work have been based on sets of calibration standards in which there is an interrelationship between the solids content at the lowest temperature and those at subsequently higher temperatures, as illustrated in Fig. 17. Based on this plot, it is not surprising that the solids data at different temperatures can be related to spectral data obtained at a single temperature. Conversely, it may be inferred that if such an interrelationship does not exist, an FTIR calibration may be difficult to obtain.

Measurements of Oxidative Status and Stability

Determination of peroxide value. Autoxidation is a major deteriorative reaction affecting the quality of fats and oils and is initiated by a variety of mechanisms (light, heat, metal ions, etc.) when oxygen is present. In the initial stages of the reaction, the main products are hydroperoxides, which subsequently undergo further degradation to form volatile short-chain oxygenated molecules, which produce characteristic rancid off-flavors (39). Since hydroperoxides are the primary products formed as autoxidation commences and serve as precursors to the subsequent formation of secondary oxidation products, their presence and rate of change are important indicators of oil quality and potential shelf life (40). The AOCS iodometric method for the determination of peroxide value (PV) is the standard method

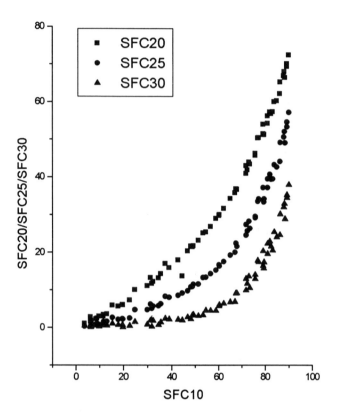

FIG. 17. Plot illustrating the interrelationship between SFC values at four temperatures, as determined by NMR, for a set of 71 partially hydrogenated soybean oils.

used in the edible oil industry to determine hydroperoxides in freshly processed oils after processing, usually after deodorization, which thermally degrades residual hydroperoxides and strips out low-molecular-weight aldehydes and free fatty acids under vacuum. High-quality edible oils usually have PV < 1.0. PV determinations are also used extensively by processors to monitor the oxidative status of bulk oils in storage, to test the efficacy of antioxidants, and, in conjunction with the active oxygen method (41), to predict the shelf life of edible oils.

The standard AOCS PV determination is based on the stoichiometric release of molecular iodine by hydroperoxides when exposed to KI in an acidic environment, the reaction converting the hydroperoxides to alcohols (42). The molecular iodine released is complexed with soluble starch, which acts as an indicator, and the iodine is quantitated by titration with sodium thiosulfate. Based on the stoichiometry of the two reactions, the hydroperoxide concentration can be calculated and is commonly expressed as milliequivalents of hydroperoxide per kilogram of fat (meq/kg). The PV test, although empirical, is relatively simple, reasonably sensitive, reliable, and reproducible when carried out under standardized conditions. On the other hand, it is labor-intensive and uses reagents and solvents (KI and chloroform or isooctane) that are increasingly difficult to dispose of and considered hazardous (43).

It has been recognized for some time that hydroperoxide functional groups can be quantitatively determined by IR spectroscopy via measurement of their O-H stretching absorption, which is observed at 3550 cm^{-1} in the spectra of fatty acid methyl ester hydroperoxides in CCl_4 solution (44) but is shifted to 3444 cm^{-1} and broadened in the spectra of oxidized oils owing to hydrogen bonding between the -OOH and the triglyceride ester C=O groups (45). Although measurement of the hydroperoxide absorption band in the spectra of neat oils can serve as a basis for the determination of PV by FTIR spectroscopy (45), the development of an FTIR method is complicated by the need to account for spectral interferences due to other OH-containing species that may be present in oils, such as alcohols, mono- and diglycerides, free fatty acids, and water, all of which exhibit O-H stretching absorptions that overlap with the hydroperoxide band. Measurements of the height or area of the hydroperoxide absorption band are also affected by overlap with a triglyceride absorption band at ~3473 cm^{-1}, assigned to the first overtone of the ester carbonyl absorption at 1748 cm^{-1}, whose intensity varies among oils of different triglyceride compositions.

Because so many potential sources of interference and variability affect the hydroperoxide absorption band, the McGill IR Group explored a PLS calibration approach in the development of an FTIR method for the determination of PV (45). Calibration standards were prepared by the addition of various amounts of *t*-butyl hydroperoxide (TBHP), a commercially available and shelf-stable hydroperoxide that is spectroscopically representative of lipid hydroperoxides, to a variety of base oils (45). Varying amounts of moisture and oleic acid were also added to the calibration standards, and the spectral contributions of mono- and diglycerides were modeled by the use of spectral coaddition techniques. Although the PLS calibration model developed using these standards accounted for these various sources of spectral interferences and spectral variability, the inherent complexity of the OH stretching region in the spectra of edible oils (due to both extensive band overlap and hydrogen bonding effects) and the need to extract a weak signal (owing to the low concentrations of hydroperoxides to be measured) from this complex spectrum make it difficult to develop a robust, "universal" calibration. Furthermore, the detection limit of this FTIR method, estimated to be 1.5 PV, is inadequate for the measurement of PV in freshly processed oils (PV < 0.5–1.0).

For these reasons, the McGill IR Group subsequently developed a simpler method that does not suffer from these limitations [Ma, K., van de Voort, F.R., Sedman, J., and Ismail, A.A. (1996) Stoichiometric Determination of Hydroperoxides in Fats and Oils by FTIR Spectroscopy, *J. Am. Oil Chem. Soc.*, submitted]. This method is based on the rapid reaction between hydroperoxides and excess triphenylphosphine (TPP), which leads to the formation of triphenylphosphine oxide (TPPO) in stoichiometric amounts:

TPP   TPPO

This reaction had previously been successfully employed by Nakamura and Maeda (46) in a microassay for lipid hydroperoxides in biological samples, using a combination of HPLC and UV detection. In our laboratory, this reaction had been used as a means of removing hydroperoxides from oil samples (45), the reaction being rapid and complete when an excess of TPP was present. The utility of this reaction for the determination of PV by FTIR spectroscopy is dependent on the ability to quantitate TPPO accurately in the presence of TPP. This proved to be readily achievable, owing to the presence of a unique and sharp band at 542 cm^{-1} in the spectrum of TPPO (dissolved in hexanol and added to canola oil; Fig. 18B), which is assigned to an X-substituent-sensitive phenyl vibration (47); the corresponding band in the spectrum of TPP is broad and shifted ~40 cm^{-1} to lower frequency (Fig. 18A). Figure 18C shows a spectrum of canola oil spiked with *t*-butyl hydroperoxide (TBHP), which has been ratioed against the spectrum of canola oil to eliminate the absorption bands of the oil; the corresponding spectrum obtained after addition of excess TPP to the sample is presented in Fig. 18D, illustrating the appearance of the sharp band at 542 cm^{-1} due to the formation of TPPO.

For this FTIR PV method, a simple univariate calibration was found to be suitable. Calibration standards were prepared by spiking canola oil (PV < 0.10, as measured by the AOCS method) with various amounts of a TPPO/hexanol stock solution (35% w/w) to produce TPPO concentrations in the oil corresponding to those that would result from the complete reaction between TPP and the hydroperoxides in oils having a PV in the range 0–15. A calibration equation relating PV to the height of the TPPO band at 542 cm^{-1}, measured relative to a single baseline point at 550 cm^{-1}, was derived by simple linear regression ($R = 0.9999$, SD = 0.056 PV), and the calibration was validated by analyzing both oxidized oils and oils spiked with TBHP. The standardized analytical protocol developed for this PV method consists of adding ~0.2 g of a TPP/hexanol stock solution (33% w/w) to ~30 g of melted fat or oil, shaking the sample, and scanning its spectrum in a 0.010-mm KCl transmission cell maintained at 80°C, the total analysis time being about 2 min/sample. With a detection limit of 0.10 PV and excellent reproducibility (±0.18 PV), this simple and rapid FTIR method is highly suited for routine quality control applications in the fats and oils industry.

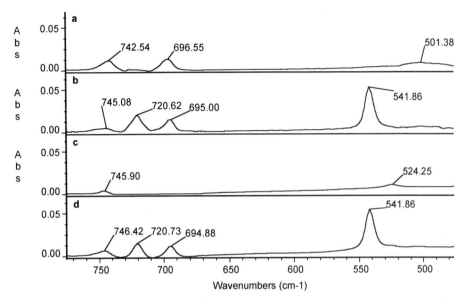

FIG. 18. Differential spectra obtained by ratioing the spectra of canola oil spiked with triphenylphosphine (A), triphenylphosphine oxide (B), t-butyl hydroperoxide (C), and t-butyl hydroperoxide + triphenylphosphine (D) against the spectrum of canola oil.

Determination of anisidine value. Oxidative rancidity in fats and oils is the result of the breakdown of hydroperoxides, the primary products of autoxidation, to a variety of secondary oxidation products, such as short-chain aldehydes, ketones, fatty acids, alcohols, and hydrocarbons (39). The anisidine value (AV) test quantitates aldehydes that are important in the development of rancid off-flavors and is a widely employed AOCS method (48). The AV test is based on UV detection of the products formed by the reaction between *p*-anisidine and aldehydes. However, all aldehydes do not contribute equally to the AV, as the molar absorptivity of the *p*-anisidine/aldehyde reaction products varies with aldehyde type (see equation 9 below). Thus, the AV test is particularly sensitive to α,β-unsaturated and especially $\alpha,\beta,\delta,\gamma$-unsaturated aldehydes (48).

Aldehydes exhibit strong bands in the 1730–1680 cm^{-1} region of the IR spectrum, owing to the high absorptivity of their C=O stretching vibrations, and the formation of aldehydes is readily observed in the FTIR spectra of oils undergoing oxidation (Fig. 19), with individual peaks due to saturated, α, β-unsaturated, and $\alpha,\beta,\delta,\gamma$-unsaturated aldehydes being discernible (49). On the basis of the ability to distinguish between these three classes of aldehydes, the McGill IR Group formulated a synthetic calibration approach to the FTIR determination of AV involving individual quantitation of hexanal, *trans*-2-hexenal, and *trans, trans*-2,4-decadienal, these compounds having been selected to represent the three aldehyde classes (50). A PLS calibration model for the prediction of the concentrations of these aldehydes in oils

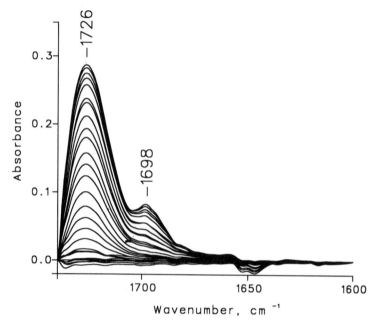

FIG. 19. Spectral overlay plot illustrating changes in the 1730–1600 cm^{-1} region of the FTIR/ATR spectrum of safflower oil heated to 76°C on the surface of a ZnSe ATR crystal as a function of time. The overlaid spectra were recorded at 30-min intervals and have been ratioed against the spectrum recorded at $t = 0$. The peak maxima at 1726 and 1698 cm^{-1} are assigned to the C=O stretching absorptions of saturated and α,β-unsaturated aldehydes, respectively; the shoulder on the low-frequency side of the 1698 cm^{-1} band is due to α,β,δ,γ-unsaturated aldehydes (from Ref. 49).

was developed by using a set of 32 synthetic calibration standards prepared by spiking canola oil with varying amounts of the three compounds, as well as with random amounts of other compounds representative of oxidation by-products, in order to model potential spectral interferences in oxidized oils. Such spectral interferences include, for example, the broadening of the triglyceride ester carbonyl absorption, centered at ~1748 cm^{-1}, by hydrogen bonding between the ester carbonyl groups and hydroperoxides or alcohols, such that this band obscures the C=O absorption band of hexanal, centered at ~1726 cm^{-1}. As a result of this band overlap, it proved necessary to base the quantitation of hexanal on its characteristic but weak aldehydic C-H stretching bands in the 2820–2700 cm^{-1} region, whereas the carbonyl absorptions of *trans*-hexenal and *trans, trans*-2,4-decadienal (1698 and 1689 cm^{-1}) are far enough away from the broadened triglyceride band to be useful for quantitation (50).

The calibration standards employed in developing the PLS calibration model were analyzed by the AOCS AV method, and the following relationship between the chemically determined AV and the gravimetrically added amounts of the three aldehydes was obtained by multiple linear regression (50)

$$AV_c = 0.34S + 2.16U + 15.36C \quad R^2 = 0.997 \quad SE = 1.37 \quad (9)$$

where
AV_c = chemical AV
S = saturated aldehydes (mol/g)
U = α,β-unsaturated aldehydes (mol/g)
C = $\alpha,\beta,\delta,\gamma$-unsaturated aldehydes (mol/g)

This equation was then employed to convert the PLS-predicted aldehyde concentrations of thermally stressed canola oil samples to "apparent" AV. Regression of the "apparent" AV values (AV_{app}) obtained against the chemically determined AV values of these samples yielded the following equation (50):

$$AV_c = 1.08 + 0.953\ AV_{app} \quad R^2 = 0.994 \quad SE = 1.65 \quad (10)$$

These results indicate that the synthetic calibration approach yields good predictive accuracy, as the SE in equation 10 is only slightly higher than that in equation 9, which is a measure of the accuracy of the chemical method. Similar predictive accuracy was obtained from an alternative calibration approach based on the use of thermally stressed oils as calibration standards (50). As such, quantitative determination of AV by FTIR spectroscopy was shown to be feasible, and the synthetic calibration approach provided additional information on the aldehyde types present in a sample. This study provided the basis for the development of a rapid, automated FTIR method for AV analysis of thermally stressed fats and oils in their neat form without the use of chemical reagents, with possible application in the monitoring of the oxidative state of frying oils.

Oil oxidation monitoring. FTIR spectroscopy has been demonstrated to provide a simple, convenient, and dynamic means of monitoring the chemical changes taking place as an oil undergoes oxidation (49). The practical utility of FTIR oil oxidation-monitoring techniques is exemplified by experiments designed to evaluate the relative effectiveness of antioxidants in retarding oxidative degradation (51). In this type of experiment, an oil sample containing an antioxidant was spread on the surface of a heated (65°C) horizontal ATR crystal, where the oil undergoes accelerated oxidation due to the combination of the elevated temperature and the relatively large surface area of the sample exposed to air (see Fig. 6). FTIR spectra were automatically recorded at preset time intervals and ratioed against the spectrum recorded at $t = 0$. This spectral ratioing technique allows one to detect very small changes in the FTIR spectrum of the oil as a function of time, as all unvarying spectral features are eliminated from the spectrum by the ratioing procedure. The spectral changes observed were primarily the growth of absorption bands associated with the formation of hydroperoxides and aldehydes and the progressive decrease in intensity of the characteristic absorption bands of *cis* double bonds. A dynamic plot of absorbance values at selected wavelengths vs. time was displayed on screen during the course of the experiment, thus allowing real-time monitoring of oil oxidation. Superposition of the time-course plots obtained in this manner for oils containing different antioxidants, or the same antioxidant at different levels, provided a graphical representation of relative antioxidant performance. As an example, Fig. 20 shows typical time-course oxidation plots obtained for menhaden oil, a highly unsaturated fish oil, containing the primary chain-breaking antioxidant butylated hydroxyanisole (BHA) at three different levels. In this figure, the peak height at 3444 cm^{-1} (corresponding to the maximum of the hydroperoxide absorption band) is plotted vs. time of heating, clearly demonstrating the prolongation of the induction period with increasing antioxidant concentration. On the basis of the results of this work, FTIR/ATR differential spectroscopy was suggested to be a simple means by which to assess the efficacy of antioxidants (51).

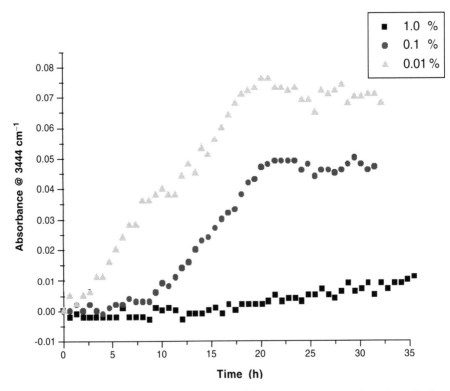

FIG. 20. Time-course oxidation plots obtained by FTIR/ATR monitoring of menhaden oil containing butylated hydroxyanisole (BHA) at levels of 0.01% (△), 0.1% (●), and 1% (■) at 65°C, showing the growth of the hydroperoxide absorption band (3444 cm^{-1}) as a function of time (from Ref. 51).

Implementation of FTIR Quantitative Analysis Methods in the Fats and Oils Industry

The fats and oils industry continues to rely largely on wet-chemical and chromatographic methods for its routine analytical requirements. However, there is a growing interest within the industry in replacing these well-established methods with automated instrumental methods in order to improve efficiency and to address increasing environmental concerns about the use of large volumes of solvents and reagents in quality control laboratories (52). In this context, FTIR spectroscopy has the potential to quickly become an important routine analytical tool in the fats and oils industry, as FTIR analysis can be performed rapidly on oils and melted fats without any sample preparation. As has been described in this chapter, FTIR spectroscopy can be used both to determine bulk properties of fats and oils and to monitor chemical changes taking place, for example, as an oil is hydrogenated or undergoes oxidation. Thus, a large number of the chemical methods that are routinely used in fats and oils analysis can potentially be replaced by FTIR methods. However, there is little familiarity with this technique in the industrial sector, and widespread application of FTIR spectroscopy for the analysis of edible fats and oils is more likely to occur if the industry is provided with practical methods that can be directly implemented and do not impose new problems or constraints or an extensive learning curve. Thus, in order to facilitate implementation of FTIR oil analysis methodology, a number of issues must be addressed, specifically in regard to sample handling, calibration stability/transfer, and automation. As a discussion of sample handling and a description of a sample handling accessory designed specifically for use in fats and oils analysis have been presented in a previous section of this chapter, only the latter two aspects will be considered here.

Calibration Stability and Calibration Transfer

Once a calibration for a particular FTIR analytical method has been developed and validated, it can be used for the prediction of unknowns, provided that two general conditions are met. These conditions may be formally stated as follows:

1. The spectra of the unknowns must be well represented by the spectra of the calibration standards within the spectral region(s) employed for prediction.
2. The spectra of the unknowns must be recorded under the same conditions as employed in the calibration step.

The first of these conditions can be met only during the method development stage and requires that the design and validation of the calibration have been based on a detailed understanding of the sample population to be analyzed. The need to ensure that condition 2 is met over time defines the problem of maintaining calibration stability; the difficulty of fulfilling this condition when a calibration developed on one instrument is transferred to one or more secondary instruments imposes severe limitations on calibration transfer.

The problems of maintaining the predictive accuracy of a calibration over time and transferring a calibration between instruments have gained attention with the increasing use of multivariate calibration techniques, particularly PLS, for several reasons. First, the need for recalibration is more problematic in multivariate calibration than in univariate calibration, because of the larger number of calibration standards and the greater amounts of time, effort, and expertise usually required. Second, because multivariate calibrations use much more of the available spectral information than simple univariate (i.e., Beer's law) calibrations, seemingly minor changes in instrument performance can have a major effect on their predictive accuracy (53,54). Similarly, multivariate calibrations are more susceptible to even minor spectral perturbations caused by the presence of interfering components in the samples to be analyzed that are not accounted for in the calibration standards; therefore, condition 1 above is a more stringent condition than in univariate calibration.

The key to maintaining the predictive accuracy of a calibration is to understand and control the factors affecting calibration stability. In FTIR spectroscopy, many potential instrumental sources of calibration instability are well controlled. Because of the high wavelength precision of FTIR spectrometers, due to the use of an internal reference laser, the possibility of wavelength drifts over time—a major concern with dispersive instruments—is eliminated. Although there are a number of other possible causes of instrumental drift (e.g., changes in the alignment of the optics, temperature fluctuations), most of these are compensated for when the single-beam FTIR spectrum of the sample is ratioed against a background spectrum in order to produce the absorbance spectrum of the sample as described earlier in this chapter. Thus, an important means of maintaining calibration stability is regular collection of the background spectrum, ideally immediately before (or after) collection of the sample spectrum. The ratioing operation is also critical in minimizing the spectral contributions of IR absorption by water vapor and carbon dioxide in the path of the infrared beam. Water vapor gives rise to two series of sharp peaks in the IR spectrum, between 3900 and 3600 cm^{-1} and 1800 and 1500 cm^{-1}, while CO_2 gives rise to a doublet with maxima at 2362 and 2340 cm^{-1}, and these absorptions, if they are not properly ratioed out of the sample spectrum, can give rise to serious errors in quantitation. These spectral contributions will be adequately removed by the spectral ratioing procedure only if the relative humidity and CO_2 levels in the air in the optical path do not change significantly between the measurement of the sample and the background spectrum, as illustrated in Fig. 21. For this reason, FTIR spectrometers are usually continually purged with dry, CO_2-free air or dry nitrogen, although some are sealed and desiccated units. In either case, it is strongly advisable to collect a background spectrum regularly for use in ratioing. This recommendation is most easily met by the use of a sample shuttle, which allows the cell to be moved out of the optical path, and recent work by Dwight Analytical (Toronto, Canada) has resulted in the redesign of the sample-handling accessory described earlier in this chapter to implement this feature (Fig. 22), making background collection a simple matter, as it does not require removal of the cell or disruption of the purge within the optical compartment.

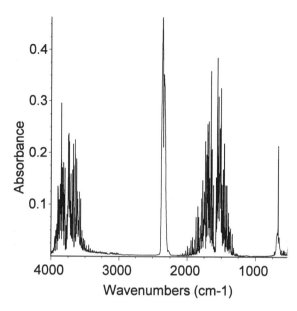

FIG. 21. Water vapor and carbon dioxide absorption bands. The spectrum was obtained by ratioing an open-beam spectrum recorded with the sample compartment open to the atmosphere against a well-purged background spectrum.

Another important consideration in maintaining calibration stability is the need to compensate for any changes in the pathlength of the IR cell that may occur over time—for example, as a result of erosion of the cell windows or buildup of a film on the surface of the windows. In the case of edible-oil analysis at elevated temperatures, such a film is readily produced, particularly when samples are left in the heated cell for prolonged periods of time, and generally cannot be removed except by disassembling

FIG. 22. Photograph of the heated sample-handling accessory placed on a sample shuttle.

the cell and polishing the windows. Since Beer's law (equation 5) is the underlying basis of any FTIR quantitative analysis method, any change in the cell pathlength from that on which the calibration is based will lead to biased predictions. Furthermore, because of the narrow pathlengths of IR cells, even a 0.5-µm change in pathlength can have a significant effect on predictive accuracy. In order to avoid the need for recalibration whenever such a change in pathlength occurs or whenever a cell has to be replaced, a pathlength correction factor may be applied to the analytical results. Such a correction factor is generated by recording the spectrum of an appropriate reference standard under the present analytical conditions and comparing it to a spectrum of the same material that was recorded when the spectra used in developing the calibration were acquired.

On the basis of the above discussion, it can be seen that calibration stability can, in principle, be maintained indefinitely on an FTIR spectrometer operated under proper conditions, with the use of software-implemented check routines and pathlength correction algorithms. The possibilities for transfer of a calibration from one FTIR spectrometer to another are less clear-cut, as they depend on a number of factors. These include the type of calibration to be transferred (e.g., univariate vs. multivariate, narrow spectral region vs. broad spectral regions), the extent to which the two instruments differ (e.g., same model, different models/same manufacturer, or different manufacturers), and the type of environment in which the instruments are operated (e.g., laboratory vs. in-plant). At least one FTIR manufacturer claims calibration transferability as a feature of their instruments, and the McGill IR Group has successfully demonstrated the transfer of both peak height–based and PLS calibrations for the prediction of the *trans* content of fats and oils between different spectrometer models produced by another FTIR manufacturer [Sedman, J., van de Voort, F. R., and Ismail, A. A. (1996) Upgrading the AOCS IR *trans* Method for Analysis of Neat Fats and Oils by FTIR Spectroscopy, *J. Am. Oil Chem. Soc.*, submitted]. On the other hand, it can be anticipated that calibration transfer between instruments made by different manufacturers will generally be problematic, owing to differences in both hardware and software; for example, although FTIR spectrometers have excellent wavelength precision, their wavelength accuracy is affected by the alignment of the source and interferometer optics, unless the manufacturer has performed a wavelength calibration and made appropriate software adjustments. Scanning a reference standard on two different instruments in the same cell under identical conditions and superimposing the spectra obtained or subtracting them from each other provides a good indication of whether calibration transfer is likely to be successful. In principle, if the difference spectrum obtained by subtracting the two spectra contains only spectral noise, it should be possible to transfer calibrations between the two instruments (54). Even with slight differences between the spectra, calibration transfer may still be possible, with the predictions obtained on the secondary instrument having a constant offset or slope, or combination thereof, relative to those obtained from the reference calibration. In this case, the calibration on the secondary instrument can be corrected mathematically to match that on the reference instrument. As such, calibration transfer between instruments is a reasonable possibility but cannot be guaranteed in all cases.

Apart from instrumental considerations, a major limitation on calibration transfer is the requirement that the calibration be representative of the samples to be analyzed. Thus, calibrations developed through the use of actual samples as calibration standards may have very limited transferability. For example, the SFI calibrations described in this chapter are based on samples of a particular type of oil that have been hydrogenated under various conditions; accordingly, these calibrations cannot necessarily be transferred to determine the SFI profile of a different oil (e.g., canola vs. soybean), or even the same oil hydrogenated under a different set of conditions. On the other hand, the IV, *cis*, *trans*, and SN PLS calibrations have the advantage of being "universal," because it proved possible to employ pure triglycerides as standards in the development of these calibrations.

Automation of the Analysis

One of the potential benefits of implementation of FTIR spectroscopic methods in the fats and oils industry is the possibility that rapid at-line analyses can be conducted routinely by plant personnel. Programming of the spectrometer to automate analysis is an essential step in achieving this goal. In this regard, one of the major advances in FTIR spectroscopy during the past five years has been the switch to the use of personal computers to drive the spectrometer, whereas formerly a dedicated computer, having its own particular (often manufacturer-specific) operating system, was an integral component of FTIR systems. This has made it possible to establish dynamic data links between FTIR software and other programs (e.g., spreadsheets and statistical packages) operating under Microsoft® Windows™ and has also enabled the creation of Windows-type user interfaces, which facilitate the development of "user-friendly" programs (55).

In developing a prototype FTIR edible oil analyzer (34,35), the McGill IR Group has accomplished the task of converting an FTIR spectrometer into a dedicated analyzer that can be operated by untrained personnel in an industrial setting. The edible-oil analyzer was constructed by equipping a Nicolet Magna 550 spectrometer with the custom-made sample-handling accessory described earlier in this chapter. Using the Nicolet Macros\Pro package and Microsoft® Visual Basic™, an edible-oil analysis software package was developed to automate collection of spectra, data processing, and output/archiving of analytical results. A simple user interface was created to allow for point-and-click operation of the system. The operator is prompted to present the sample at the input spigot of the sample-handling accessory and enter a sample ID; a description of the sample, or other information, can also be entered and will be archived with the spectrum of the sample. The system is precalibrated for IV, *cis*, *trans*, and SN analyses, and the software incorporates automatic check routines and an instrument/cell standardization routine to ensure that the predictive accuracy of the calibrations is maintained. In addition, an automatic calibration update routine is provided to compensate for changes in cell pathlength and to facilitate calibration transfer between instruments. The basic programming/automation protocol described here is an integral part of all our completed analytical packages, some of which require no calibration and are simply implemented directly from a floppy disk, making it simple and convenient to set up and operate the method.

Conclusion

In this chapter, the basic principles and advantages of FTIR quantitative analysis have been described, and the potential utility of FTIR spectroscopy as an analytical tool in the quality control of fats and oils has been discussed. As exemplified by the IV/SN/*cis*/*trans* method described herein, an FTIR spectrometer can, in effect, perform a number of common oil analyses on the basis of a single spectral measurement. Thus, FTIR analysis can provide substantial savings, in terms of time and labor, for quality control laboratories in the fats and oils industry, with the additional benefit of reducing the use of solvents and chemical reagents. In order to facilitate practical implementation of FTIR oil analysis methodologies in the industry, a sample-handling accessory equipped with a heated flow-through transmission cell and heated input and output lines has been designed for operation at 80°C (a temperature suitable for handling both fats and oils) and has been employed in all of the methods developed to date by the McGill IR Group. In addition, the McGill IR Group has developed calibrations, basic algorithms for data processing, and a simple user interface in order to produce analytical packages that are readily implemented, requiring no knowledge of IR spectroscopy on the part of the end user. The methodology development work to date has been targeted toward providing alternatives to the more common AOCS official methods used in routine quality control applications (*trans* and *cis* content, IV, SN, PV, AV, SFI, etc.), with other FTIR methods still slated for development. It is expected that some of the major FTIR instrument manufacturers will commercialize these methods and provide technical support, making them more readily available to the industry. The implementation of FTIR methodology is only viable where the volume of analyses; improved quality control through more frequent analyses; or savings in reagent, reagent disposal, and labor costs provide a rapid payback. For example, hydrogenation/deodorization operations could benefit substantially from investing in FTIR technology, as it would allow *cis*/*trans*/IV/SN, SFI, and PV determinations all to be carried out on a single instrument, eliminating a variety of chemical and physical procedures. Although there is significant potential for savings and improved quality control at the plant level, FTIR quantitative analysis is relatively sophisticated and new to a largely conservative industrial sector, and it will take some time for it to become accepted as routine procedure. To facilitate its acceptance, the McGill IR Group will be working with the American Oil Chemists' Society and the Association of Official Analytical Chemists to standardize selected methods and the group plans to carry out collaborative studies with industrial and research laboratories to validate and ruggedize the methodology, thereby diffusing FTIR technology to the industry.

Acknowledgments

The authors would like to acknowledge the assistance of Nicolet Instrument Corporation, who supplied the instrumentation used in this work, and Dr. Stephen Dwight of Dwight Analytical, who constructed the heated sample handling accessory. We also acknowledge the Natural Sciences and Engineering Research Council of Canada for financial support of this research.

References

1. van de Voort, F.R., and Ismail, A.A. (1991) Proximate Analysis of Food by Mid–FTIR Spectroscopy, *Trends Food Sci. Technol. 2*, 13–17.
2. Ismail, A.A., van de Voort, F.R., Emo, G., and Sedman, J. (1993) Rapid and Quantitative Determination of Free Fatty Acids by FTIR Spectroscopy, *J. Am. Oil Chem. Soc. 70*, 335–341.
3. Thomas, E.V. (1994) A Primer on Multivariate Calibration, *Anal. Chem. 66*, 795A–804A.
4. Haaland, D.M., and Thomas, E.V. (1988) Partial Least-Squares Methods for Spectral Analyses. 1. Relation to Other Quantitative Calibration Methods and the Extraction of Qualitative Information, *Anal. Chem. 60*, 1193–1202.
5. Fuller, M.P., Ritter, G.L., and Draper, C.S. (1988) Partial Least-Squares Quantitative Analysis of Infrared Spectroscopic Data. Part I, Algorithm Implementation, *Appl. Spectrosc. 42*, 217–227.
6. Haaland, D.M. (1988) Quantitative Infrared Analysis of Borophosphosilicate Films Using Multivariate Statistical Methods, *Anal. Chem. 60*, 1208–1217.
7. Fuller, M.P., Ritter, G.L., and Draper, C.S. (1988) Partial Least-Squares Quantitative Analysis of Infrared Spectroscopic Data. Part II, Application to Detergent Analysis, *Appl. Spectrosc. 42*, 228–236.
8. van de Voort, F.R., Sedman, J., Emo, G., and Ismail, A.A. (1992) Assessment of Fourier Transform Infrared Analysis of Milk, *J. Assoc. Off. Anal. Chem. 75*, 780–785.
9. Heise, H.M., Marbach, R., Koschinsky, T., and Gries, F.A. (1994) Multicomponent Assay for Blood Substrates in Human Plasma by Mid-Infrared Spectroscopy and Its Evaluation for Clinical Analysis, *Appl. Spectrosc. 48*, 85–95, and references therein.
10. De Lène Mirouze, F., Boulou, J.C., Dupuy, N., Meurens, M., Huvenne, J.P., and Legrand, P. (1993) Quantitative Analysis of Glucose Syrups by ATR/FT-IR Spectroscopy, *Appl. Spectrosc. 47*, 1187–1191.
11. Miller, R.G.J., and Stace, C. (1979) *Laboratory Methods in Infrared Spectroscopy*, Heyden and Sons, London.
12. Ferraro, J.R., and Krishnan, K. (1990) *Practical FT-IR Spectroscopy, Industrial and Laboratory Chemical Analysis*, Academic Press, New York.
13. Coleman, P.B. (1993) *Practical Sampling Techniques for Infrared Analysis*, CRC Press, Boca Raton, Florida.
14. Harrick, N.J. (1967) *Internal Reflection Spectroscopy*, Wiley-Interscience, New York.
15. van de Voort, F.R., Sedman, J., Emo, G., and Ismail, A.A. (1992) Rapid and Direct Iodine Value and Saponification Number Determination of Fats and Oils by Attenuated Total Reflectance/Fourier Transform Infrared Spectroscopy, *J. Am. Oil Chem. Soc. 69*, 1118–1123.
16. *Official Methods and Recommended Practices of the American Oil Chemists' Society, 1995–1996 Additions and Revisions,* American Oil Chemists' Society, Champaign, Illinois, AOCS Official Method Cd 14-95.

17. Gurr, M. (1996) A Fresh Look at Dietary Recommendations, *INFORM 7*, 432–435.
18. Anon. (1995) Controversy, Three Nations Wrestle with *trans* Issue, *INFORM 6*, 1148–1149.
19. Anon. (1995) Some Food-Labeling Questions Still Unresolved, *INFORM 6*, 335–340.
20. Swern, D., Knight, H.B., Shreve, O.D., and Heether, M.R. (1950) Comparison of Infrared Spectrophotometric and Lead Salt-Alcohol Methods for Determination of *trans* Octadecenoic Acids and Esters, *J. Am. Oil Chem. Soc. 27*, 17–21.
21. Firestone, D., and LaBouliere, P. (1965) Determination of Isolated *trans* Isomers by Infrared Spectrophotometry, *J. Am. Oil Chem. Soc. 48*, 437–443.
22. Firestone, D., and Sheppard, A. (1992) Determination of *trans* Fatty Acids, in *Advances in Lipid Methodology—One* (Christie, W.W., ed.), The Oily Press, Alloway, Scotland, pp. 273–322.
23. Sleeter, R.T., and Matlock, M.G. (1989) Automated Quantitative Analysis of Isolated (Nonconjugated) *trans* Isomers Using Fourier Transform Infrared Spectroscopy Incorporating Improvements in the Procedure, *J. Am. Oil Chem. Soc. 66*, 121–127.
24. Mossoba, M., Yurawecz, M.P., and McDonald, R.E. (1996) Rapid Determination of the Total *trans* Content of Neat Hydrogenated Oils by Attenuated Total Reflection Spectroscopy, *J. Am. Oil Chem. Soc. 73*, 1003–1009.
25. Mossoba, M.M., and Firestone, D. (1996) New Methods for Fat Analysis in Foods, *Food Testing and Analysis 2(2)*, 24–32.
26. Huang, A., and Firestone, D. (1971) Determination of Low Level Isolated *trans* Isomers in Vegetable Oils and Derived Methyl Esters by Differential Infrared Spectrophotometry, *J. Assoc. Off. Anal. Chem. 54*, 47–51.
27. Firestone, D. (1996) General Referee Reports; Fats and Oils, *J. Assoc. Off. Anal. Chem. Int. 79*, 216–220.
28. van de Voort, F.R., Ismail, A.A., and Sedman, J. (1995) A Rapid, Automated Method for the Determination of *cis* and *trans* Content of Fats and Oils by Fourier Transform Infrared Spectroscopy, *J. Am. Oil Chem. Soc. 72*, 873–880.
29. Ratnayake, W.M.N. (1995) Determination of *trans* Unsaturation by Infrared Spectrophotometry and Determination of Fatty Acid Composition of Partially Hydrogenated Vegetable Oils by Gas Chromatography/Infrared Spectrophotometry: Collaborative Study, *J. Assoc. Off. Anal. Chem. Int. 78*, 783–802.
30. Ratnayake, W.M.N., and Pelletier, G. (1996) Methyl Esters from a Partially Hydrogenated Vegetable Oil Is a Better Infrared External Standard than Methyl Elaidate for the Measurement of Total *trans* Content, *J. Am. Oil Chem. Soc. 73*, 1165–1169.
31. Anderson, B.F., Miller, R., and Pallansch, M.J. (1974) Measuring Unsaturation in Milkfat and Other Oils by Differential Infrared Spectroscopy, *J. Dairy Sci. 57*, 156–159.
32. Arnold, R.G., and Hartung, T.E. (1971) Infrared Spectroscopic Determination of Degree of Unsaturation of Fats and Oils, *J. Food Sci. 36*, 166–168.
33. Afran, A., and Newbery, J.E. (1991) Analysis of the Degree of Unsaturation in Edible Oils by Fourier Transform–Infrared/Attenuated Total Reflectance Spectroscopy, *Spectroscopy 6*, 31–34.
34. van de Voort, F.R., Sedman, J., and Ismail, A.A. (1996) Edible Oil Analysis by FTIR Spectroscopy, *Lab. Robotics Automation 8*, 205–212.
35. van de Voort, F.R., Sedman, J., Ismail, A.A., and Dwight, S. D. (1996) Moving FTIR Spectroscopy into the Quality Control Laboratory. 2. Applications, *Lipid Technol.*, 117–119.

36. van de Voort, F.R., Memon, K.P., Sedman, J., and Ismail, A.A. (1996) Determination of Solid Fat Index by Fourier Transform Infrared Spectroscopy, *J. Am. Oil Chem. Soc. 73*, 411–416.
37. *Official Methods and Recommended Practices of the American Oil Chemists' Society,* 4th edn., American Oil Chemists' Society, Champaign, Illinois, AOCS Official Method Cd 10-57.
38. *Official Methods and Recommended Practices of the American Oil Chemists' Society,* 4th edn., American Oil Chemists' Society, Champaign, Illinois, AOCS Official Method Cd 16-81.
39. Paquette, G., Kupranycz, D.B., and van de Voort, F.R. (1985) The Mechanisms of Lipid Oxidation I. Primary Oxidation Products, *Can. Inst. Food Sci. Technol. J. 18*, 112–118.
40. Gray, J.I. (1978) Measurement of Lipid Oxidation. A Review, *J. Am. Oil Chem. Soc. 55*, 539–546.
41. *Official Methods and Recommended Practices of the American Oil Chemists' Society,* 4th edn., American Oil Chemists' Society, Champaign, Illinois, AOCS Official Method Cd 12-57.
42. *Official Methods and Recommended Practices of the American Oil Chemists' Society,* 4th edn., American Oil Chemists' Society, Champaign, Illinois, AOCS Official Method Cd 8-53.
43. Berner, D.L. (1996) Two Methods Offer New Solvent, Catalyst, *INFORM 1*, 884–886.
44. Fukuzumi, K., and Kobayashi, E. (1972) Quantitative Determination of Methyl Octadecadienoate Hydroperoxides by Infrared Spectroscopy, *J. Am. Oil Chem. Soc. 49*, 162–165.
45. van de Voort, F.R., Ismail, A.A., Sedman, J., Dubois, J., and Nicodemo, T. (1994) The Determination of Peroxide Value by Fourier Transform Infrared (FTIR) Spectroscopy, *J. Am. Oil Chem. Soc. 71*, 921–926.
46. Nakamura, T., and Maeda, H. (1991) A Simple Assay for Lipid Hydroperoxides Based on Triphenylphosphine Oxidation and High-Performance Liquid Chromatography, *Lipids 26*, 765–768.
47. Deacon, G.B., and Green, J.H.S. (1968) Vibrational Spectra of Ligands and Complexes—II. Infrared Spectra (3650–375 cm^{-1}) of Triphenylphosphine, Triphenylphosphine Oxide and Their Complexes, *Spectrochim. Acta 24A*, 845–852.
48. *Official Methods and Recommended Practices of the American Oil Chemists' Society,* 4th edn., American Oil Chemists' Society, Champaign, Illinois, AOCS Official Method Cd 18-90.
49. van de Voort, F.R., Ismail, A.A., Sedman, J., and Emo, G. (1994) Monitoring the Oxidation of Edible Oils by FTIR Spectroscopy, *J. Am. Oil Chem. Soc. 71*, 243–253.
50. Dubois, J., van de Voort, F.R., Sedman, J., Ismail, A.A., and Ramaswamy, H.R. (1996) Quantitative Fourier Transform Infrared Analysis for Anisidine Value and Aldehydes in Thermally Stressed Oils, *J. Am. Oil Chem. Soc. 73*, 787–794.
51. Sedman, J., Ismail, A.A., Nicodemo, A., Kubow, S., and van de Voort, F.R., Application of FTIR/ATR Differential Spectroscopy for Monitoring Oil Oxidation and Antioxidant Efficiency, in *Natural Antioxidants* (Shahidi, F., ed.) AOCS Press, Champaign, Illinois, 1996, pp. 358–378.
52. Steiner, J. (1993) Efforts to Eliminate Toxic Solvents, *INFORM 4*, 955.
53. Wang, Y., Veltkamp, D.J., and Kowalski, B.R. (1991) Multivariate Instrument Standardization, *Anal. Chem. 63*, 2750–2756.

54. Adhihetty, I.S., McGuire, J.A., Wangmaneerat, B., Niemczyk, T.M., and Haaland, D.M. (1991) Achieving Transferable Multivariate Spectral Calibration Models: Demonstration with Infrared Spectra of Thin-Film Dielectrics on Silicon, *Anal. Chem.* 63, 2329–2338.
55. Brereton, R.G. (1994) Object-Oriented Programming on Personal Computers, *Analyst* 119, 2149–2160.

Chapter 15
Recent Applications of Iatroscan TLC-FID Methodology

R.G. Ackman[a] and H. Heras[b]

[a]Canadian Institute of Fisheries Technology, Technical University of Nova Scotia, Halifax, Nova Scotia Canada B3J 2X4, and [b]INIBIOLP, Universidad Nacional de La Plata, 60 y 120, (1900) La Plata, Argentina

Introduction

TLC-FID (thin-layer chromatography with flame ionization detection) has been "around" for two decades (1,2). The technology and operating principles have changed only slightly in that period and are well described in a recent review (3). In that time this technique has become widely used in marine research, especially in oceanographic laboratories for reporting and quantifying lipid classes, because of its extreme (µg-level) sensitivity (4,5). One special advantage is that the basic separation is carried out in one operation on ten Chromarods, enabling rapid screening of many samples of unknown mixtures, ranging trials if the concentrations of analytes are not known, or replication in any desired combination of two to ten.

The Chromarod has changed little since the original design. It consists of a pencil lead–thin (1 mm diameter) quartz rod with a thin coating of silica gel held in place with a soft glass frit. Alumina-coated Chromarods are also available, and for the common lipid classes they offer somewhat similar separations to those obtained on silica gel (6). Separations of other types of analytes are discussed elsewhere (1,7). However, since the silica gel Chromarod is by far the most widely used basis for lipid separations, this brief review will focus on some recent applications of silica gel Chromarods S-III. These are machine-made and more uniform in characteristics than the Chromarods S-II. They improve on the reproducibility of the latter, which they have superseded without particularly changing the separations possible.

The FID electronic signal is, of course, handled exactly as is the corresponding output from GLC (gas-liquid chromatography) with FID detection. An important difference is that analyte volatility is necessary for GLC, and the FID recognizes only that part of the sample (or of its degradation products) exiting the column. TLC-FID can lose slightly volatile materials, such as short-chain hydrocarbons, from radiant heating of the silica gel on the Chromarod as the sample band approaches the flame, but with molecular weights above hexadecane (MW 226) there is usually no problem among hydrocarbons or other lipid classes. The addition of polar groups (hydroxyl, carbonyl functions, nitrogen, or sulfur) to organic molecules obviously reduces volatility and generally improves FID responses; however, as in GLC-FID, the oxygenated carbon may reduce molecular response in the flame (8), so highly oxygenated simple sugars are not good candidates for a high response in TLC-FID. However as early as 1980, plant chloroplast lipids were analysed by TLC-FID (9), and recent work with glycoglycerolipids shows what can be done in this respect (10). The latter study confirms an earlier comparison of planar TLC and TLC-FID in analysis of galactolipids (11). This review will focus on selected actual lipid samples as examples of difficult analyses in this field and give a few other examples for different chemical classes.

Gastric Digestate Lipids

Digestion is superficially well known to everybody but is, in fact, a complex process, discussed in a detailed review (12). What is not understood is that this process begins with "lingual lipases." Since those from the saliva promptly reach the stomach, they are usually lumped together as gastric lipases (13). Especially important and active in infants (14), they cease to be effective in the older human (15). Pups of marine mammals such as seals receive milk with up to 50% fat (16). The digestive status of this milk is therefore of considerable interest, and the digestate is a complex mixture not easily analyzed.

Sample Preparation

The samples were taken from the gastric contents of nursing hooded seal (*Cystophora cristata*) pups by gastric intubation with a stomach tube and syringe. Samples were stored in chloroform and stabilized with BHT from the time of collection. Lipids were extracted by a modified Folch method (17) and kept at −35°C in $CHCl_3$ under a nitrogen atmosphere until analysis.

Quantitation of lipid classes and the general TLC-FID procedure broadly followed that of Parrish and Ackman (18). Lipid standards were supplied by Serdary Research (London, ON). All solvents were ACS grade and were glass-distilled. Standard mixtures and samples were spotted on Chromarods S-III (Iatron Laboratories, Inc., Tokyo) with a Hamilton syringe fitted to a repeating dispenser (P6100, Hamilton, Reno, NV). To minimize rod-to-rod variability in response, each sample was spotted onto five of the rods, that is, two samples were developed in quintuplicate on each run. Each set of five rods was calibrated for response independently. Spotted rods were placed in a 76% constant-humidity chamber for 7 min before development in hexane-based solvent systems. Chambers (lined with filter paper) were allowed to saturate for 7 min with the eluting solvent. The solvent mixture was changed every two runs at a room temperature of 24°C, but with temperatures of 27°C or more it was necessary to change it after each run. Following separation of lipid classes, the Chromarods were dried for 3 min at 100°C prior to scanning. Rods were scanned at 0.4 cm/s in an Iatroscan Mark III analyzer (Iatron

Laboratories, Tokyo) with a hydrogen flow rate of 160 mL/min and an air flow of 2000 mL/min. Data were acquired with an SP4200 integrator (Spectra-Physics) via the analog output.

Quantification of Lipid Standards and Properties of the Solvent System

Quantitative mixtures of cholesteryl stearate, trilinolenin, α-linolenic acid, 1,2-dipalmitoylglycerol, cholesterol (CHO), and dioleoylphosphatidylcholine were dissolved in chloroform and used as standards for calibration. Each lipid was calibrated at six load levels between 0.08 and 5 µg, except that for the dominant lipid triacylglycerol (TG), which was calibrated up to 20 µg, with five analyses per level on each set. l-tetracosanol was chosen as an IS (internal standard) because it is stable, and has a high response in the FID, and also because, with our solvent system, the R_f lies between those of free CHO and free fatty acids (FFA), in the center of the chromatogram. It is reasonably well separated from other peaks. Van Tornout et al. (19) used 1-octadecanol for much the same reasons. Their petroleum ether–based solvent system was essentially similar to hexane-based systems.

Quadratic equations were found to be adequate to fit the data for quantification, with r^2 values ranging from 0.98 to 0.99 when plotting peak area ratios for each lipid standard with respect to the IS. The y-intercepts were not significantly different from zero for all standard curves except those for phospholipids.

Hexane, or a related hydrocarbon, must be the basic component of any mobile phase designed to resolve nonpolar lipids by normal-phase TLC. Polar modifiers from one or another of the solvent selectivity groups defined by Rutan et al. (20) must be added to affect elution of lipids such as CHO and diacylglycerols (DG). In this project we found that a quaternary solvent system composed of hexane, ethyl acetate, ethyl ether, and formic acid has unique advantages for the resolution of lipids that are based on saturated and unsaturated fatty acids of animal tissues. The separations that can be achieved with this solvent mixture are illustrated by Fig. 1. Cholesterol esters (CE) and wax esters (WE) are eluted together (a not uncommon problem in TLC on silica gel) but are well separated from TG; a reasonable separation of the CE and WE by planar TLC on silica gel with the solvent system hexane : benzene : acetic acid (50 : 50 : 0.5, by volume) has recently been published (21). The FFA fraction elutes next after TG, followed by free sterols and DG. Monoacylglycerols (MG) are well resolved from polar lipids, essentially phospholipids (PL) in the standards.

The use of solvent mixtures is always problematic, as a relatively small change in the composition can lead to a large change in the selectivity and thus also in the reproducibility of the analysis. The resolution of some individual marine lipid classes such as TG and FFA into two distinct groups (Fig. 1) based on total fatty acid unsaturation (22–25) is peculiar to marine lipids, where some fatty acids are highly unsaturated, with five and six ethylenic bonds. When the TG and FFA are resolved into two peaks, the first peak consists of groups of molecules of primarily saturated acids and monoethylenic acids (which may match common standards of such fatty acids), and

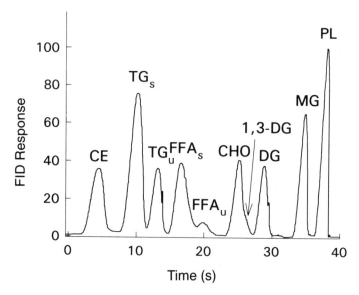

Fig. 1. TLC-FID separation of a range of standard lipids. The mobile phase was hexane : ethyl acetate : diethyl ether : formic acid (91:6:3:1, by volume). Time refers to scanning time for the Chromarod. PL = phosphatidylcholine; MG = 1-monoacylglycerol; DG = 1,2-diacylglycerol; 1,3-DG = 1,3-diacylglycerol; CHO = cholesterol; FFA_u = highly unsaturated free fatty acid; FFA_s = less unsaturated free fatty acid; TG_u = highly unsaturated triacylglycerol; TG_s = saturated triacylglycerol; CE = cholesterol ester. Tripalmitin and palmitic acid were used to complement the trilinolenin and linolenic acid.

the second consists of less-mobile groups of molecules with highly unsaturated fatty acids which are not usually included in standards). These separations, which can be seen in planar TLC as well as on Chromarods (24), could be very useful in some circumstances (25). If they are not desired, total lipid hydrogenation overcomes any subfractionation problems and presents other benefits; not only is peak sharpness, improved, but there is consequently a gain in FID response of the order of 7 to 45%, depending on the lipid class (26).

Most solvent systems for single-step chromatography of simple lipids are based on combinations of n-hexane with diethyl ether, with small amounts of low-molecular-weight organic acids, especially formic (usually called the HDF system). Such systems have been widely used for nearly two decades (27–29). By adding ethyl acetate, we obtained better results than with any equivalent HDF mixture. Although ethyl ether (bp 34°C) and ethyl acetate (bp 77°C) have similar elution strengths on silica, fully replacing one by the other to obtain a ternary mixture gave poor results (not shown). This might be due to the different boiling points, which may lead to changes in the composition of the solvent mixture inside the chamber, and also to the different rates at which the mixed solvents rise on the thin silica gel layer of the Chromarod.

The properties of ethyl acetate in a mobile phase for TLC-FID and its effects on the selectivity of separation of lipids have not been reported before for hexane-based systems. An increase of ethyl acetate primarily increases the mobility of TG, but the selectivity of the separation also depends on secondary effects that are a function not only of the solvent composition, but also of the solvent and solute. It is very difficult to quantify these dependencies; thus, an experimental approach remains the best way to choose the optimal eluent composition.

Figure 2 shows the lipid classes recovered from the gastric contents of a hooded seal pup. The actual (w/w%) figures represented are, for five Chromarods: TG, 77.18 ± 1.96; DG, 11.41 ± 1.13; MG, 1.83 ± 0.27; FFA, 9.57±2.57; internal calibration standard (IS), 1-tetracosanol.

The necessity of calibration of the Chromarods is probably one reason for the lack of enthusiasm for TLC-FID in a world of analytical methods based on push-buttons and electronic digital readout. However, the use of an internal standard is advantageous (27,30), and the Chromarods S-III can be used several hundred times with little change in separations. When such changes do occur, they usually are

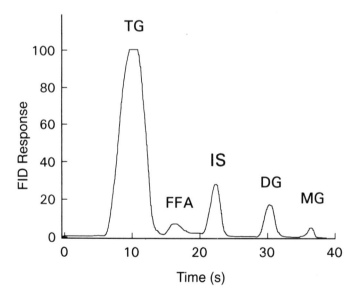

Fig. 2. TLC-FID separation of lipids recovered from the gastric contents of a hooded seal pup. The mobile phase was hexane:ethyl acetate:diethyl ether:formic acid (91:6:3:1, by volume). Time refers to scanning time for the Chromarod. MG = monoacylglycerol; DG = 1,2-diacylglycerol; IS = Internal standard; FFA = free fatty acid; TG = triacylglycerol.

sudden and result from changes in laboratory humidity, which should always be suspected before more complex reasons are considered.

Use of Multiple Development for Phospholipids of Fish Muscle

Although limited use of multiple development in different solvent systems has been applied in planar TLC, the FID scan of a Chromarod totally cleans and reactivates the silica gel. In a difficult case of fish flesh lipid analysis (31), it was necessary to do two things with one sample. The neutral lipids had to separate low levels of FFA distinctly from a large TG component, and the phospholipids also had to be quantified in order to test their influence on the customary AOCS Method Ca 5a-40 for FFA (32) by titration with alkali. In this case three partial scans were conducted with a development sequence of three different solvent systems. The extracted lipids were dissolved in chloroform at an appropriate concentration, and this lipid solution was then spotted onto Chromarods S-III in 1 µL volumes from glass Microcap disposable pipettes (Drummond Scientific Co., Broomall, PA). The Chromarods were then conditioned in a constant-humidity chamber for 5 min. The first development was carried out for 55 min in hexane:chloroform:isopropanol:formic acid (80:14:1:0.2, by volume). The Chromarods were then dried at 100°C for 1.5 min and partially scanned from the top to a point just below the DG peak (Fig. 3-I). The Chromarods were then redeveloped in acetone for 15 min, dried at 100°C for 1.5 min and partially scanned to below the acetone-mobile polar lipid (AMPL) position (Fig. 3-II). Finally the Chromarods were again developed in chloroform:methanol:water (70:30:3, by volume) for 60 min, dried at 100°C for 3 min, and completely scanned to reveal different phospholipids (Fig. 3-III). The quantification calibration of the system was conducted under the same conditions with authentic standards. Each sample was spotted on 10 rods treated as one lot, and each was analyzed in triplicate. The natural FFA is clearly seen in Fig. 3, although it constitutes only 0.8% ± 0.2 of a total lipid content of 2.4% of the fish muscle. The total phospholipids averaged 20.2 ± 1.8% of the lipids and appreciably affected the titration results when the FFA were a relatively low proportion of total lipids in fresh fish muscle.

AMPL and Polar Plant Lipids

Acetone-mobile polar lipid was described in some detail for aquatic lipid classes by Parrish (30). We have, however, also found AMPL lipid classes in fish meal lipid extracts (33). The polar plant lipids were examined by TLC-FID by Banerjee and Ackman (11), who found that monogalactosyl diacylglycerol (MGDG) and digalactosyl diacylglycerol (DGDG) could be easily resolved in several solvent systems, but not necessarily without interference from phospholipids. Originally, Parrish (30) did not clarify the chemical nature of AMPL. Phospholipids are almost by definition not soluble in acetone, so AMPL could refer to any material mobile in this solvent, such

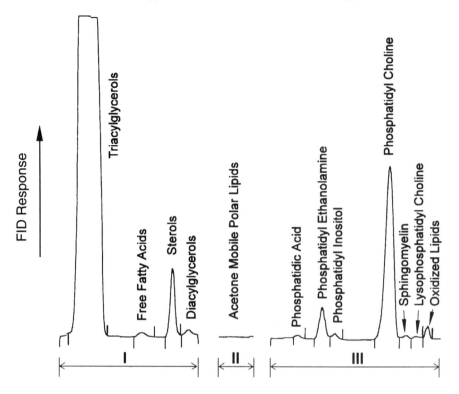

Fig. 3. Sequential Iatroscan thin-layer chromatography–flame ionization detector profiles of the lipid classes extracted from silver hake muscle tissue. I, II, and III represent partial chromatograms from the three-stage development sequence of total lipids on Chromarods S-III as described in the text. Reproduced with permission from Ref. 31.

as a natural mixture of monoacylglycerols and pigments (10). Figure 4 shows how an isopropanol extract of spinach—a traditional natural reference material for plant lipids—can be resolved into galactosyl lipids by an appropriate TLC-FID approach. A system of two steps was found by Parrish et al. (10) to be optimal; first a 40-min development in $CHCl_3$: acetone (3 : 2, by volume), then a partial scan from the top of the Chromarod to just above the MGDG peak, followed by redevelopment for 30 min in acetone : formic acid (49 : 1, by volume), followed by a full scan.

Qualitatively, all of the polar lipid glycoglycerolipids found by others in the photosynthetic system of plants could be found by TLC-FID in spinach and in the lipids of the unicellular marine alga *Isochrysis galbana*. The proportions of various lipid classes closely matched those reported by other workers using other forms of chromatography with gravimetric recovery. The TLC-FID totals for recovered lipids were, however, slightly lower (5 to 10%) than those obtained by more conventional techniques. The equipment was a Mark IV Iatroscan; a previous study by the same

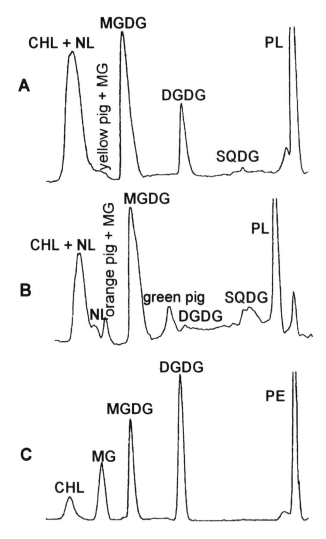

Fig. 4. Chromatograms of isopropanol extract of spinach (A), of *Isochrysis galbana* (Clone T-Iso) (B) and standards (C) showing the separation of galactosyl lipids in a two-stage development as described in the text. CHL = chlorophyll a; MG = monoacylglycerol; MGDG = monogalactosyl diacylglycerol; DGDG = digalactosyl diacylglycerol; PE = phosphatidyl ethanolamine; NL = neutral lipid; PL = polar lipid; SQDG = sulfoquinovosyl diacylglycerol. Adapted from Ref. 10.

author with a Mark II unit had also shown the same magnitude of shortfall for total lipids (30). The reason is obscure, but it could arise from the slightly wider peaks from the mixtures of fatty acids in natural lipids, compared to those of the usually homogeneous reference material used for calibration. A small percentage of the natural-lipid band on the Chromarod would be very thinly distributed on one or both sides of the main distribution band, perhaps not at a high enough concentration to contribute to the ion current. This hypothetical thin material layer could also be subject to premature vaporization from the heat of the approaching flame, as proposed by Parrish (10). The Mark IV was an interim model derived from the Mark III with an improved FID detector and could be fitted with a flame thermionic detector (34). It has been superseded by the fully redesigned Mark 5.

Other Materials

A particularly thorough investigation of operating parameters of TLC-FID in the analysis of MG, DG, and TG and of free oleic acid has recently been published (29), confirming that it is an effective method of analysis for lipid classes of interest in the food industry.

In early work comparing TLC-FID with planar TLC it was usually possible to recover traces of any material mobile on a silica gel plate from the silica gel below a streak of that material. In effect, a "memory" of the passage of the mg-scale mobile material would not be noticed on staining or charring a TLC plate, because it was spread over a large area, from the streaking position at the origin to the much higher concentration in the observed band. The sensitivity of the TLC-FID system, however, could record this material recovered from the TLC plate silica gel, because it operated on a µg scale. Possibly, quantitation in TLC of any type would suffer from this "memory" effect, although with TLC-FID and an internal standard the process should be self-correcting. In the case of the fairly simple mixture published by Peyrou et al. (29) mean results were 5% too low for oleic acid and 5% too high for α-monoolein, with cholesterol used as an internal standard in ten Chromarods. The actual responses in area units/µg appeared different, the α-monoolein always giving a 10% higher response. This result contradicts one of the basic tenets of FID responses in GLC: that the molar response of a molecule reflects the proportion of active (i.e., non-oxygen-bearing) carbon in the molecule (8). Possibly two hydroxyl groups of the α-monoolein bind that molecule to the silica or make it less volatile, either process being capable of retaining some extra material for the FID combustion process.

Cholesterol is a common component of animal tissues, and simple transesterification permits analysis of membrane lipid total cholesterol, fatty acids, and plasmalogens on Chromarods (35). This interesting idea of transesterification of a lipid mixture with BF_3-MeOH (instead of alkali) should be compared with many saponification methods to recover cholesterol for GLC, such as that of Kovacs et al. (36). However, for cholesterol in seafoods (37) an Iatroscan method gave an excellent correlation between weight applied and TLC-FID peak area (Fig. 5). There is also a novel method for rapid quantification of cholesterol, total bile salts, and phospholipids by TLC-FID (38), showing similar linearity. More advanced research with a different and newer model of Iatroscan shows linear calibration lines (39) for most lipid classes.

There have been reports in the older Iatroscan literature of difficulties with quantitative cholesterol determinations by TLC-FID. It has also been suggested that some of the many methods used in food analysis of cholesterol are of doubtful value when compared with a modern and elegant approach consisting of gas chromatography–mass spectrometry (GC-MS) of *tert*-butyldimethylsilyl derivatives (40). Possibly solvent focusing (41) could reduce problems with cholesterol determination by TLC-FID. An example of the benefits of solvent focusing recently noted in our laboratory is shown in Fig. 6 for caffeine. An aqueous system to be analyzed for caffeine could be applied directly to Chromarods and scanned without development. The slow evaporation of the water led to band spreading and a double peak from concentration of the caffeine at each side of the band (Fig. 6A). Development with methanol, in which caffeine is not very soluble, to above the spot concentrated the caffeine into a narrow band, resulting in the superior chromatogram of Fig. 6B. Some workers have resorted to spotting the Chromarods above a warm surface to accelerate solvent evaporation and produce narrower peaks. This band spreading on the Chromarods is a vari-

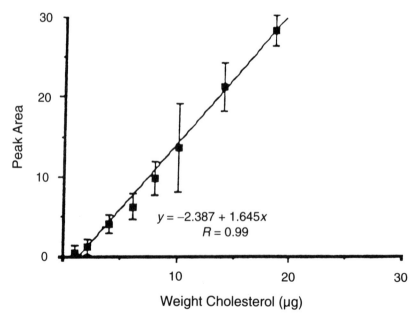

Fig. 5. Calibration curve of Iatroscan peak area versus weight of cholesterol. Bars indicate standard deviation.

ant on an old problem in paper chromatography and silica gel planar TLC, where the concentration of analyte on the outside ring of a spot can give "W-shaped" spots on development.

Caffeine contains nitrogen, but the results of Fig. 6 were achieved with the standard TLC-FID operation. To increase sensitivity in selected analyses a nitrogen-sensitive detector based on flame thermionic technology (DETector Engineering & Technology, Inc., Walnut Creek, CA) can be fitted on the standard Iatroscan equipment (34). Figure 7 shows how this increases the Iatroscan sensitivity in respect to the nitrogen of the nonlipid complex shellfish toxins (for structures see Ref. 42). It functions at a 100-fold increased sensitivity, compared to the ordinary TLC-FID detector. These results were obtained with the Mark 5 Iatroscan, in which the Chromarods are scanned in a horizontal position, rather than the sloped position of all older models (1).

Since reviews of Iatroscan TLC-FID include many nonlipid materials analyzed with manual spotting techniques, it is worth mentioning that automated spotting equipment is available and yet may not solve all problems, as recently shown for coal tar pitch (43). This report has some parallels with the foregoing case in which an aqueous solution was spotted. Since all the coal tar material is combusted, it is suggested that it is possible to go directly to mass analysis of the total sample with-

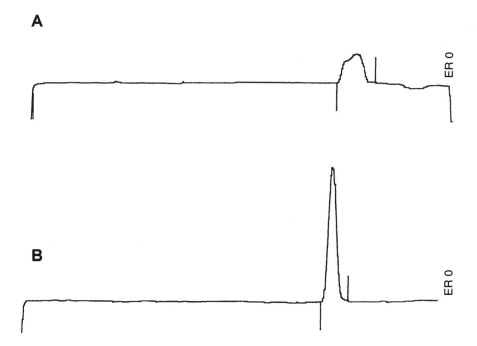

Fig. 6. Caffeine deposited on a Chromarod S-III from an aqueous solution (A) and impact of solvent focusing with methanol (B).

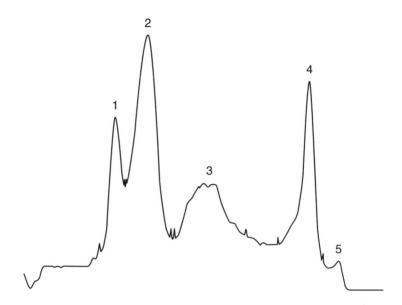

Fig. 7. Chromatogram of shellfish toxins containing nitrogen when resolved on a Chromarod S-III and scanned with a nitrogen-sensitive thermionic FID. 1 = C toxins, 2 and 3 = gonyautoxins; 4 = neosaxitoxin; 5 = saxitoxin.

out calibration. TLC-FID is not a technique for the faint-hearted or impatient, but it rewards a thoughtful approach.

Use of TLC-FID in the food industry has been limited, as sample amounts are usually not a limitation. For serial studies of lipids of meat samples (44) it is ideal. The fact that oxidative polymers tend to stay at the origin has considerable potential in frying operations or edible-oil research (45,46).

In Fig. 3 a very small peak at the origin is marked "oxidized lipids." Whether these components are polymeric or contain hydroxy or hydroperoxy groups is usually not known, but in the hexane : diethyl ether : formic acid systems actual oxidation products from fish oils cannot be distinguished from phospholipids (47). In the marine lipid and animal biochemistry work the phospholipids are usually highly unsaturated and susceptible to oxidation, so the designation of a small amount of "polar lipid" is not misleading, as it may well include oxidized phospholipids. Kaitaranta and Ke (47) were able to show that in oxidizing fish oils both the thiobarbituric acid values and actual weight gain from oxygen uptake could be calculated with the polar lipid peak. Nonlipid material, such as dissolved proteins and particulate matter, will also remain at the origin (30), so some care in assigning this peak to any type of lipid material is needed.

It is not possible to conclude a review of Iatroscan TLC-FID without mentioning two names. Mr. John Newman in the United Kingdom was for many years the world distribution agent for Iatroscans outside of Japan and personally did much to further the growth of diverse applications for this equipment. He is now retired to Wales. Mojmír Ranný of Prague wrote the definitive book on the subject of TLC-FID (48) and has particularly reviewed the basic principles and illustrated numerous industrial chemistry applications.

Acknowledgments

Dr. S.J. Iverson, Dalhousie University, is gratefully acknowledged for providing the gastric digestate of the hooded seal pup.

References

1. Ackman, R.G. (1981) Flame Ionization Detection Applied to Thin Layer Chromatography on Coated Quartz Rods, in *Methods in Enzymology,* Lowenstein, J.M., Academic Press, New York, Vol. 72, pp. 205–252.
2. Sipos, J.C., and Ackman, R.G. (1978) Automated and Rapid Quantitative Analysis of Lipids with Chromarods, *J. Chromatogr. Sci. 16,* 443–447.
3. Shantha, N.C. (1992) Thin-Layer Chromatography–Flame Ionization Detection Iatroscan System, *J. Chromatogr. 624,* 21–35.
4. Volkman, J.K. and Nichols, P.D. (1991) Applications of Thin Layer Chromatography–Flame Ionization Detection to the Analysis of Lipids and Pollutants in Marine and Environmental Samples, *J. Planar Chromatogr. 4,* 19–26.
5. Parrish, C.C., McKenzie, C.H., MacDonald, B.A., and Hatfield, E.A. (1995) Seasonal Studies of Seston Lipids in Relation to Microplankton Species Composition and Scallop Growth in South Broad Cove, Newfoundland, *Mar. Ecol. Prog. Ser. 129,* 151–164.
6. Indrasena, W.M., Parrish, C.C., Ackman, R.G., and Paulson, A.T. (1990) Separation of Lipid Classes and Carotenoids in Atlantic Salmon Feeds by Thin Layer Chromatography with Iatroscan Flame Ionization Detection, *Bull. Aquaculture Assoc. Canada 90-4,* 36–40.
7. Ackman, R.G., McLeod, C.A., and Banerjee, A.K. (1990) An Overview of Analyses by Chromarod-Iatroscan TLC-FID, *J. Planar Chromatogr. 3,* 450–490.
8. Ackman, R.G. (1968) The Flame Ionization Detector: Further Comments on Molecular Breakdown and Fundamental Group Responses, *J. Gas Chromatogr. 6,* 497–501.
9. Hirayama, O., and Morita, K. (1980) A Simple and Sensitive Method for the Quantitative Analysis of Chloroplast Lipids by Use of Thin Layer Chromatography and Flame Ionization Detector, *Agric. Biol. Chem. 44,* 2217–2219.
10. Parrish, C.C., Bodennec, G., and Gentien, P. (1996) Determination of Glycoglycerolipids by Chromarod Thin-Layer Chromatography with Iatroscan Flame Ionization Detection, *J. Chromatogr. A 741,* 91–97.
11. Banerjee, A.K., and Ackman, R.G. (1990) A Comparative Study of the Efficacies of TLC/FID and Plate-TLC Techniques in the Separation of Phospholipids and Galactolipids, *J. Indian Chem. Soc. 67,* 932.

12. Nelson, G.J., and Ackman, R.G. (1988) Absorption and Transport of Fat in Mammals with Emphasis on n-3 Polyunsaturated Fatty Acids, *Lipids 23*, 1005–1014.
13. Armand, M., Borel, P., Cara, L., Senft, M., Chautan, M., Lafont, H., and Lairon, D. (1990) Adaptation of Lingual Lipase to Dietary Fat in Rats, *J. Nutr. 120*, 1148–1156.
14. Armand, M., Hamosh, M., Mehta, N.R., Angelus, P.A., Philpott, J.R., Henderson, T.R., Dwyer, N.K., Lairon, D., and Hamosh, P. (1996) Effect of Human Milk or Formula on Gastric Function and Fat Digestion in the Premature Infant, *Pediatric Res. 40*, 429–437.
15. Moreau, H., Gargouri, Y., Bernadal, A., Pieroni, G., and Verger, R. (1988) Étude Biochimique et Physiologique des Lipases Préduodénales d'Origines Animale et Humaine: Revue, *Rev. Franç. Corps Gras 35*, 169–176.
16. Iverson, S.J., Sampugna, J., and Oftedal, O.T. (1992) Positional Specificity of Gastric Hydrolysis of Long-Chain n-3 Polyunsaturated Fatty Acids of Seal Milk Triglycerides, *Lipids 27*, 870–878.
17. Folch, J., Lees, M., and Sloane Stanley, G.H. (1957) A Simple Method for the Isolation and Purification of Total Lipids from Animal Tissues, *J. Biol. Chem. 226*, 497–509.
18. Parrish, C.C., and Ackman, R.G. (1985) Calibration of the Iatroscan-Chromarod System for Marine Lipid Class Analyses, *Lipids 20*, 521–530.
19. Van Tornout, P., Vercaemst, R., Caster, H., Lievens, M.J., De Keersgieter, W., Soetewey, F., and Rosseneu, M. (1979) Use of 1-Octadecanol as an Internal Standard for Plasma Lipid Quantitation on Chromarods, *J. Chromatogr. 164*, 222–227.
20. Rutan, S.C., Carr, P.W., Cheong, W.J., Park, J.H., and Snyder, L.R. (1989) Re-Evaluation of the Solvent Triangle and Comparison to Solvatochromic Based Scales of Solvent Strength and Selectivity, *J. Chromatogr. 463*, 21–37.
21. Ichihara, K., Shibahara, A., Yamamoto, K., and Nakayama, T. (1996) An Improved Method for Rapid Analysis of the Fatty Acids of Glycerolipids, *Lipids 31*, 535–539.
22. Kramer, J.K.G., Fouchard, R.C., and Farnworth, E.R. (1985) The Effect of Fatty Acid Chain Length and Unsaturation on the Chromatographic Behavior of Triglycerides on Iatroscan Chromarods, *Lipids 20*, 617–619.
23. Ohshima, T., Ratnayake, W.M.N., and Ackman, R.G. (1987) Cod Lipids, Solvent Systems and the Effect of Fatty Acid Chain Length and Unsaturation on Lipid Class Analysis by Iatroscan TLC-FID, *J. Am. Oil Chem. Soc. 64*, 219–223.
24. Shantha, N.C., and Ackman, R.G. (1991) Silica Gel Thin-Layer Chromatographic Method for Concentration of Longer-Chain Polyunsaturated Fatty Acids from Food and Marine Lipids, *Can. Inst. Food Sci. Technol. J. 24*, 156–160.
25. Parrish, C.C., Bodennec, G., and Gentien, P. (1992) Separation of Polyunsaturated and Saturated Lipids from Marine Phytoplankton on Silica Gel–Coated Chromarods, *J. Chromatogr. 607*, 97–104.
26. Shantha, N.C., and Ackman, R.G. (1990) Advantages of Total Lipid Hydrogenation Prior to Lipid Class Determination on Chromarods-SIII, *Lipids 25*, 570–574.
27. Farnworth, E.R., Thompson, B.K., and Kramer, J.K.G. (1982) Quantitative Determination of Neutral Lipids on Chromarods, *J. Chromatogr. 240*, 463–474.
28. Parrish, C.C., and Ackman, R.G. (1983) Chromarod Separations for the Analysis of Marine Lipid Classes by Iatroscan Thin-Layer Chromatography–Flame Ionization Detection, *J. Chromatogr. 262*, 103–112.

29. Peyrou, G., Rakotondrazafy, V., Mouloungui, Z., and Gaset, A. (1996) Separation and Quantitation of Mono-, Di- and Triglycerides and Free Oleic Acid Using Thin-Layer Chromatography with Flame-Ionization Detection, *Lipids 31,* 27–32.
30. Parrish, C.C. (1987) Separation of Aquatic Lipid Classes by Chromarod Thin-Layer Chromatography with Measurement by Iatroscan Flame Ionization Detection, *Can. J. Fish. Aquat. Sci. 44,* 722–731.
31. Zhou, S., and Ackman, R.G. (1996) Interference of Polar Lipids with the Alkalimetric Determination of Free Fatty Acid in Fish Lipids, *J. Am. Oil Chem. Soc. 73,* 1019–1023.
32. AOCS Official Method Ca 5a-40 (1989) *Official Methods and Recommended Practices of the American Oil Chemists' Society,* 4th edn. *1995–1996 Additions and Revisions,* American Oil Chemists' Society, Champaign, Illinois.
33. Gunnlaugsdóttir, H., and Ackman, R.G. (1993) Three Extraction Methods for Determination of Lipids in Fish Meal: Evaluation of a Hexane/Isopropanol Method as an Alternative to Chloroform-Based Methods, *J. Sci. Food Agric. 61,* 235–240.
34. Parrish, C.C., Zhou, X., and Herche, L.R. (1988) Flame Ionization and Flame Thermionic Detection of Carbon and Nitrogen in Aquatic lipid and Humic-Type Classes with an Iatroscan Mark IV, *J. Chromatogr. 435,* 350–356.
35. Beaumelle, B.D., and Vial, H.J. (1986) Total Cholesterol, Fatty Acids, and Plasmalogens Can Be Reliably Quantitated by Analysis on Chromarods After the Methylation Step Required for Fatty Acid Analysis by Gas-Liquid Chromatography, *Anal. Biochem. 155,* 346–351.
36. Kovacs, M.I.P., Anderson, W.E., and Ackman, R.G. (1979) A Simple Method for the Determination of Cholesterol and Some Plant Sterols in Fishery-Based Food Products, *J. Food Sci. 44,* 1299–1201, 1205.
37. Walton, C.G., Ratnayake, W.M.N., and Ackman, R.G. (1989) Total Sterols in Seafoods: Iatroscan TLC/FID Versus the Kovacs GLC/FID Method, *J. Food Sci. 54,* 793–795, 804.
38. Rigler, M.W., Leffert, R.L., and Patton, J.S. (1983) Rapid Quantification on Chromarods of Cholesterol, Total Bile Salts and Phospholipids from the Same Microliter Sample of Human Gallbladder Bile, *J. Chromatogr. 277,* 321–327.
39. Ohshima, T., and Ackman, R.G. (1991) New Developments in Chromarod/Iatroscan TLC-FID: Analysis of Lipid Class Composition, *J. Planar Chromatogr. 4,* 27–34.
40. Stewart, G., Gosselin, C., and Pandian, S. (1992) Selected Ion Monitoring of *tert*-Butyldimethylsilyl Cholesterol Ethers for Determination of Total Cholesterol Content in Foods, *Food Chem. 44,* 377–380.
41. Parrish, C.C., and Ackman, R.G. (1983) The Effect of Developing Solvents on Lipid Class Quantification in Chromarod Thin Layer Chromatography/Flame Ionization Detection, *Lipids 18,* 563–565.
42. Concon, J.M. (1988) *Food Toxicology, Part B: Contaminants and Additives,* Marcel Dekker, Inc., New York.
43. Cebolla, V.L., Vela, J., Membrado, L., and Ferrando, A.C. (1996) Coal-Tar Pitch Characterization by Thin-Layer Chromatography with Flame Ionization Detection, *Chromatographia 42,* 295–299.
44. St. Angelo, A.J., and James, C., Jr. (1993) Analysis of Lipids from Cooked Beef by Thin-Layer Chromatography with Flame-Ionization Detection, *J. Am. Oil Chem Soc. 70,* 1245–1250.

45. Hara, K., Cho, S.-Y., and Fujimoto, K. (1989) Measurement of Polymer and Polar Material Content for Assessment of the Deterioration of Soybean Oil Due to Heat Cooking, *Yakagaku 38*, 463–470.
46. Sébédio, J.L., Astorg, P.O., Septier, C., and Grandgirard, A. (1987) Quantitative Analyses of Polar Components in Frying Oils by the Iatroscan Thin-Layer Chromatography–Flame Ionization Detection Technique, *J. Chromatogr. 405*, 371–378.
47. Kaitaranta, J.K., and Ke, P.J. (1981) TLC-FID Assessment of Lipid Oxidation as Applied to Fish Lipids Rich in Triglycerides, *J. Am. Oil Chem. Soc. 58*, 710–713.
48. Ranný, M. (1987) *Thin-Layer Chromatography with Flame Ionization Detection*, D. Reidel Publishing Company, Dordrecht, Germany.

Chapter 16

Natural Antioxidants in Lipids

T. Rathjen and H. Steinhart

Institute of Biochemistry and Food Chemistry, University of Hamburg, Grindelallee 117, D-20146 Hamburg, Germany

Introduction

One of the major changes that occurs during processing, final preparation, and storage of foods especially rich in fat is lipid oxidation. The spontaneous reaction of atmospheric oxygen with lipids initiates changes in the food system that affect its nutritional quality, wholesomeness, safety, color, flavor, and structure. Hence, antioxidants are used to improve the oxidative stability of food lipids. Antioxidants are substances that retard the above-mentioned changes by inhibiting the lipid autoxidation reactions. According to their mode of action, antioxidants can be classified as free radical terminators, chelators of transition metal ions (which are capable of catalyzing lipid oxidation), or oxygen scavengers. The primary antioxidants are radical scavengers and function as hydrogen atom donors by a one-electron transfer reaction with the high-energy lipid radicals formed during the autoxidation reactions. The antioxidants are converted to radicals that are so stable that they break the autoxidation chain reaction. Preventive antioxidants (so-called secondary antioxidants) reduce the rate of initiation reactions, for instance, by inactivating metal ions. Overall, they exhibit an effect analogous to chain-breaking antioxidants.

A variety of synthetic substances have been designed to function as antioxidants, but because of the rigorous toxicological testing required for food additives, only a very few are permitted for use in foods. As public health awareness has increased, natural antioxidants have gained from the belief that natural food ingredients are better and safer than synthetics. It is important to realize very clearly that the reactive chemical moiety in most natural antioxidants is the same as that in synthetic antioxidants. According to their origin, natural antioxidants can be classified into four groups:

1. Substances formed by *in situ* reactions, such as fermentation or Maillard reaction
2. Enzyme systems, such as glucose oxidase/catalase
3. Defined chemical substances, such as Vitamin E or ascorbic acid
4. Plant extracts, including spices, herbs, and seeds

Only a few of these are commercially utilized. Several defined chemical substances (Vitamin E, ascorbic acid) and plant extracts (rosemary, sage) are especially used; others may gain importance in the future. Table 1 summarizes the most important of these.

TABLE 1
Important Natural Antioxidants

Type of antioxidant	Conventional applications
Tocopherols	Stabilization of edible oils and processed foods
Rosemary extracts	Stabilization of processed foods
Flavonoids	Potential natural antioxidants for application in processed foods
Ascorbic acid	Synergist to tocopherols, stabilization of beverages and juices
Citric acid	Metal chelator

Determination of Natural Antioxidants in Foods

Because of the great variety of relevant compounds, it is increasingly difficult to form a clear overview of feasible determination methods for natural antioxidants. It is therefore not possible to employ a universal method. Generally speaking, most analytical procedures are based on high-pressure liquid chromatography (HPLC) with adequate detection. Almost every method requires considerable sample preparation, such as saponification and extraction, before separation and determination of individual antioxidants. Oxidation may cause loss of natural antioxidants that is exacerbated by exposure of samples to heat, light, atmospheric oxygen, or peroxide-containing chemicals, so great attention has to be paid to appropriate sample preparation. Hence, it should be pointed out that care must be taken with the purity of solvents used and the analytical techniques applied. For some identification and quantification procedures, attention must also be paid to the elimination of potential interferences by appropriate extract purification, preparation, or derivatization. To ensure comparable analytical results, tested and approved analytical methods should be used (Table 2).

Determination of Tocopherols and Tocotrienols

Eight different homologs that have Vitamin E activity occur in nature. They belong to two families, tocopherols and tocotrienols, having the prefix α, β, γ and δ, depending on number and position of methyl groups attached to a 6-hydroxychroman ring. The tocotrienols differ from the corresponding tocopherols by the presence of three nonconjugated double bonds in the side chain (Fig. 1). With respect to Vitamin E activity, α-tocopherol is the most potent member of the family, whereas the antioxidant activity decreases from δ to α.

Oilseeds, vegetable oils, nuts, and cereals are rich sources of tocopherols, with varying composition of homologs. Remarkable amounts of tocotrienols have been found in cereals, palm oil, coconut oil, rapeseed oil, and rice oil. The different homologs may be found in esterified or free forms; seed oils predominantly contain the free tocopherols. In animal and human fat, α-tocopherol is present in trace quantities.

Many methods for the determination of vitamin E have been described in the literature (1–4). The early coloric methods based on complex formation of Fe^{2+} and

TABLE 2
Determination of Tocopherols and Tocotrienols: Selected Methods

Sample	Sample preparation	Homolog	HPLC-conditions Stationary Phase Mobile Phase (v/v)	Detection	Reference no.
Vegetable oils	Dil. in mobile phase	α-,β-,γ-,δ-T, α-,β-,γ-,δ-T3	R sil (10 μm) n-hexane/ethyl acetate (97/3)	Fluorescence ex. 303 nm, em. 328 nm	12
Vegetable oils	Dil. in n-hexane	α-,β-,γ-,δ-T	LiChrosorb Si 60 n-hexane/diethyl ether (95:5)	UV 292 nm, fluorescence ex. 290 nm, em 330 nm	13
Vegetable oils	Dil. in n-hexane	α-,β-,γ-,δ-T	Partisil 5 n-hexane/dibutyl ether (90:10)	UV 292, fluorescence ex. 290 nm, em 330 nm	14
Milk products	Sap. + extr. with petroleum ether/diethyl ether (1/1)	α-,β-,γ-,δ-T	Partisil Si 60 n-hexane/2-propanol (98,5:1,5)	UV 292 nm	15
Vegetable oils	Dil. in n-hexane	α-,β-,γ-,δ-T, α-,β-,γ-,δ-T3	Partisil PAC (5 μm) n-hexane/tetrahydrofuran (94:6)	Fluorescence ex. 210 nm, em 325 nm	16
Vegetable oils, margarine, butter	Dil. in n-hexane	α-,β-,γ-,δ-T, α-,β-,γ-,δ-T3	LiChrosorb Si 60 (5 μm) n-hexane/1,4-Dioxan (94:6)	Fluorescence ex. 295 nm, em 330 nm	5,17
Vegetable oils	Dil. in n-hexane	α-,β-,γ-,δ-T	Corasil II n-hexane/diisopropyl ether (95:5)	Fluorescence ex. 295 nm, em 340 nm	9
Vegetable oils	Dil. in n-hexane	α-,β-,γ-,δ-T	Jascopack-WC-03-500 n-hexane/diisopropyl ether (98:2)	UV 295, fluorescence ex. 298 nm, em 325 nm	18
Vegetable oils, milk, cereals	Extr. by 2-propanol and acetone, n-hexane and water; sap. for α-TAc	α-,β-,γ-,δ-T, α-,β-,γ-,δ-T3 α-TAc	LiChrosorb Si 60 (5 m) n-hexane/diethyl ether (95:5) or n-hexane/2-propanol	UV 290, fluorescence ex. 290 nm, em 330 nm	6
Vegetable oils	Direct injection	α-,β-,γ-,δ-T, α-,β-,γ-,δ-T3 α-TAc	Merckosorb Si 60 (10 μm)	Fluorescence ex. 290 nm, em 330 nm	19
Juices, infant formulas, human milk, cereals	Extr. by ethanol and tert-butyl methyl ether	α-,β-,γ-,δ-T, α-,β-,γ-,δ-T3	LiChrospher 100 diol (5 μm) n-hexane/tert-butyl methyl ether (gradient)	Fluorescence ex. 295 nm, em 330 nm	7,20
Meat	Sap. + extr. by n-hexane	α-T	LiChrosorb Si 60 (5 μm) n-hexane/ethyl acetate (95:5)	Fluorescence ex. 295 nm, em 330 nm	8

Abbreviations used: dil. = dilution; sap. = saponification; extr. = extraction; T = Tocopherol; T3 = Tocotrienol; TAc = Tocopheryl acetate.

Tocopherols

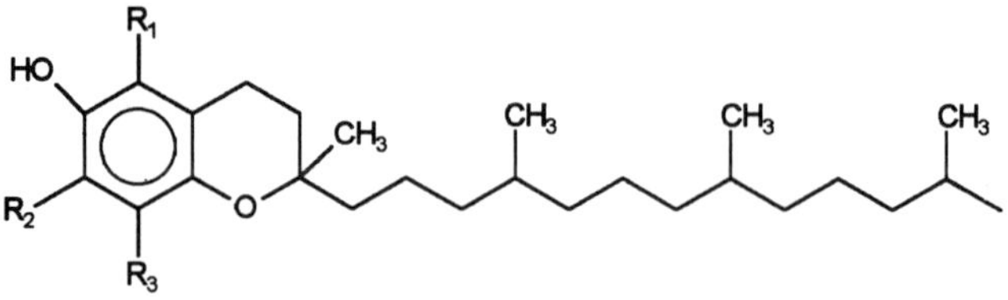

	R₁	R₂	R₃

		R_1	R_2	R_3
5,7,8-Trimethyl tocol	(α-Tocopherol)	CH_3	CH_3	CH_3
7,8-Dimethyl tocol	(β-Tocopherol)	H	CH_3	CH_3
5,8 Dimethyl tocol	(γ-Tocopherol)	CH_3	H	CH_3
8-Methyl tocol	(δ-Tocopherol)	H	H	CH_3

Tocotrienols

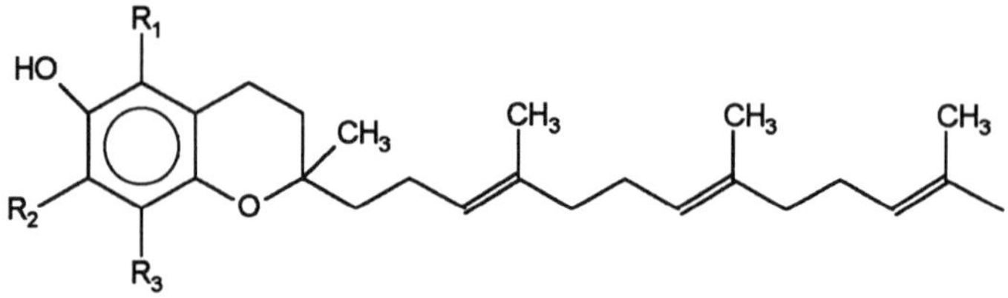

		R_1	R_2	R_3
5,7,8-Trimethyl tocotrienol	(α-tocotrienol)	CH_3	CH_3	CH_3
7,8-Dimethyl tocotrienol	(β-tocotrienol)	H	CH_3	CH_3
5,8 Dimethyl tocotrienol	(γ-tocotrienol)	CH_3	H	CH_3
8-Methyl tocotrienol	(δ-tocotrienol)	H	H	CH_3

FIG. 1. Structure of tocopherols and tocotrienols.

2,2′-bipyridyl are very sensitive but are not selective. Thus, chromatographic methods such as HPLC and GC are the methods of choice.

As already mentioned, the analysis of tocopherols requires sample preparation prior to HPLC or GC analysis. In the case of vegetable oils, except for dissolving the sample in *n*-hexane, no pretreatment is required (5). These dissolved samples with mainly unesterified tocopherols may be injected directly on the HPLC column. Extraction methods for lipid-containing foodstuffs have been published for a variety of food samples, including spinach, milk, beef, and cereal products (6). Balz et al. (7) describe a simple procedure for extraction of all homologs and α-tocopheryl acetate from infant formulas, human milk, breakfast cereals, multivitamin juices, and isotonic beverages by the addition of ethanol, *tert*-butyl methyl ether, and petroleum ether. Recoveries obtained varied between 98 and 100%.

It is in any case necessary to saponify the samples for subsequent determination of tocopherols by means of GC (in contrast to HPLC). However, saponification also offers some advantages for application in HPLC analysis; removal of the fatty acids from the sample and possible concentration of the tocopherols by evaporation of the extraction solvent. Moreover, saponification could be indicated to free tocopherols from the sample matrix or if tocopherols are esterified. Because saponification is usually carried out with potassium hydroxide in aqueous ethanol under nitrogen at about 80°C for 30–40 min, it is vitally important to protect tocopherols against oxidation by the addition of powerful antioxidants such as ascorbic acid. The saponification is followed by solvent extraction of tocopherols with *n*-hexane, *n*-hexane/ethyl acetate mixture, or solvents of similar polarity. An alternative to these time- and chemical-consuming preparation steps is reported by Pfalzgraf et al. (8), who combined all steps into one method, including saponification and single-step extraction, without sample transfer.

Since the publication of the first HPLC method for the determination of free tocopherols in vegetable oils by Van Niekerk (9), many separation systems have been proposed for application. Commonly, separation of the tocopherol homologs is carried out by normal-phase HPLC using silica gel, which offers some important advantages, such as relative insensitivity to high amounts of injected lipids. Only normal-phase HPLC with a mixture of two or three solvents (such as *n*-hexane, *n*-heptane, or diethyl ether) as major components and more polar solvents as modifiers makes it possible to separate all eight tocopherol homologs. With a flow rate of 1 to 2 mL/min and a 5-µm column packing, the analysis time will be 10 to 15 min. Commonly, detection is accomplished by UV set at 290 to 295 nm or, by reason of an improved selectivity and sensitivity, by fluorescence with excitation between 292 and 295 nm and emission between 320 and 330 nm. The elution order is

1. α-tocopherol
2. α-tocotrienol
3. β-tocopherol
4. γ-tocopherol
5. β-tocotrienol
6. γ-tocotrienol
7. δ-tocopherol
8. δ-tocotrienol

A typical chromatogram of normal-phase HPLC is shown in Fig. 2. For a selection of normal-phase HPLC methods employed, see Table 2. Official methods [IUPAC 2.432 (10) and DGF F-II 4 (11)] are based on silica gel and fluorescence (5–9,12–20).

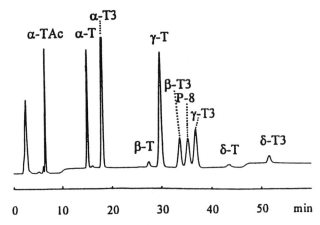

FIG. 2. HPLC chromatogram of α-tocopheryl acetate (α-TAc), tocopherols (T), tocotrienols (T3) and plastochromanol-8 (P-8). For conditions, see Ref. 7.

In contrast to normal phase, in reverse-phase HPLC, using a mobile phase of methanol and water with gradient elution, separation of all homologs—especially β-tocopherol and γ-tocopherol—is impossible. The order of elution is reversed. Moreover, a nonpolar column does not tolerate heavy loads of triglycerols and other nonpolar compounds and thus requires additional cleanup steps. Therefore, reverse-phase is used for determination of supplemental tocopherols and for the simultaneous determination of tocopherols and other fat-soluble vitamins.

Further increase in sensitivity for detection in HPLC demands the use of an electrochemical detector. As reported by Murphy and Kehrer (21), tocopherols and their quinones in chicken liver and muscle are detected 2–3 times more sensitively than by regular fluorescence detection. However, this improvement requires more extensive sample preparation prior to HPLC as well as solvents of great purity. On account of these considerations, this system cannot compete with the established HPLC separation systems.

The great significance and quality of HPLC in the analysis of tocopherols has led to a decrease in the employment of GC. As mentioned above, prior to GC analysis, sample preparation such as saponification is required. In contrast to reverse-phase HPLC, β-tocopherol and γ-tocopherol are difficult to separate with GC. However, the use of a 30-m apolar capillary column (DB-5) allows baseline separation of all four tocopherol homologs in vegetable oils within 8 min. A silylation of free hydroxy groups shows no increase in performance (22). Similar methods are the IUPAC (10) and AOCS (23) reference methods. Overall, HPLC and GC allow analysis of tocopherols in lipids and oils, but with regard to sample preparation and performance, HPLC is by far superior.

Determination of Spice Extracts

In addition to the tocopherols, extracts obtained from spices and herbs are an important source of natural antioxidants. A great variety of food spices and herbs have been shown to possess antioxidant properties (24,25).

However, for many reasons including cost, availability, and antioxidant activity, among this diversity only a few are used as potential sources of natural antioxidants. Great commercial importance has been achieved by rosemary-based preparations. Sage and oregano have also been found to constitute potent antioxidant spices (26).

As investigations have revealed, spice extracts contain a great number of substances, mainly phenolics, that have some antioxidant efficacy. The compounds mainly responsible for these antioxidant properties are the phenolic diterpenes carnosol and carnosic acid, as well as rosmanol, epirosmanol, and isorosmanol (Fig. 3) (27–32).

As use of rosemary extracts is becoming increasingly important, it is of great interest to possess analytical methods for the determination of rosemary constituents in foods, on the one hand to control the quality of the extracts and on the other hand to enable the examination of food labels. However, to date no official methods have been established. Investigations in this field of interest during the last few years offer a variety of possible methods. Most of them have been developed for the determination of phenolic compounds in rosemary extracts.

Generally, extraction of phenolic compounds from rosemary and sage leaves is based on methanol extraction (32,33); other solvents may be n-hexane, diethyl ether, chloroform, or dioxane. Yet methanol shows the highest yields of antioxidant compounds (34), so it should be used for the extraction from other foodstuffs.

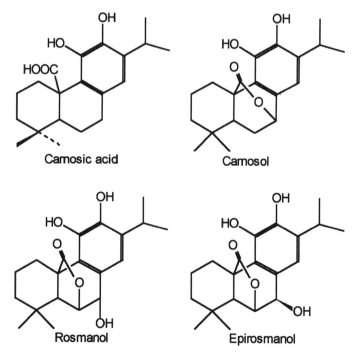

FIG. 3. Structure of phenolic diterpenes.

Because of the fairly low stability of phenolic diterpenes, sample preparation should be accelerated (35) or under protection by a strong antioxidant.

Schwarz et al. (36–38) reported a sample preparation for rosemary and sage extracts as well as for liquid fats. The lipids are extracted in a centrifuge tube with methanol and centrifuged. The methanol layer is decanted into a further centrifuge tube, kept at −20°C for precipitation of methanol-soluble fat components, and then centrifuged. The methanol layer is decanted and diluted with methanol. Recoveries for rosmanol, epirosmanol, 7-methyl-epirosmarol, carnosol, and carnosic acid varied between 90 and 93%; even naturally occurring tocopherols were coextracted (recoveries 54–59%).

Separation of these compounds and the tocopherol homolog has been realized by reverse-phase HPLC (ODS Hypersil 5 µm, 250 × 4 mm) using a methanol/water/citric acid/tetraethylene ammonium hydroxide gradient. For determination, an electrochemical detector (glassy carbon working electrode, Ag/AgCl reference electrode) was employed. A typical chromatogram is shown in Fig. 4.

Cuvelier et al. (33) describe a method for the identification and determination of natural antioxidants in sage. Methanol extracts are analyzed by reverse-phase HPLC (Shandon C18 Hypersil) with a gradient between acetonitrile/water with 1% citric acid (15:85, v/v) and methanol, and UV detection at 284 nm. In contrast to the above-mentioned method, interferences with the decomposition products of fats and rosemary extracts in foodstuffs may affect the determination.

LC–MS (liquid chromatography–mass spectrometry) may be capable of characterizing natural antioxidants. An application of LC–MS with atmospheric-pressure chemical ionization is reported by Maillard et al. (39), enabling characterization of diterpenes from rosemary and sage. These studies demonstrate that separation and determination of rosemary and sage antioxidants in foods are possible; nevertheless, applications to further foodstuffs with additional rosemary extracts have to be developed.

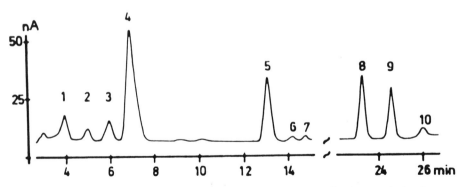

FIG. 4. HPLC chromatogram of (1) rosmanol, (2) epirosmanol, (3) 7-methyl-epirosmanol, (4) carnosol, (5) carnosic acid, (8) δ-tocopherol, (9) β-γ-tocopherol, and (10) α-tocopherol (from Ref. 36).

Determination of Other Natural Antioxidants and Synergists

Further naturally occurring compounds furnish antioxidant activity or synergism to other antioxidants.

Flavonoids and related compounds. Certain flavonoids and related compounds possess antioxidant activities in lipid-aqueous and lipid-food systems. Flavones, flavonols, flavenones, flavanones, and cinnamic acid derivatives can be assigned to this group. The flavonoids are ubiquitous and widespread in plant material, especially in leaves and fruits, whether as glycosides or aglycones (40,41). The effectiveness of flavonoids as antioxidants depends on the position and the degree of hydroxylation in the chemical structure (42). Rather, synergistic effects between α-tocopherol and flavonoids are observed. The very low solubility of these compounds in lipids is often considered a disadvantage, but suspension in the aqueous phase of a lipid-aqueous system shows acceptable protection against lipid oxidation. Unless flavonoids obtain importance as supplements, they are remarkable as naturally occurring antioxidants or synergists in foodstuffs which could influence oxidation stability.

A great variety of methods for the determination of flavonoids is available. Generally, sample preparation is based on extraction with methanol/water (4:1, v/v). A subsequent cleanup may be solid-phase extraction [e.g., Sep-pak C18 cartridges (43)] or centrifugation. Ferreres et al. (44) reported a combination of chromatography on Amberlite XAD-2 and Sephadex LH-20 columns for the purification of flavonoid fractions from honeys; samples thus prepared were analyzed by reverse-phase HPLC. Ooghe (45) used a Novapak RP18 and a mobile phase gradient (partially replacing an aqueous phosphate buffer solution by acetonitrile), and photodiode array detection to analyze flavone glycosides of oranges; 22 flavonoids and cinnamic acid derivatives from different foods were separated on a Novapak C18 column (150 mm × 3.9 mm i.d., 4 µm) with a ammonium phosphate buffer/acetonitrile/orthophosphoric acid buffer gradient. Detection was performed with a diode array UV-visible detector (46). Engelhardt et al. (47) described a method for the detection of flavonol glycosides in tea. Samples were extracted with pure methanol followed by 70% aqueous methanol, and the extracts were cleaned up by polyamide column chromatography. The flavonol glycosides were then analyzed by HPLC on a Nucleosil C18 guard column and a Hypersil-ODS (5 µm) main column with 2% acetic acid/acetonitrile (85:15, v/v) and 2% acetic acid/1,4-dioxane/methanol (77:13:10, v/v/v) mobile phases. A multichannel detector was used. Separation of 14 flavonol glycosides was possible, with recovery rates of 88.3–112.1% and detection limit of 15 mg/kg.

A good analytical tool for identifying polyphenols is the LC-thermospray-MS (48).

Phospholipids. The phospholipids phosphatidylcholine, phosphatidylserine, phosphatidylinositol, and phosphatidylethanolamine are widely distributed in foods, and pro- and antioxidant effects are attributed to them (49,50). It should, however, be emphasized that phospholipids, which are good emulsifiers, act more as synergists (51–53) or chelating agents than as actual primary antioxidants.

In addition to various methods to determine the phospholipid content by determination of phosphorus, there are suitable HPLC applications for the qualification and quantification of phospholipids. Blunck and Steinhart (54) have developed a rapid extraction method for phospholipids from bovine muscle. Phospholipids are extracted by dichloromethane/methanol, followed by a cleanup using silica gel solid-phase extraction to separate phospholipids from glycerol, cholesterol, and free fatty acids. Determination of phosphatidylcholine, phosphatidylserine, phosphatidylinositol, phosphatidylethanolamine, lysophosphatidylcholine, and lysophosphatidylethanolamine is carried out by HPLC on a Zorbax column (250 mm × 4.6 mm i.d., 7 µm) with a gradient of isooctane, tetrahydrofuran, isopropanol, chloroform, and water (55). An evaporative light scattering detector is used for detection, as also reported by Christie (56).

Christie (57) combined isocratic reverse-phase HPLC (Hypersil, 5 µm, acetonitrile/methanol/water eluent, 50:45:6.5, v/v/v) with refractometric detection. Another possibility for detection may be the use of a UV detector, but within the range of 203–210 nm it is difficult to obtain accurate results. Hence, normal-phase HPLC combined with an evaporative light-scattering detector seems to be most suitable.

Ascorbic acid. L-ascorbic acid is a substance widespread in nature, though important quantities are manufactured by chemical synthesis. L-ascorbic acid helps to improve the quality and increases the shelf life of many food products; it destroys or terminates free radical-mediated chain reactions, and it shows great synergism to tocopherols, where it acts as a transformer of tocopheryl radicals.

In addition to the enzymatic estimation after treatment of the sample with metaphosphoric acid (58), Zapata and Dufour (59) described an HPLC method for the determination of ascorbic acid along with isoascorbic acid and dehydroascorbic acid. The procedure uses a Bondapak C18 column, a mobile phase of methanol/water (5:95, v/v) containing cetrimide and potassium dihydrogen phosphate, and UV detection set at 261 nm (348 nm for dehydroascorbic acid).

A further use of the reverse-phase HPLC in the separation and quantification of ascorbic acid along with other antioxidant synergists in fatty pharmaceuticals, cosmetics and foods is described by Irache et al. (60). The samples were extracted with hot water, centrifuged, filtered, and directly injected on a Spherisorb-ODS column (150 mm × 4.6 mm i.d., 3 µm). Elution was carried out with a mobile phase of double distilled water acidified by sulfuric acid. The UV detector was set at 210 nm for antioxidant synergists and 254 nm for ascorbic acid. This method allows parallel determination of ascorbic acid and citric acid at a detection limit less than 50 mg/kg sample. Citric acid acts as a chelating reagent; it can even be determined by GC after silylation (61).

Evaluation of Natural Antioxidants

Because the use of antioxidants in edible oils and food emulsions is increasing, it is important not only to have methods for the determination of natural antioxidants in foods but also to have efficient methods to analyze the activity of antioxidants. Only in this way is it feasible to obtain information about interactions of antioxidants with the matrix, the surrounding emulsion system, and the concentration of antioxidants for optimum effectivity. There is extensive literature on this field of analysis. However, the results of the various investigations are difficult to compare on account of the different oxidation conditions and the determination of the stage of oxidation.

As reported by Frankel (62), high-temperature tests such as oxygen uptake, the active oxygen method (AOM), or the Rancimat method are not suitable for reproducing normal storage conditions, because many parameters are influenced by the elevated temperature [e.g., change in mechanism of lipid oxidation, dependence of oxidation on oxygen concentration, or decomposition of phenolic antioxidants and the food system (emulsions) (2)].

The thiobarbituric acid–reactive substances method (TBARS) is a sensitive method for the determination of oxidation products of polyunsaturated lipids. However, other components of the foodstuff may interfere with the TBARS color complex.

The spectrophotometric method (63) for carotene bleaching by co-oxidation of linoleic acid is simple and very sensitive. However, interferences from oxidizing and reducing agents present in crude extracts may occur.

Further assays use manifold methods (2,64); however they all have the disadvantage of nonspecificity. They are practical for a simple and fast screening of a large sample number. If a closer examination of the interaction of natural antioxidants with the food system during the lipid oxidation is necessary, other methods are required.

It is suggested that standardized oxidation and determination methods be developed to obtain reliable information on the activity of natural antioxidants in different food lipid systems. Testing should be carried out at storage temperatures of 40–60°C to have accelerated lipid oxidation but avoid high-temperature effects on the mechanism of lipid oxidation. During oxidation, samples have to be shaken efficiently (62). The degree of oxidation should be determined at suitable time intervals (depending on temperature and lipid system used) by more than one method, measuring initial oxidation products (hydroperoxides) and decomposition products (carbonyl compounds) (65). Hydroperoxides can be measured by determining the peroxide value or the level of conjugated dienes by spectrophotometry at 234 nm (66). The decomposition products as carbonyl compounds can be analyzed by use of static headspace GC (67) or HPLC after derivatization (68). α-tocopherol, or BHA and BHT at the same molar concentrations as the tested antioxidants, should be used as a reference standard to enable a comparison with other results.

References

1. Parrish, D.B. (1980) Determination of Vitamin E in Foods—A Review, *CRC Crit. Rev. Food Sci. Nutr. 13,* 161–186.
2. Kochhar, S.P., and Rossell, J.B. (1990) Detection, Estimation and Evaluation of Antioxidants in Food Systems, in *Food Antioxidants,* Hudson, B.J.F., Elsevier Applied Science, London, pp. 19–64.

3. Bourgeois, C. (1992) *Determination of Vitamin E: Tocopherols and Tocotrienols*, Elsevier Applied Science Publishers Ltd., Barking, United Kingdom.
4. Bourgeois, C. (1992) Determination of Vitamin E in Oils and Fats: Tocopherols and Tocotrienols, *Lipid Technol. 4*, 143–145.
5. Coors, U. (1991) Anwendung des Tocopherolmusters zur Erkennung von Fett- und Ölmischungen, *Fat Sci. Technol. 90*, 519–526.
6. Thompson, J.N., and Hatina, G. (1979) Quantitative Determination of Tocopherols and Tocotrienols in Foods and Tissues by HPLC, *J. Liq. Chromatogr. 2*, 327–344.
7. Balz, M.K., and Thier, H.P. (1993) Simultaneous Determination of α-Tocopheryl Acetate, Tocopherols and Tocotrienols by HPLC with Fluorescence Detection in Foods, *Fat Sci. Technol. 95*, 215–220.
8. Pfalzgraf, A., Steinhart, H., and Frigg, M. (1995) Rapid Determination Of α-Tocopherol in Muscle and Adipose Tissues of Pork, *Z. Lebensm. Unters. Forsch. 200*, 190–193.
9. Van Niekerk, P.J. (1973) Direct Determination of Free Tocopherols in Plant Oils by Liquid-Solid Chromatography, *Anal. Biochem. 52*, 533–537.
10. Paquot, C., and Hautfenne, A., (1986) *Standard Methods for the Analysis of Oils, Fats and Derivatives,* 7th edn., International Union of Pure and Applied Chemistry, Blackwell/Oxford.
11. Deutsche Gesellschaft für Fettwissenschaft (DGF) (1987) *Deutsche Einheitsmethoden zur Untersuchung von Fetten, Fettprodukten und verwandten Stoffen,* Wissenschaftliche Verlagsgesellschaft, Stuttgart.
12. Deldime, P., Lefebvre, G. Sadin, Y., and Wybauw, M. (1980) Analysis of Tocopherols in Vegetable Oils by HPLC, *Rev. Franç. Corps Gras 27*, 279–282.
13. Barnes, P.J., and Taylor, P. (1980) The Composition of Acyl Lipids and Tocopherols in Wheat Germ Oils from Various Sources, *J. Sci. Food Agric. 31*, 997–1006.
14. Taylor, P.W., and Barnes, P.J. (1981) Analysis for Vitamin E in Edible Oils by HPLC, *Chem. Ind. 20*, 722–726.
15. Reynolds, S.L. (1985) Determination of Tocopherols in Milk Products, *Proc. Inst. Food Sci. Technol. 18*, 43–47.
16. Rammell, C.G., and Hoogenboom, J.J.L. (1985) Separation of Tocols by HPLC on an Amino-Cyano Polar Phase Column, *J. Liq. Chromatogr. 8*, 707–717.
17. Coors, U., and Montag, A. (1985) Untersuchungen zum Tocopherolgehalt in Milchfett, *Milchwissenschaft 40*, 470–473.
18. Abe, K. Yuguchi, Y., and Katsui, G. (1975) Qualitative Determination of Tocopherols by HPLC, *J. Nutr. Sci. Vitaminol. 21*, 183–188.
19. Van Niekerk, P.J., and du Plessis, L.M. (1976) The Application of Liquid Chromatography to Food Analysis, *S. Afr. Food Rev. 3*, 167–171.
20. Balz, M.K., Schulte, E., and Thier, H.P. (1992) Trennung von Tocopherolen durch HPLC, *Fat Sci. Technol. 94*, 209–213.
21. Murphy, M.E., and Kehrer, J.P. (1987) Simultaneous Measurement of Tocopherols and Tocopheryl Quinones in Tissue Fractions Using HPLC with Redox-Cycling Electrochemical Detection, *J. Chromatogr. 421*, 71–82.
22. Ulberth, F., Reich, H., and Kneifel, W. (1992) Zur Analytik von Tocopherolen—Ein Methodenvergleich zwischen HPLC und GC, *Fat Sci. Technol. 94*, 51–54.
23. *Official and Tentative Methods of the American Oil Chemists' Society,* (1986) American Oil Chemists' Society, Champaign, Illinois.
24. Chipault, J.R., Mizuno, G.R., Hawkins, J.M., and Lundberg, W.O. (1952) The Antioxidant Properties of Natural Spices, *Food Res. 17*, 46–55.

25. Chipault, J.R., Mizuno, G.R., and Lundberg, W.O. (1955) Antioxidant Properties of Spices in Oil-In-Water Emulsions, *Food Res. 20,* 443–448.
26. Lölinger, J. (1991) Natural Antioxidants, *Lipid Technol. 3,* 58–61.
27. Brieskorn, C.H., and Dömling, H.J. (1969) Carnolsäure, der wichtige antioxidativ wirksame Inhaltsstoff des Rosmarin- und Salbeiblattes, *Z. Lebensm. Unters. Forsch. 141,* 10–16.
28. Inatani, R., Nakatani, N., and Fuwa, H. (1983)Two Antioxidative Diterpenes from Rosemary and a Revised Structure for Rosmanol, *Agric. Biol. Chem. 48,* 2081–2085.
29. Chang, S.S., Ostric-Matijasevic, B., Hsieh, O.A.L., and Huang, C.L. (1977) Natural Antioxidants from Rosemary and Sage, *J. Food Sci. 42,* 1102–1106.
30. Houlihan, C.M., Ho, C.-T., and Chang, S.S. (1984) Elucidation of the Chemical Structure of a Novel Antioxidant, Rosmarindiphenol, Isolated from Rosemary, *J. Am. Oil Chem. Soc. 61,* 1036–1039.
31. Houlihan, C.M., Ho, C.-T., and Chang, S.S. (1985) The Structure of Rosmariquinone— A New Antioxidant Isolated from *Rosmarinus officinalis* L., *J. Am. Oil Chem. Soc. 62,* 96–99.
32. Wu, J.W., Lee, M.-H., Ho, C.-T., and Chang, S.S. (1982) Elucidation of the Chemical Structure of Natural Antioxidants Isolated from Rosemary, *J. Am. Oil Chem. Soc. 59,* 339–345.
33. Cuvelier, M.E., Berset, C., and Richard, H. (1994) Separation of Major Antioxidants in Sage by HPLC, *Sci. Aliments 14,* 811–815.
34. Chang, S.S., Ostric-Matijasevic, B., Hsieh, O.A.L., and Huang, C.L. (1977) Natural Antioxidants from Rosemary and Sage, *J. Food Sci. 42,* 1102–1106.
35. Brieskorn, C.H., and Dömling, H.J. (1969) Natürliche und synthetische Derivate der Carnolsäure, *Arch. Pharm. 302,* 641–649.
36. Schwarz, K., and Ternes, W. (1992) Antioxidative Constituents of *Rosmarinus officinalis* and *Salvia officinalis* I. Determination of Phenolic Diterpenes with Antioxidative Activity Amongst Tocochromanols Using HPLC, *Z. Lebensm. Unters. Forsch. 195,* 95–98.
37. Schwarz, K., and Ternes, W. (1992) Antioxidative Constituents of *Rosmarinus officinalis* and *Salvia officinalis* II. Isolation of Carnosic Acid and Formation of other Phenolic Diterpenes, *Z. Lebensm. Unters. Forsch. 195,* 99–103.
38. Schwarz, K., Ternes, W., and Schmauderer, E. (1992*)* Antioxidative Constituents of *Rosmarinus officinalis* and *Salvia officinalis* III. Stability of Phenolic Diterpenes of Rosemary Extracts Under Thermal Stress as Required for Technological Processes*, Z. Lebensm. Unters. Forsch. 195,* 104–107.
39. Maillard, M.N., Giampaoli, P., and Cuvelier, M.E. (1996) Atmospheric Pressure Chemical Ionization Liquid Chromatography–Mass Spectrometry: Characterization of Natural Antioxidants, *Talanta 43,* 339–347.
40. Herrmann, K.J. (1976) Flavonols and Flavones in Food Plants: A Review, *Food Technol. 11,* 433–448.
41. Vekari, S.A., Oreopulou, C.T., and Thomopoulos, C.D. (1993) Oregano Flavonoids as Lipid Antioxidants, *J. Am. Oil Chem. Soc. 70,* 483–487.
42. Nieto, S., Garrido, A., Shanhuza, J., Loyola, L.A., Morales, G., Leighton, F., and Valenzuela, A. (1993) Flavonoids as Stabilizers of Fish Oil: An Alternative to Synthetic Antioxidants, *J. Am. Oil Chem. Soc. 70,* 773–778.
43. Terada, H., and Miyabe, M. (1993) Determination of Rutin and Quercetin in Processed Foods by Fast Semi-Micro High-Performance Liquid Chromatography, *J. Food Hyg. Soc. Jpn. 34,* 385–391.

44. Ferreres, F., Andrade, P., and Tomas-Barberan, F.A. (1994) Flavonoids from Portuguese Heather Honey, *Z. Lebensm. Unters. Forsch. 199*, 32–37.
45. Ooghe, W.C., Ooghe, S.J., Detavernier, C.M., and Huyghebaert, A. (1994) Characterization of Orange Juice (*Citrus sinensis*) by Flavanone Glycosides, *J. Agric. Food Chem. 42*, 2183–2190.
46. Lamuela-Raventos, R.M., and Waterhouse, A.L. (1994) A Direct HPLC Separation of Wine Phenolics, *Am. J. Enol. Viti. 45*, 1–5.
47. Engelhardt, U.H., Finger, A., Herzig, B., and Kuhr, S. (1992) Determination of Flavonol Glycosides in Black, *Dtsch. Lebensm. Rundsch. 88*, 69–73.
48. Kiehne, A., and Engelhardt, U.H. (1996) Thermospray–LC–MS Analysis of Various Groups of Polyphenols in Tea. I. Catechins, Flavonol O-Glycosides and Flavone C-Glycosides, *Z. Lebensm. Unters. Forsch. 202*, 48–54.
49. Kaur, N., Sukhija, P.S., and Bhatia, I.S. (1982) A Comparison of Seed Phosphatides and Synthetic Compounds as Antioxidants for Cow and Buffalo Ghee (Butter Fat), *J. Sci. Food Agric. 33*, 576–578.
50. Olcott, H.S., and Van der Veen, J. (1963) Role of Individual Phospholipids as Antioxidants, *J. Food Sci. 28*, 313–315.
51. Dziedic, S.Z., and Hudson, B.J.F. (1984) Phosphatidylethanolamine as a Synergist for Primary Antioxidants in Edible Oils, *J. Am. Oil Chem. Soc. 61*, 1042–1045.
52. Kwon, T.W., Snyder, H.E., and Brown, H.G. (1984) Oxidative Stability of Soybean Oil at Different Stages of Refining, *J. Am. Oil Chem. Soc. 61*, 1843–1846.
53. Hamzawi, L.F. (1990) Role of Phospholipids and α-Tocopherol as Natural Antioxidants in Buffalo Butterfat, *Milchwissenschaft 45*, 95–97.
54. Blunk, H.C., and Steinhart, H. (1990) Separation of Phospholipids in Bovine Tissue with Disposable Silica Gel Extraction Columns, *Z. Lebensm. Unters. Forsch. 190*, 123–125.
55. Melton, S.L. (1992) Analysis of Soybean Lecithins and Beef Phospholipids by HPLC with an Evaporative Light-Scattering Detector, *J. Am. Oil Chem. Soc. 69*, 784–788.
56. Christie, W.W. (1986) Separation of Lipid Classes by HPLC with the "Mass Detector," *J. Chromatogr. 361*, 396–399.
57. Christie, W.W., and Hunter, M.L. (1984) HPLC in the Analysis of the Products of Phospholipase: A Hydrolysis Of Phosphatidylcholine, *J. Chromatogr. 294*, 489–493.
58. Boehringer Mannheim GmbH, Biochemica (1989) *Methoden der biochemischen Analytik und Lebensmittelanalytik*, Mannheim, pp. 24–26.
59. Zapata, S., and Dufour, J.P. (1992) Ascorbic Acid, Dehydroascorbic Acid and Isoascorbic Acid Simultaneous Determinations by Reverse-Phase Ion Interaction HPLC, *J. Food Sci. 57*, 506–511.
60. Irache, J.M., Ezpeleta,I., and Vega, F.A. (1993) HPLC Determination of Antioxidant Synergists and Ascorbic Acid in Some Fatty Pharmaceuticals, Cosmetics and Foods, *Chromatographia 35*, 232–236.
61. Kim, K.R., Zlatkis, A., Horning, E.C., and Middleditch, B.S. (1989) Gas Chromatography of Volatile and Nonvolatile Carboxylic Acids as *tert*-Butyldimethylsilyl Derivatives, *J. Chromatogr. Commun. 10*, 289–301.
62. Frankel, E.N. (1993) In Search of Better Methods to Evaluate Natural Antioxidants and Oxidative Stability in Food Lipids, *Trends Food Sci. Technol. 4*, 220–225.
63. Mallet, J.F., Cerrati, C., Ucciani, E., Gamisans, J., and Gruber, M. (1994) Antioxidant Activity of Plant Leaves in Relation to Their α-Tocopherol Content, *Food Chem. 49*, 61–65.

64. Madsen, H.L., Nielsen, B.R., Bertelsen, G., and Skibsted, L.H. (1996) Screening of Antioxidative Activity of Spices. A Comparison Between Assays Based on ESR Spin Trapping and Electrochemical Measurement of Oxygen Consumption, *Food Chem. 57*, 331–337.
65. Frankel, E.N. (1995) Natural and Biological Antioxidants in Foods and Biological Systems. Their Mechanism of Action, Applications and Implications, *Lipid Technol.*, 77–80.
66. Huang, S.W., Frankel, E.N., Schwarz, K., Aeschbach, R., and German, J.B. (1996) Antioxidant Activity of Carnosic Acid and Methyl Carnosate in Bulk Oils and in Oil-In-Water Emulsions, *J. Agric. Food Chem. 44*, 2951–2956.
67. Frankel, E.N. (1993) Formation of Headspace Volatiles by Thermal Decomposition of Oxidized Oils vs. Oxidized Vegetable Oils, *J. Am. Oil Chem. Soc. 70*, 767–772.
68. Reindl, B., and Stan, H.J. (1982) Determination of Volatile Aldehydes in Meat as 2,4-Dinitrophenylhydrazones Using RP–HPLC, *J. Agric. Food Chem. 30*, 849–854.

Chapter 17

Coffee Lipids: Analysis of the Diterpene 16-*O*-methylcafestol as an Indicator of Admixing of Coffees

N. Sehat[a] and G. Niedwetzki[b]

[a]Food and Drug Administration, Center for Food Safety and Applied Nutrition, 200 C Street, S.W., HFS-175, Washington, DC 20204, USA, and [b]Handels und Umweltschutzlaboratorium, Stenzelring 14b, 21107 Hamburg, Germany

Introduction

The coffee plant, which originated in the highlands of Abyssinia, belongs to the botanical genus *Cofea*. Coffee has been consumed in Europe since the 17th century (1). At the present time, coffee is grown in many countries in South America and Asia between the latitudes 20°N and 20°S. The most important botanic species are *arabica*, including the varieties *typical, bourbon maragogype,* and others; *canephora*, with the varieties *robusta, quillou, Uganda,* and others; *Liberia;* and *mocca* (2). The most favorable average annual temperature range for growing coffee is between 15 and 25°C. Unlike the *robusta* types, *arabica* coffee is sensitive to high heat and high atmospheric humidity and thrives only in higher elevations, where temperatures of about 20°C are typically found. Robusta coffees are grown in lower elevations in the tropics.

Crude coffee beans are about 1 cm long and yellowish-green to blue-green in color. The *arabica* beans are rather slim and elongated, whereas *robusta* beans are smaller and round or irregularly shaped.

C. arabica coffee is characterized sensorially by a delicate (or subtle) acidity and full flavor and is preferred to *robusta* coffee in the Scandinavian countries, Germany, and the United States. Mixtures (blends) containing *robusta* coffees are consumed to a greater degree in Southern European countries, in Turkey, and in North Africa. Depending upon their types, the tastes of these coffees vary from very tart (sharp) to bitter, and they are less aromatic than *arabica* coffee.

For consumers, the most important criteria for the choice of a coffee are its taste and quality. In addition, personal likes and dislikes are based on individual responses to, for example, caffeine and other irritating substances in coffees.

Economic Importance of Coffee

World coffee production in 1991 was 4.5 million tons. In world commerce, international trade in coffees is second only to that of crude oil (3). For this reason, the adulteration of premium coffees with coffees of lesser quality is of considerable economic importance.

Analytically verified species-specific differences between *Cofea arabica* and *C. canephora robusta* include their amounts of fat, caffeine, trigonelline, and chlorogenic acid (2). *C. arabica* and *robusta* coffees also differ significantly in their amounts of saccharose and reducing sugars (e.g., glucose, fructose, galactose) (4). However, these latter variations are not suitable for use in the identification of admixtures of coffees because of their lack of specificity.

Coffee Lipids

The total lipid content of coffee varies with the species from about 7 to 17%, with average values of about 15% for *arabica* and 10% for robusta types (1,5–7). The largest fraction consists of liquid coffee oil, which is found in the endosperm. About 0.24% (i.e., between 1–2% of the total lipid) is referred to as "coffee wax." This component forms a thin, firm layer on the surface of the coffee bean (2). Most investigations have been carried out on coffee wax, which is obtained by partial extraction of whole coffee beans with lipophilic solvents or on total coffee lipids (including coffee wax), following extraction of the ground beans with petroleum ether.

Total lipids. The lipid composition of green coffee beans is as follows: triacylglycerol, 75.2%; diterpene esters, 18.5%; free diterpenes, 0.4%; sterol esters, 3.2%; free sterols, 2.2%; phosphatides, 0.1–0.5%; carboxylic acid-5-hydroxytryptamides, 0.6–1.0%; and tocopherols, 0.04–0.06% (1). The physical characteristics (indices) of the petroleum ether extract are summarized by Maier (1). The large deviations in the unsaponifiable material are due to the use of different solvents (8). For example, much less coffee lipid is dissolved in petroleum ether than in diethyl ether.

Fatty acids and triacylglycerides. As can be seen from the low acid index (1), only small amounts of the fatty acids in green coffee beans are present in their free form. The largest fraction is esterified with glycerol. The predominant components of coffee lipids are triacylglycerides at 75.2%. The triacylglycerides are esters of long-chain unbranched carboxylic acids with the trihydric alcohol glycerol. In addition to the fatty acids bound to glycerol, fatty acids in green coffee beans also occur bound to diterpenes (10%), to sterols (1.5%), and to 5-hydroxytryptamine (0.7%), as well as in the free form (1). The free fatty acids in extracts of coffee lipids can be quantitated by conventional methods.

A higher acid index correlates with the length of storage of green coffee (9) and with sensory alterations in brews prepared from the roasted coffee beans.

Total fatty acid composition has been thoroughly investigated by Pokorny and Forman (10), Roffi, *et al.* (11), Streuli (2), Vitzthum (12), and Wurziger (13) and has been summarized by Maier (1). Data from Kaufmann and Hamsagar (14) regarding the composition of fatty acids from triacylglycerides and diterpene esters are also included (1). The composition of fatty acids from diterpene esters is as follows: palmitic acid, 42.5%; stearic acid. 17.5%; oleic acid, 11.0%; linoleic acid, 20.5%; arachidic acid, 6.0%; and behenic acid, 2.5% (1).

Coffee beans are covered by a thin film of solid lipids (coffee wax), which comprises 1–2% of the total coffee lipids. The distribution of fatty acids in coffee wax differs from that in coffee oil. The long-chain fatty acids arachic, behenic, and lignoceric, which together comprise only about 4% of the fatty acids of coffee oil, make up 40% of the fatty acid content of coffee wax. Twenty-one percent consists of behenic acid (8). This difference in distribution of fatty acids is shown clearly with the carboxylic acid-5-hydroxytryptamides, a group of compounds found only in coffee wax. Only arachic, behenic, and lignoceric acids are found in the carboxylic acid-5-hydroxytryptamides, with behenic acid making up almost 60% of the total fatty acids (15,16).

Sterols. The most important sterols occurring in coffee lipids are sitosterol, stigmasterol, campesterol, and δ-5-avenasterol. Altogether, 17 sterols have been detected in coffee, 13 of which have been positively identified (1,17,18). These are all sterols that also occur in other plants. The occurrence of lanosterol and dihydrolanosterol has not been confirmed. The amounts of the most important sterols in a sample of *arabica* coffee analyzed by Nagasampagi, et al. (17) are listed as percentages of the total sterols, which make up 5.4% of the lipids. Those researchers also found 0.4% stigmastenol, 0.1% campestanol, and cycloartenol as well as traces of cholesterol and cholestanol. Probably 4-α, 24R-dimethyl-5-α–cholest-8-ene-3-β-ol, 4-α; 24R-dimethyl-5-α–cholest-7-ene-3-β-ol; and 4-α–methyl-5-α–stigmast-5-ene-3-β-ol are also present.

Diterpenes and the significance of diterpene 16-O-methylcafestol (16-OMC). Terpenes are formed from isoprene units (C_5H_8). Mono-, di-, tri-, tetra-, sesqui-, and polyterpenes can be distinguished, based on the number of isoprene units. Monoterpenes contain two, and diterpenes four, isoprene units (19). The diterpenes occurring in the coffee lipid are the three pentacyclic diterpenes kahweol, cafestol and 16-O-methylcafestol. These diterpenes have not been detected in foods other than coffee. A survey of the chemistry of coffee diterpenes and their esters is given by Kaufmann and Hamsagar (14,20). Because of their high content of diterpenes, which are physiologically unwholesome and cannot be removed by normal refining processes, coffee lipids are considered unfit for human consumption or for use as cooking oil (14).

In the 1930s, two diterpenes were identified in the unsaponifiable fraction of coffee oil. Bengis and Anderson (21) were the first to isolate kahweol. Slotta and Neisser (22) identified cafestol, which they initially named cafesterol, because they mistook it for a steroid.

Between 1942 and 1960, working groups under Wettstein (23–28), Chakravorty (29–31), Djerassi (32–37), and Haworth (38–40) dealt with identifying the structure of the two coffee diterpenes. Kaufmann and Sen Gupta (41,42) prepared kahweol as a pure substance and discovered that kahweol differs from cafestol only by an additional double bond, a finding that was later confirmed by mass spectrometry (43).

Cafestol is the primary diterpene component in *arabica* coffee, with kahweol making up to 15% of the cafestol (1). Although Wurziger (44) reported that kahweol does not occur in *robusta* coffee, Pettitt (45) and Speer and Mischnik (46) subsequently confirmed the presence of small amounts of this diterpene in *robusta* coffees.

The diterpenes occur largely as fatty acid esters, esterified at the primary hydroxyl group. Surveys of the chemistry of the diterpenes and their esters, their chromatography, and their IR spectra are given by Kaufmann and Hamsagar (14,20,47). Free cafestol forms colorless crystals with a melting point of 158–160°C. It is optically active (α-95° in chloroform). In comparison with the other coffee lipids, it is relatively poorly soluble in petroleum ether and is more soluble in slightly more polar organic solvents such as ether, chloroform, acetone, and benzene. The purification is best carried out by recrystallization from diethyl ether. The fatty acid esters (for preparation, see Ref. 14) are liquids to waxy crystalline masses that are soluble in nearly all organic solvents, are insoluble in water, and, insofar as they are solid, recrystallizable from acetone. They are optically active (α-30° to α-60°); like free cafestol they are very sensitive to acids and tend to polymerize easily.

Kahweol has two additional double bonds in conjugation to the furan ring and therefore shows strong UV absorption at 290 nm. It is more strongly levo (left) rotatory than cafestol. Kahweol's other physical and chemical characteristics are very similar to those of cafestol, but it is more sensitive to acids, heat, and light. If heated with potassium iodide/acetic acid, a blue color results (44,48). The amounts of kahweol decrease with the commercial treatment of green coffee and also with steam treatment and heating.

Cafestol, kahweol, and their esters give color reactions with antimony trichloride (2) and (on paper) with phosphormolybdenic acid (14). A tricyclic diterpene was found by Nagasampagi et al. (17) in small amounts. According to Wahlberg et al. (49), this tricyclic compound is 19-hydroxy-(–)-kaur-(16/17)-ene. Other diterpenes occur in the form of glycosides. However, these compounds are not considered to be lipids.

Because of economic considerations, coffee researchers have long been interested in distinguishing *arabica* and *robusta* coffees. When unground, the coffee species can be distinguished by the shape of their beans. However, once ground and roasted, the addition of *robusta* coffee to *arabica* coffee can be detected by sensory investigations only if more than 10–15% *robusta* has been added. Therefore, the need for chemical methods for identifying admixtures is increasingly important.

For example, Wurziger (44,48) used a color reaction in attempts to identify coffee admixtures. Many researchers have looked for differences in the composition or distribution of specific components. Nurok et al. (50) investigated highly volatile sulfur compounds. Tiscornia et al. (51), Duplatre et al. (52), Picard et al. (53) and Mariani and Fedeli (54) studied the occurrence and distribution of sterols. Nackunstz (55) and Nackunstz and Maier (56) studied the diterpenes cafestol and kahweol. The addition of *arabica* coffee to *robusta* mixtures can be identified by measuring the diterpenes cafestol and kahweol, but this measurement cannot detect the addition of *robusta* coffee to *arabica* coffee.

Dickhaut (1985, unpublished communication) discovered a substance in the unsaponifiable part of *robusta* coffee oil that does not occur in *arabica* coffee oil; it was detected in the thin-layer chromatogram, after detection with vanillin/sulfuric acid, as a red spot. Pettitt (45) detected a diterpene in *robusta* coffee oil, which he identified as 16-methoxycafestol, in his investigations on the distribution of diterpene fatty acid esters in *arabica* and *robusta* coffees. However, its structural formula was established by Speer and Mischnik (46) by use of high-resolution mass spectrometry as 16-*O*-methylcafestol (16-OMC). Structure confirmation by synthesis of this new diterpene was first carried out by Speer and Mischnik (46). The structural formulas of the coffee diterpenes kahweol, cafestol, and 16-OMC are shown in Fig. 1.

An HPLC method for determining the amounts of diterpene in the unsaponifiable fraction of coffee lipids was developed by Speer (57). The diterpene 16-OMC has been found in every *robusta* coffee investigated, whereas it has not been detected in *arabica* coffee. The amount of 16-OMC in *robusta* coffee dry matter is between 0.6 and 1.8 g/kg. The arithmetic mean and the median value for 20 samples investigated is 1.1 g/kg (58). Determination of 16-OMC is suitable for the detection of small amounts of *robusta* coffee (as low as 2%) in *arabica* coffee. Further investigations showed that only 1.0 to 3.4% of 16-OMC occurs in the free form (59), whereas the largest part is esterified with fatty acids.

The presence of 16-OMC in *robusta* but not in *arabica* coffees makes it possible to identify admixtures of these coffees.

16-*O*-Methylcafestol (16-OMC)

After the discovery and synthesis of 16-OMC, a method for isolating it from natural sources in order to use it as a standard substance was developed by Speer and Montag (60). For this purpose, *robusta* green coffee lipid was saponified and extracted with tertiary butyl methyl ether (*t*-BME). After solid-phase extraction, the

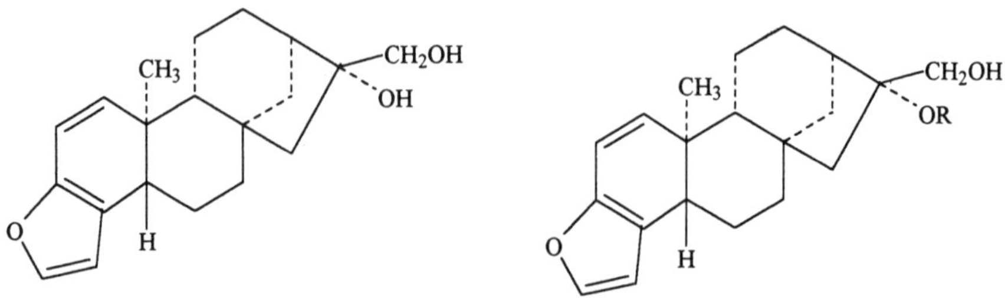

FIG. 1. Structural formulas of the coffee diterpenes.

16-OMC was isolated from the diterpene fraction by semipreparative HPLC. This method was very time- and labor- intensive and could not be automated.

A simpler, automatable method was developed by Sehat (61). The 16-OMC is no longer prepared from the total unsaponifiable fraction of the coffee lipid, but rather by fractionation of the diterpene esters. In this way, the liquid–liquid extraction of the saponified solution and subsequent manual solid-phase extractions can be avoided. The preparation of the diterpene ester fraction by gel permeation chromatography (GPC) can be automated.

In the following section, the preparation of 16-OMC, outlined in Fig. 2, is described in detail.

Extraction and Purification from Natural Sources

Green *robusta* coffee beans are ground in a cutting mill with addition of dry ice to avoid heating, followed by partial roasting of the green coffee (56). The coffee powder is stored at −20°C under nitrogen. The method of Weibull-Stoldt (62,63), which is frequently used for the extraction of lipid from foods, cannot be applied to the extraction of total coffee lipid because the digestion with hydrochloric acid would destroy the diterpenes (55). Therefore, a Soxhlet extraction is performed. The ground green coffee is first mixed with anhydrous sodium sulfate and then extracted with *t*-BME. *t*-BME is used in place of petroleum ether (20) because the diterpene esters, which are not totally nonpolar, are better solubilized. *t*-BME is used increasingly in place of diethyl ether in laboratory applications, because in contrast to diethyl ether it does not tend to form peroxides, and thus the danger of explosions is reduced. Following extraction, the solvent is distilled, remaining solvent is removed under nitrogen, and the residue (coffee lipid) is dried in a desiccator and weighed.

Gel permeation chromatography (GPC). GPC is used to isolate the diterpene fatty acid esters from the coffee lipid. Small molecules in the mixture can enter pores in the gel and move slowly down the column; large molecules, which cannot enter the pores, move more quickly. Thus, mixtures of molecules can be separated on the basis of their size. For this reason, GPC is also referred to as "size exclusion chromatography." In addition to the molar mass, the steric construction, degree of solvation, and degree and swelling of the molecule in the solvent influence the separation. For adequate separation of two substances, the difference in their molar masses should be at least 10% (64).

The diterpene fatty acid esters are separated from the triacylglycerol, the sterol esters, and the free fatty acids by GPC. The stationary phase is a granulated polystyrol gel (Bio Beads S-X_3) where the designation "3" indicates that the polystyrol gel is esterified with 3% of divinyl benzene.

In previous studies, a mixture of ethyl acetate and cyclohexane (1:1, v/v), a combination of solvents frequently used in pesticide analysis, was used for the GPC separation of coffee lipids. Use of this mixture resulted in long chromatography

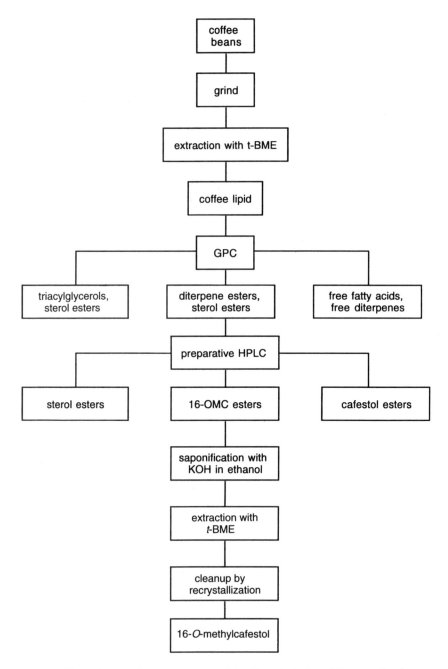

FIG. 2. Flow diagram for preparation of 16-O-methylcafestol from coffee beans (Ref. 61).

runs (49 min), high consumption of nonreusable solvents, and incomplete separation of fractions. An improved method was developed by Sehat (61) that used only dichloromethane as the solvent and required only 35 min for GPC separation. A verification of the composition of the fractions is carried out regularly by thin layer chromatography (TLC).

TLC rapid method for the identification of lipid fractions. TLC can be used to confirm the purity and composition of the GPC eluate fractions. The TLC stationary phase was silica gel (Alugram Sil G/UV); the mobile phase was a mixture of n-hexane, diethyl ether and acetic acid (90:10:1, v/v/v) by this method. The sterol esters, 16-OMC esters, free fatty acids, cafestol esters, free 16-OMC and free cafestol can be separated, whereas the triacylglycerol has the same R_f value as the 16-OMC esters. The R_f values of free fatty acids and cafestol esters are very close to one another. If the TLC plate is sprayed with vanillin/sulfuric acid in the first detection, the diterpene esters and free diterpenes appear as red spots, while the other components remain invisible. Therefore, there are no difficulties in detecting the 16-OMC esters and cafestol esters. By a second detection with 20% molybdophosphoric acid and subsequent heating in a drying oven at 80°C, the free and the esterified sterols, the triacylglycerol and the free fatty acids are made to appear as dark blue spots.

Isolation of 16-OMC fatty acid esters by semipreparative HPLC. The 16-OMC esters can be isolated from the GPC diterpene esters fraction by semipreparative HPLC. After the eluent dichloromethane has been removed under vacuum, the collected diterpene ester fractions from several GPC runs are dissolved in n-hexane:ethyl acetate (8:2, v/v), filtered through a 0.2-μm membrane filter, and injected on a semipreparative HPLC (61). UV detection of the diterpenes at 220 nm is not possible, because the n-hexane in the eluent is not transparent to UV light at this wavelength.

Preparation of 16-OMC fatty acid esters: Saponification and extraction. The 16-OMC fatty acid ester fractions collected during the semipreparative HPLC runs are evaporated, and residual solvent is removed under nitrogen. The saponification is based on the same principle as that for the determination of diterpenes by the Speer method (57) with 10% ethanolic potassium hydroxide solution (about 1 g ester per 100 mL). Sodium ascorbate is added before saponification to avoid oxidation. After two hours of reflux boiling, the alcohol is evaporated and the residue dissolved in 100 mL of warm water (75°C) and 40 mL of 10% sodium chloride solution.

After cooling to room temperature, the saponification solution is extracted with t-BME. The emulsions that form are broken by addition of sodium chloride and ethanol. The combined t-BME phases are washed with 2% sodium chloride solution and filtered over anhydrous sodium sulfate.

Clean-up and tests for purity and identity. The volume of the t-BME phase is reduced to about 20 mL. Because diterpenes are not highly soluble in nonpolar solvents such as n-hexane, the polarity of the t-BME solution is decreased by slow addition of n-hexane until the 16-OMC precipitates. t-BME can be evaporated while some n-hexane remains. This phase contains small amounts of sterols, as the TLC control shows. Therefore, n-hexane is added to the 16-OMC; the mixture is brought to the boiling point and refrigerated, and the supernatant solution containing the sterols is removed. This procedure, during which some loss of 16-OMC occurs, is repeated about four times.

The purity of the 16-OMC was tested by the following methods: TLC, HPLC with diode array UV detection, capillary gas chromatography (CGC) with FID detection, and mass spectrometry. The HPLC was carried out by the method of Speer (57). However, instead of a simple UV detector, a diode array detector was used, taking UV spectra between 190 and 350 nm during the entire run. The 16-OMC peak proved to be very pure in the peak purity control. No other peaks were detected.

Gas chromatographic detection provides an advantage over HPLC with UV detection because substances such as sterols are detected with high sensitivity by CGC/FID. However, analysis of a derivative is necessary for GC of coffee diterpenes. By silylation of the free hydroxy groups, the 16-OMC reacts as its O-trimethyl silyl derivate.

The mass spectrometry of the prepared 16-OMC gives the same EI mass spectrum as that of synthesized 16-OMC (46).

Synthesis and Purification of 16-OMC Fatty Acid Esters

16-OMC fatty acid esters can by obtained by chemical synthesis for use as analytical standard substances. 16-OMC as an alcohol is esterified with fatty acids such as palmitic, stearic, oleic, linolic, linoleic, arachic, and behenic acids (61).

The esterification of alcohols is far easier via reaction with the acid chloride or acid anhydride than with the underivatized carboxylic acid. This is especially true for sterically hindered alcohols such as 16-OMC. In this instance, the fatty acid chlorides are chosen because of their reactivity.

Acylation of alcohols can be accelerated by acidic or basic catalysts. 16-OMC is sensitive to acids, so acidic catalysis cannot be applied. Strong bases are not suited, because they would partially saponify the esters formed. Kaufmann and Hamsagar (20) used the Einhorn reaction for the production of cafestol fatty acid esters. In this reaction, the alcohol is dissolved in pyridine, and the acid chloride is added in excess, using ice for cooling. The reaction is stopped by adding water. Pyridine serves both as solvent and as a basic catalyst. For 16-OMC, in which the free hydroxy group is more sterically hindered than the free hydroxy group of cafestol due to its methoxy group, the acylation reaction is probably very slow. Therefore, a stronger catalyst than pyridine is used, such as 4-dimethylaminopyridine (DMAP) (65). Pyridine is used as a solvent for 16-OMC and for DMAP. The fatty acid chlorides are dissolved in n-hexane.

The synthesis of 16-OMC fatty acid esters can be carried out as follows in an Erlenmeyer flask: 16-OMC is dissolved in pyridine that has been previously dried for

48 h over a molecular sieve. The fatty acid chloride, which is dissolved in *n*-hexane (75% of the volume of the pyridine solution), is added dropwise in two- to threefold excess to the 16-OMC solution at ambient temperature. The catalyst DMAP is dissolved in pyridine, then added dropwise in 1.3-fold excess. In order to increase the solubility of the reaction mixture, dichloromethane in a volume of 25% of the volume of the reaction mixture is added. The reaction mixture is stirred in a closed flask for 15 min at room temperature on a magnetic stirrer. After 48 h of standing in the dark at room temperature, the reaction is stopped by addition of about the twice the volume of water in relation to the volume of the reaction solution. The degree of esterification can be estimated by TLC during and at the end of the esterification.

After the reaction, the pyridine is evaporated at 30°C (at which time foaming might occur), and the aqueous reaction residue is transferred with *t*-BME following sonication to a brown glass separatory funnel. The ether phases are extracted with a 10% aqueous sodium carbonate solution in order to remove unreacted fatty acids. If gelatinous clots form, the aqueous phase, including the clots, is centrifuged under nitrogen after removal of the ether phase. The ether phases are combined and filtered over anhydrous sodium sulfate and diluted with acetone. The volume of the solvent mixture is again reduced to about 5 mL, then placed in a freezer at −20°C for 24 h to allow the crystallization to proceed. Finally, residual solvent is removed in a nitrogen stream.

Solid-phase extraction (SPE) is applied for the separation of nonesterified (free) 16-OMC. According to Speer (57), the coffee diterpenes which are applied on a silica column in a solution of *n*-hexane:ethyl acetate (9:1, v/v), are adsorbed so strongly that they are not eluted with *n*-hexane:ethyl acetate (8:2, v/v) but only with ethyl acetate. In contrast, diterpene esters are not adsorbed on silica gel but pass through the column unhindered. The SPE extraction procedure is described by Sehat (61). The yields of the 16-OMC fatty acid esters are between 56 and 82%.

Considerations in the Analysis of Coffee Lipids

Methods. The concept for the analysis of coffee lipids presented here is based on my own research (61) (see Fig. 3). The coffee lipid, extracted from the ground coffee beans with *t*-BME, is entirely separated into the fractions of triacylglycerol, diterpene esters, and free fatty acids by use of GPC. The fraction of the diterpene esters is further subdivided by SPE into sterol esters, 16-OMC esters, and cafestol esters. The 16-OMC esters are then analyzed by HPLC, whereas the fatty acid analysis of the triacylglycerol fractions and the identification of free fatty acids are carried out by CGC.

Gel permeation chromatography (GPC). GPC is applied for the isolation of the diterpene fatty acid ester fraction of *robusta* coffee lipid for the purpose of obtaining 16-OMC. The presence of small amounts of free fatty acids in the diterpene fatty acid ester fraction is acceptable, because these components are separated during the subsequent semipreparative HPLC step. However, a total separation of the three lipid fractions (triacylglycerol, diterpene esters, and free fatty acids) is essential for analytical purposes. The amount of coffee lipid injected per run is reduced from 300 to

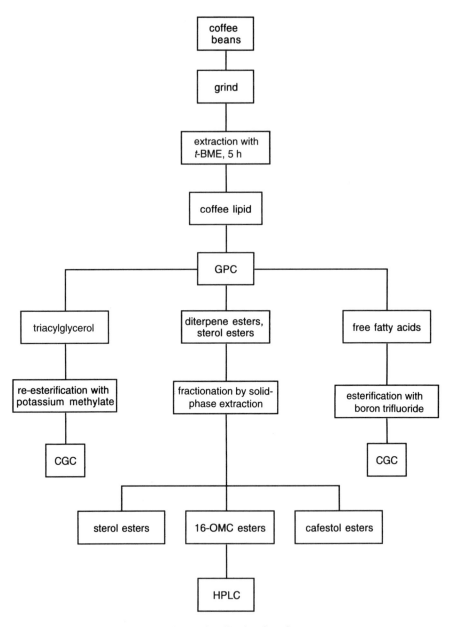

FIG. 3. Working scheme for analysis of coffee lipid (Ref. 61).

25 mg, and the time windows are altered, in comparison to the preparative method. The injection volume is 200 µL. The stationary phase, mobile phase, and flow rate remain the same as for the preparative method.

Solid-phase extraction (SPE). In analytical chemistry, SPE is used for the selective extraction of the component of interest from a complex mixture for purification or for enrichment of substances from dilute solutions prior to analysis. As cafestol esters are more polar than 16-OMC esters, because of their additional free hydroxy group, their hydrophilic interactions with a sorbent can be used for separation (e.g., cafestol esters are more strongly adsorbed on silica gel than are 16-OMC esters). In contrast, sterol esters do not possess an ether group and therefore are still more nonpolar than 16-OMC esters. The SPE eluate fractions are tested by TLC and HPLC at frequent intervals.

Chromatographic determination. Cafestol fatty acid esters of *arabica* coffee were first investigated by Kaufmann and Hamsagar (20) by paper chromatography. Lam et al. (43) analyzed the acetic acid esters of cafestol and kahweol by use of a preparative column with a silver-impregnated silica gel filling. Nackunstz (55) analyzed the free diterpenes kahweol and cafestol on a reversed-phase (RP) C-18-HPLC column (particle size 5 µm), with an acetonitrile/water eluent and UV detection at 220 nm. Five kahweol and cafestol fatty acid esters were also analyzed by RP–HPLC (45), using RP C-8 material on an analytical (250 × 4.6 mm i.d.) and a preparative (250 × 10 mm i.d.) column. The mobile phase for this separation was a linear gradient of 80 to 100% acetonitrile in water.

An HPLC method for the determination of the total amounts of 16-OMC in the unsaponified fraction of the extracted coffee lipids was developed by Speer (57).Further investigation showed that only 1.0–3.4% of 16-OMC occurs in the free form (59).

An analytical method for the determination of 16-OMC esters by HPLC was developed by Sehat (61) using synthesized standard substances. A mixture of acetonitrile and water was not used for the mobile phase, because the 16-OMC esters are less soluble in it than are the kahweol and cafestol esters. Instead, mixtures of acetonitrile and isopropanol were tested during method development. By the use of acetonitrile and isopropanol (65:35, v/v) a baseline separation of nearly all investigated 16-OMC fatty acid esters was achieved (61).

The method for the determination of 16-OMC esters (GPC and HPLC) was validated with a series of experiments. The relative standard deviations for HPLC injections of standard solutions of 16-OMC esters were generally between 0.8 and 1.6%. However, the relative standard deviation for HPLC injections of standard solutions for 16-OMC palmitate was 2%, due to incomplete separation from the preceding oleate peak. Reproducibility and recoveries were investigated by running a 16-OMC standard solution through GPC and subsequent HPLC analyses. Recoveries were found to be 94–99%. In order to simulate matrix effects, the 16-

OMC standard mixture was also added to *arabica* coffee lipid, which is naturally free of 16-OMC compounds. The recoveries were between 80 and 89%. The relative standard deviations of the measured 16-OMC ester concentrations ranged from 2.3 to 5.0% in a fourfold determination.

Determination of Fatty Acid Composition of Triacylglycerides and Identification of the Free Fatty Acids

The amounts and distributions of fatty acids in the coffee lipid fractions of the triacylglycerol (TAG) and the free fatty acids (FFA) obtained by GPC were determined by gas chromatography.

Previous Investigations

Early investigations of coffee lipids focused primarily on the total fatty acid compositions rather than on fatty acids in specific lipid fractions (2,10–13,66–69). Linoleic is the commonest of the fatty acids, representing more than 40% of the total. Linoleic acid is followed by palmitic, representing about 35%. Stearic and oleic acids are present in amounts between 7 and 12%, according to most authors. Arachic acid is present at levels from 0.7 to 5%, linolenic and behenic acid between 0.3 and 3%, and myristic acid in very small amounts. Information about the composition of the fatty acids from the triacylglycerol and diterpene ester fractions can be found in the work of Kaufmann and Hamsagar (20), Folstar et al. (8), and Folstar (70). Kaufmann and Hamsagar (20) performed paper chromatography, a color reaction, and photometric evaluation with a single sample of Brazilian green coffee oil. Folstar et al. (8) investigated coffee oil (free from coffee wax) of Colombian *arabica*. The coffee wax was removed from the whole green coffee beans by chloroform treatment; then the coffee oil was extracted from the ground coffee beans. The fatty acid compositions of the TAG and the diterpene esters were analyzed after a column chromatographic separation. The results of the investigations in 1962 and 1975 were similar: in the TAG, linoleic acid was definitely predominant; palmitic acid was the commonest fatty acid in the diterpene esters; all unsaturated fatty acids were found at higher percentages in the TAG than in the diterpene esters; the reverse was found for the saturated fatty acids.

Prior to the work of Sehat (61), the free fatty acids in the coffee lipid had been investigated only by the titrimetric acid index method, described in the following section. Kaufmann and Hamsagar (47) found acid indices of 3.1 to 4.2 in *arabica* green coffee; Calzolari and Cerma (66), 4.5 to 7.3; and Carisano and Gariboldi (67), 1.0 to 3.9. Given an average molar mass of the fatty acids of 275 g/mL (71), these acid indices are equivalent to 0.5 to 3.6% free fatty acids. In roasted coffee, Carisano and Gariboldi (67) found acid indices of 4.5 to 7.9; and Kaufmann and Hamsagar (47) found 4.2 to 8.7. These figures are equivalent to 2.1 to 4.3% of free fatty acids.

Determination of the Acid Index

The acid index is the amount of potassium hydroxide in mg needed to neutralize the free acids present in 1 g of lipid. Because this method is nonspecific, it does not allow a distinction to be made among mineral acids, FFA, and other organic acids.

For the determination of the acid index, the weighed amount of lipid is dissolved in ethanol:toluene (1:1, v/v) or 96% ethanol and titrated with 0.1 M KOH, using phenolphthalein as an indicator. For dark-colored fats, thymolphthalein is used for better visibility of the titration end point (2).

The color of green coffee lipid obtained by t-BME extraction is strongly greenish to yellow, whereas the color of roasted coffee lipid darkens toward brown. Because of this, the color change of thymolphthalein is barely recognizable for the determination of the acid index. Therefore, the endpoint (pH 9.3) is measured potentiometrically. The coffee lipid is dissolved in a mixture of n-hexane and ethanol (1:1, v/v) that has been adjusted to pH 9.3. The solution is titrated with 0.1 M KOH solution to an end point of pH 9.3. The acid index is calculated from the consumption of KOH solution.

Gas Chromatography of Fatty Acid Methyl Esters

Gas chromatography of the fatty acid methyl esters has been used in lipid chemistry for many years (71,72). For reesterification, the lipid mixture is reacted with sodium or potassium methoxide so that all fatty acid esters (e.g., the triacylglycerol) are converted to the fatty acid methyl esters. Free fatty acids are also esterified by the method of esterification with boron trifluoride in methanol (73).

Determination of Coffee Lipids in Various Coffee Brews

In order to investigate the transfer behavior of coffee lipids from the coffee powder to the coffee brew, it is necessary to isolate the lipids from the coffee brews. The isolated lipid can then be analyzed by the methods already described to determine the amounts and the fatty acid distribution of 16-OMC, the free fatty acids and the triacylglycerol. Various parameters influencing the transfer of coffee lipids into the coffee brew have been investigated by Sehat (61) following the method described subsequently.

The commonest grinding grades for coffee powder are coarse-ground for boiled coffee, medium- to fine-ground for filtered coffee, and very finely ground (pulverized) for espresso.

The coffee-brewing procedures are performed as follows:

- Boiled coffee (Scandinavian style): Hot water is added to the coffee powder in a pot (500 mL for 25 g), the mixture is boiled and stirred, and the boiling is continued for 2 to 20 min. The finished brew is decanted through a fine metal sieve.

- Espresso (steam extraction under pressure): In a commercial espresso machine, the water (100 mL per 15 g) is pressed through the coffee powder at a temperature above 100°C.
- Filtered coffee: 25 g of coffee powder are placed in a paper filter in a commercial coffee machine and 500 mL of boiling water are slowly poured over the powder. The filtrate is the coffee brew.

Isolation of Lipids from Coffee Brews

Three methods can be applied for the isolation of lipids from coffee brews: liquid–liquid extraction, filtration, and freeze drying.

In the liquid–liquid extraction procedure, the coffee brew is cooled to room temperature and is transferred to a separatory funnel. In order to have the free fatty acids in their undissociated form, a tartaric acid/tartrate buffer is added to give a pH value of 3. The brew is saturated with sodium chloride and then extracted with t-BME (20% of the volume of the brew). Emulsions can be broken by addition of ethanol and by allowing the brew to stand overnight. The extraction is repeated three times with the same volume of t-BME. The combined ether phases are dehydrated over anhydrous sodium sulfate, filtered, and evaporated nearly to dryness. Remaining solvent is removed under a nitrogen stream. The remaining black, hard residue is sonicated in 20 mL dichloromethane three times (10 min. each). The dichloromethane solution is collected in a weighed 100-mL round-bottom flask and evaporated. Residual solvent is removed under a nitrogen stream, and the residue is dried for one day in a desiccator and weighed.

In the filtration procedure, the coffee brew is cooled to room temperature and is applied to a membrane filtration device equipped with a glass fiber prefilter. First, the upper part of the coffee brew, which contains relatively few solid particles, must be carefully decanted. The suction of the vacuum must be adjusted to avoid boiling and foaming of the filtrate. The filter residue obtained is dried in a desiccator for one day (together with the glass fiber prefilter), and is then mixed with a fourfold amount of anhydrous sodium sulfate and placed in a Soxhlet apparatus for lipid extraction with t-BME. Finally, the amount of extraction residue is determined gravimetrically.

For the freeze-drying procedure outlined in Fig. 4, the coffee brew is quantitatively transferred to a round-bottom flask and carefully reduced by evaporation to about 50 mL. This volume reduction is carried out in a first step with a water bath temperature of 62°C until a pressure of 240 bar is reached. Then, after a wait of 3 min while the coffee brew warms, the pressure is slowly reduced to 65 bar. The volume reduction procedure takes 30 to 90 min depending on the type of coffee brew. Attention must be paid to prevent evaporation to dryness, because the resulting hard residue cannot be redissolved.

Before freeze drying, the round-bottom flask is prechilled in a deep freeze, and the remaining liquid is frozen by rotating the flask in a Dewar flask containing solid

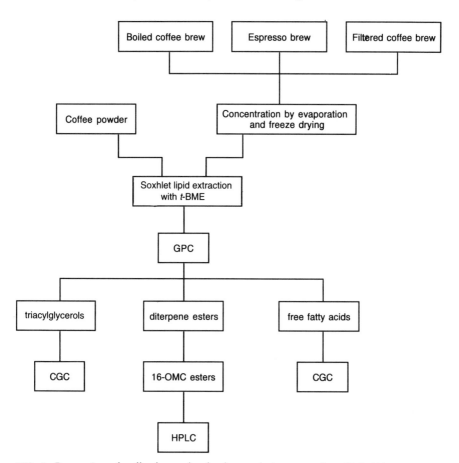

FIG. 4. Processing of coffee brews by the freeze-drying procedure (Ref. 61).

carbon dioxide. Then, during a period of 8 h, the water is completely removed by freeze drying. The dry, porous extract obtained in this way can be very easily removed from the flask. The extract is quantitatively transferred to a mortar, mixed with twice the amount of desiccated sodium sulfate and sea sand, and then placed in a Soxhlet apparatus for lipid extraction with t-BME. The amount of lipid obtained is determined gravimetrically.

The best extraction yields are obtained by the freeze-drying and the filtration procedures. The extraction yields are much lower by liquid–liquid extraction; the reproducibility is poor and the method is very time-consuming. Despite the high extraction yields, filtration procedures are very labor-intensive. The freeze-drying procedure has practical advantages, such as partial automation and parallel treatment of up to 10 samples. Therefore, the freeze-drying procedure is the method of choice for the isolation of lipids from large numbers of coffee brews.

The reproducibility of the method for the determination of coffee lipids in coffee brews was tested by Sehat (61) with the Scandinavian brew of a roasted coffee mixture from the French market. The coefficients of variation of the sample results were 4.4% for the total lipid content, 4.5% for the 16-OMC esters, 1.4% for the free fatty acids, and 3.7% for the triacylglycerides. The recovery of the coffee lipid groups in the coffee brew and the brewing residue in comparison with the coffee powder was 100% for the 16-OMC esters and about 90% for the free fatty acids and triacylglycerides. The fatty acid distributions of the three lipid substance classes are the same in the brew, the brewing residue and the coffee powder, respectively. In the brewing residue, the percentage of free fatty acids in the lipid is slightly higher, and in the brew slightly lower, than in the coffee powder.

Results of Analysis of Coffee Samples, from Unroasted Coffee Beans to the Coffee Brew

In agreement with older references (1,5–7), Sehat (61) found lower total lipid amounts in *robusta* green coffee (between 9 and 11% in the dry matter) than in *arabica* green coffee (about 15%). Whereas the *arabica* samples were free of 16-OMC esters, *robusta* green coffee contained 1 to 2 g of 16-OMC esters per 100 g of lipid.

The amounts of free fatty acids determined by the method described were distinctly lower than the amounts calculated from the acid index. *Robusta* green coffee contained higher amounts of free fatty acids (about 2 g/100 g lipid) than *arabica* green coffee (about 1 g/100 g lipid).

The distributions of fatty acids in the groups of triacylglycerol and free fatty acid published by Speer et al. (74) and Sehat (61) were in good agreement with the values found by Folstar (70). The only difference between *arabica* and *robusta* coffee in this respect was the higher content of oleic acid in *robusta* samples. In the fraction of free fatty acids, the amount of linoleic acid was lower, but the amounts of arachic and behenic acid were higher than in the triacylglycerol fraction.

In contrast to the similarity of the fatty acid distribution in triacylglycerol and free fatty acids, the 16-OMC esters show distinctly higher amounts of palmitic acid (48% vs. about 30%) and lower amounts of linoleic acid (12% vs. 40 to 50%). Sehat et al. (75) and Sehat (61) investigated the influence of the degree of roasting on the total lipid concentration, and the 16-OMC esters, FFA, and TAG in a series of samples roasted at different temperatures (242°C, 250°C, 253°C, 264°C). The percentages of all lipid groups with respect to the coffee powder dry matter increased significantly with increasing roasting temperature. With respect to the effects of increased roasting temperatures on specific fractions of lipids, a slight decrease in linoleic acid and an increase in the amount of stearic and oleic acids were found in the FFA fraction.

A series of commercial roasted coffee samples from Germany, France, and the Netherlands was analyzed. Indications for *robusta* admixtures were the total lipid content, the FFA content and the proportion of oleic and stearic acids in the FFA.

Samples containing *robusta* coffee had a tendency to have lower total lipid, higher FFA content, and a higher ratio of oleic to stearic acid in the FFA (61,74). Determinative evidence for *robusta* admixture, as well as an estimation of the *robusta* fraction, was produced by determination of the 16-OMC esters using the method developed by Sehat (61) as well by determination of the total 16-OMC content after saponification (57). Both methods proved to be comparable.

Roasted coffee from the French market tended to have higher amounts of FFA in comparison to coffee from the German market, even when pure *arabica* mixtures were analyzed. The instant coffees investigated contained 0.3–0.7% of lipid in the dry matter, and all showed measurable amounts of 16-OMC.

The coffee brews investigated by Sehat et al. (75) and Sehat (61) were boiled coffee (Scandinavian style), espresso, and filtered coffee. The influence of the use of demineralized water and tap water, the duration of boiling, and the coffee powder dosage were investigated in boiled coffee brews.

Comparison of the brewing methods showed that about 22% of coffee lipids are transferred to the brew by the boiling procedure, whereas about 2.5% and 0.6% are transferred by the espresso and filter procedures, respectively. These results are in good agreement with the values reported by Viani (76) and Van Dusseldrop et al. (77) for lipids transferred by the boiling and filter procedures, respectively. The distributions of fatty acid were not influenced by the brewing methods.

The influence of the degree of roasting was investigated in all three types of brews. Under the chosen roasting conditions, the degree of lipid extraction by the boiling procedure was not influenced by the roasting temperature; the amounts of lipid in the brews increased in parallel with the amounts of lipid in the roasted coffee powder with increasing roasting temperature.

In contrast, the influence of the grinding grade on lipid transfer into the coffee brews is relevant. In brews prepared by the boiling procedure, the degree of lipid extraction increases with decreasing particle size. In contrast, by the espresso procedure, the lipids are less efficiently extracted from finer coffee powder than from the more coarsely ground powder. In filtered coffee, the degree of lipid extraction increases from coarse to medium grinding and decreases again for fine grinding (61,75).

Much of the early research work on coffee emphasized the possible carcinogenic potential of coffee and caffeine. Currently, interest has turned to the effect of coffee on the cardiovascular system. According to several studies, the consumption of Scandinavian-style coffee brew or coffee lipid is correlated with an elevated serum cholesterol level in coffee consumers (78–92). In contrast, other researchers have not found correlations between serum cholesterol and consumption of filtered coffee or coffee lipid (93–98). In a review of numerous studies, Thelle et al. (99) concluded that a positive correlation between coffee intake and serum cholesterol level was present in approximately two-thirds of the studies. Stavric (100) reviewed the literature from 1989–1990 on the relationship between coffee consumption and coronary heart disease. His overall assessment of the safety of drinking coffee and the effect of coffee on human health indicates that certain significant issues are still unresolved.

A problem with many of the studies cited above was that the investigators did not report the lipid composition of the coffee consumed. Hence, in some studies there are uncertainties in how the coffee brews were prepared and the amounts and types of coffee components (including lipids) actually consumed. This may explain some of the seemingly inconsistent findings of the studies.

Acknowledgments

Dr. Sehat gratefully acknowledges Prof. Dr. A. Montag, University of Hamburg, for providing support for the research; Prof. Dr. A. Montag and Prof. Dr. K. Speer, Technical University of Dresden, for helpful discussions throughout the Ph.D. thesis work; Prof. Dr. K. Speer for generously providing the coffee samples; and Dr. J. Rader, U.S. Food and Drug Administration, for her assistance in proofreading the chapter.

References

1. Maier, H.G. (1981) *Kaffee*, Paul Parey, Hamburg, Germany.
2. Streuli, H. (1970) in *Handbuch der Lebensmittelchemie* (Schormueller, J., ed.), Vol. VI, Springer Verlag, Berlin, Heidelberg, New York, pp. 1–95.
3. Deutscher Kaffee-Verband, e.V. (1992) *Kaffee-Text 2/92*, Hamburg, Germany.
4. Tressl, R., Holzer, M., and Kamperschroer, H. (1983) Formation of Aromatic Substances in Roasted Coffee in Relation to the Levels of Free Amino Acids and Reducing Sugars, *Association Scientifique Internationale sur le Café (ASIC)*, Paris, France, pp. 279–283.
5. Streuli, H. (1973) Der heutige Stand der Kaffee-Chemie, *Association Scientifique Internationale sur le Café (ASIC)*, Paris, France pp. 179–200.
6. Clifford, M.N. (1975) Composition of Green and Roasted Coffee Beans, *Proc. Biochem.*, 3–8.
7. Clifford, M.N., and Wilson, K.C. (1985) *Coffee; Botany, Biochemistry and Production of Beans and Beverage*, Croom-Helm London.
8. Folstar, P., Pilnik, W., De Heus, J.G., and Van der Plas, H.C. (1975) The Composition of Fatty Acids in Coffee Oil and Wax, *Lebensm.-Wiss. und Technol. 8*, 286–288.
9. Wajda, P., and Walczyk, D. (1978) Relationship Between Acid Value of Extracted Fatty Matter and Age of Green Coffee Beans, *J. of Science and Food Agriculture 29*, 377–380.
10. Pokorny, J., and Forman, L. (1970) Pflanzenlipide, 2. Mitt. Kaffee-Lipide, *Nahrung 14*, 631–632.
11. Roffi, J., Corte dos Santos, A., Mexia, J.T., Busson, F., and Maigrot, M. (1971) Café Verts et Torrefies de l'Angola, Étude chimique, *Association Scientifique Internationale sur le Café (ASIC)*, Paris, France, pp. 179–200.
12. Vitzthum, O.G. (1976) *Chemie und Bearbeitung des Kaffees; Kaffee und Coffein* (Eicher, O., ed.), Springer Verlag, Berlin, New York, pp. 3–64.
13. Wurziger, J. (1963) L'Huile du Café Vert et du Café Torrefie, *Café, Cacao, Thé 7*, 331–340.

14. Kaufmann, H.P., and Hamsagar R.S. (1969) Über die Lipoide der Kaffeebohne, *Forschungsbericht des Landes Nordrhein-Westfalen Nr. 2062*, Köln, Westdeutscher Verlag, Opladen.
15. Wurziger, J. (1974) Carbonsäurehydroxytryptamide und Alkalifarbzahlen in Rohkaffees als analytische Hilfsmittel zur Beurteilung von Röstkaffee-Genusswert und Bekömmlichkeit, *Association Scientifique Internationale sur le Café (ASIC)*, Paris, France, pp. 332–342.
16. Wurziger, J., Capot, J., and Vincent, J.C. (1977) Über Untersuchungen an Arabusta-Rohkaffees, *Kaffee und Tee Markt 27(12)*, 3–8.
17. Nagasampagi, B.A., Rowe, J.W., Simpson, R., and Goad, L.J. (1971) Sterols of Coffee, *Phytochem. 10*, 1101–1107.
18. Alcaide, A., Devys, M., Barbier, M., Kaufmann, H.P., and Sen Gupta, A.K. (1971) Triterpenes and Sterols Of Coffee Oil, *Phytochem. 10*, 209–210.
19. Liebscher, W. (1979) *Handbuch zur Anwendung der Nomenklatur Organisch-Chemischer Verbindungen*, Berlin Akademie.
20. Kaufmann, H.P., and Hamsager, R.S. (1962) Zur Kenntnis der Lipoide der Kaffeebohne. I. Über Fettsäure-Ester des Cafestols, *Fette, Seifen, Anstrichmittel 64*, 206–213.
21. Bengis, R.O., and Andersen, R.J. (1932) The Chemistry of the Coffee-Bean. I. Concerning the Unsaponifiable Matter of the Coffee-Bean Oil. Preparation and Properties of Kahweol, *J. Biol. Chem. 97*, 99–113.
22. Slotta, K.H., and Neisser, K. (1938) Determination of Trigonelline, Isolation of Cafesterol and other Components from Unsaponifiable Portion of Coffee Oil, *Ber. Dtsch. Chem. Ges. 71*, 1991–1994.
23. Wettstein, A., Fritsche, H., Hunziger, F., and Miescher, K. (1941) Steroids. XXXII. Constitution of Cafesterol, *Helv. Chim. Acta 24*, 332–358.
24. Wettstein, A., and Miescher, K. (1942) Zur Konstitution des Cafestols, IV. Über das inerte Sauerstoffatom, *Helv. Chim. Acta 25*, 718–731.
25. Wettstein, A., and Miescher, K. (1943) Zur Konstitution des Cafestols, V. Cafestol und Kahweol, *Helv. Chim. Acta 26*, 631–641.
26. Wettstein, A., and Miescher, K. (1943) Zur Konstitution des Cafestols, VII. Nachweis eines Furankerns im Cafestol, *Helv. Chim. Acta 26*, 788–800.
27. Wettstein, A., and Miescher, K. (1943) Zur Konstitution des Cafestols, VIII. Die Verknuepfung des Furanringes mit dem Rest der Molekel, *Helv. Chim. Acta 26*, 1197–1218.
28. Wettstein, A., Spilmann, M., and Miescher, K. (1945) Zur Konstitution des Cafestols, XI. Cafestol und Kahweol, *Helv. Chim. Acta 28*, 1004–1013.
29. Chakravorty, P.N., Wesner, M.M., and Reed, G. (1942) On the Chemical Behavior of Cafesterol, *J. Am. Chem. Soc. 64*, 2235.
30. Chakravorty, P.N., Levin, R.H., Wesner, M.M., and Reed, G. (1943) Cafesterol. II., *J. Am. Chem. Soc. 65*, 929–932.
31. Chakravorty, P.N., Levin, R.H., Wesner, M.M., and Reed, G. (1943) Cafesterol. III. Isocafesterol, *J. Am. Chem. Soc. 65*, 1325–1328.
32. Djerassi, C., Wilfred, E., Visco, L., and Lemin, A.J. (1953) Terpenoids II. Experiments in the Cafestol Series, *J. Org. Chem. 18*, 1449–1460.
33. Djerassi, C., and Bendas, H. (1955) Constitution of Cafestol, *Chemistry and Industry*, 1481–1482.

34. Djerassi, C., Bendas, H., and Sen Gupta, P. (1955) Terpenoids. XIX. The Pentacyclic Skeleton of Cafestol, *J. Org. Chem. 20*, 1046–1055.
35. Djerassi, C., Cais, M., and Mitscher, L.A. (1958) Terpenoids. XXXIII. The Structure and Probable Absolute Configuration of Cafestol, *J. Am. Chem. Soc. 80*, 247–248.
36. Djerassi, C., Cais, M., and Mitscher, L.A. (1959) Terpenoids. XXXVII. The Structure of the Pentacyclic Diterpene Cafestol. On the Absolute Configuration of Diterpenes and Alkaloids of the Phyllocladene Group, *J. Am. Chem. Soc. 81*, 2385–2398.
37. Djerassi, C., and Finnegan, R.A. (1960) Terpenoids. XLV. Further Studies on the Structure and Absolute Configuration of Cafestol, *J. Am. Chem. Soc. 82*, 4342–4344.
38. Haworth, R.D., Jubb, A.H., and McKenna, J. (1955) Cafestol. Part I, *J. Chem. Soc.*, 1983–1989.
39. Haworth, R.D., and Johnstone, R.A.W. (1956) The Structure of Cafestol, *Chem. Ind.*, 168.
40. Haworth, R.D., and Johnstone, R.A.W. (1957) Cafestol. Part II, *J. Chem. Soc. (London)*, 1492–1496.
41. Kaufmann, H.P., and Sen Gupta, A.K. (1963) Zur Kenntnis der Lipoide der Kaffeebohne III. Die Reindarstellung des Kahweols, *Fette, Seifen, Anstrichmittel 65*, 529–532.
42. Kaufmann, H.P., and Sen Gupta, A.K. (1963) Terpene als Bestandteile des Unverseifbaren von Fetten. Zur Konstitution des Kahweols, *Chem. Ber. 96*, 2489–2498.
43. Lam, L.K.T., Sparnis, V.L., and Wattenberg, L.W. (1982) Isolation and Identification of Kahweol Palmitate and Cafestol Palmitate as Active Constituents of Green Coffee Beans that Enhance Glutathione S-Transferase Activity in the Mouse, *Cancer Research 42*, 1193–1198.
44. Wurziger, J. (1976) Viridinsaure zum Nachweis von Chlorogensäuren, *Kaffee und Tee Markt 26(19)*, 3–5.
45. Pettitt, B.C., Jr. (1987) Identification of the Diterpene Esters in *Arabica* and *Canephora* Coffees, *J. Agric. Food Chem. 35*, 549–551.
46. Speer, K., and Mischnik, P. (1989) 16-*O*-Methylcafestol-ein neues Diterpen im Kaffee. Entdeckung und Identifizierung, *Z. Lebensm. Unters. Forsch. 189*, 219–222.
47. Kaufmann, H.P., and Hamsager, R.S. (1962) Zur Kenntnis der Lipoide der Kaffeebohne II. Die Veraenderung der Lipoide bei der Kaffee-Röstung, *Fette, Seifen, Anstrichmittel 64*, 734–738.
48. Wurziger, J. (1976) Über Nachweis und Bewertung von Rohkaffee-Bearbeitungseinflüssen, *Kaffee und Tee Markt 26(17)*, 3–12.
49. Wahlberg, E.C.R., and Rowe, J.W. (1975) Ent-16-kauren-19-ol from Coffee, *Phytochem. 14*, 1677.
50. Nurok, M.B., Anderson, J.W., and Zlatkis, A. (1978) Profiles of Sulfur Containing Compounds Obtained from *Arabica* and *Robusta* Coffees by Capillary Column Gas Chromatography, *Chromatographia 11*, 188–192.
51. Tiscornia, E., Centi–Grossi, M., Tassi-Micco, C., and Evangelisty, F. (1979) The Sterol Fraction of the Oil Extracted from Coffee, *Riv. Ital. Sostanze Grasse 61*, 283–292.
52. Duplatre, A., Tisse, C., and Estienne, J. (1984) Identification of *Arabica* and *Robusta* (coffee) Species by Studying the Sterol Fraction, *Ann. Fals. Exp. Chim. 828*, 259–270.
53. Picard, H., Guyot, B., and Vincent, J.C. (1984) Study of the Sterol Compounds of *Coffea canephora* Oil, *Café Cacao Thé 28*, 47–62.

54. Mariani, C., and Fedeli, E. (1991) The Sterol Fraction of the Coffee Oil, *Riv. Ital. Sostanze Grasse 68*, 111–115.
55. Nackunstz, B. (1986) Diterpene im Kaffee, Ph.D. Thesis, Institute of Food Chemistry, University of Braunschweig, Germany.
56. Nackunstz, B., and Maier, H.G. (1987) Diterpenoide im Kaffee. II. Cafestol und Kahweol, *Z. Lebensm. Unters. Forsch. 184*, 494–499.
57. Speer, K. (1989) 16-*O*-Methylcafestol-ein neues Diterpen im Kaffee. Methoden zur Bestimmung des 16-*O*-Methylcafestol in Rohkaffees und in behandelten Kaffees, *Z. Lebensm. Unters. Forsch. 189*, 326–330.
58. Speer, K. (1992) Neue Untersuchungen der Lipidfraktion des Kaffees, *Habilitationsschrift*, Institute of Biochemistry and Food Chemistry, University of Hamburg, Germany.
59. Speer, K., Tewis, R., and Montag, A. (1991) 16-*O*-Methylcafestol-ein neues Diterpen im Kaffee. Freies und gebundenes 16-*O*-Methylcafestol, *Z. Lebensm. Unters. Forsch. 192*, 451–454.
60. Speer, K., and Montag, A. (1989) 16-*O*-Methylcafestol-ein neues Diterpen im Kaffee. Erste Ergebnisse in Roh- und Roestkaffees, *Deutsche Lebensmittel-Rundschau 85*, 381–384.
61. Sehat, N. (1994) Contribution to the Knowledge of Coffee Lipids, Ph.D. Thesis, Institute of Biochemistry and Food Chemistry, University of Hamburg, Germany.
62. Stoldt, W. (1949) Fettbestimmung in Lebensmitteln, *Deutsche Lebensmittel-Rundschau 45*, 41–46.
63. Stoldt, W. (1952) Vorschlag zur Vereinheitlichung der Fettbestimmung in Lebensmitteln, *Fette, Seifen, Anstrichmittel 54*, 206–207.
64. Meyer, V. (1988) Praxis der Hochleistungs-Fluessigchromatographie- (HPLC), 5th edn., Diesterweg Verlag, Frankfurt, Germany.
65. Höfle, G., Steglich, S., and Vorbruggen, H. (1978) 4-Dialkylaminopyridine als hochwirksame Acylierungskatalysatoren, *Angew. Chem. 90*, 602–615.
66. Calzolari, C., and Cerma, E. (1963) Über die Fettsubstanzen des Kaffees, *Riv. Ital. Sostanze Grasse 40*, 176–180.
67. Carisano, A., and Gariboldi, L. (1964) Gas Chromatographic Examination of the Fatty Acids of Coffee Oil, *J. Agric. Food Chem. 15*, 619–622.
68. Hartman, L. Lago, R.C.A., Tango, J.S., and Teixeira, C.G. (1968) The Effect of Unsaponifiable Matter on the Properties of Coffee Seed Oil, *J. Am. Oil Chem. Soc. 45*, 577–579.
69. Chassevent, F., Dalger, G., Gerwig, S., and Vincent, J.C. (1974) Lipid and Nonsaponifiable Fraction. Possible Relation Between Caffeine and Chlorogenic Acid Contents, *Cafe Cacao The 18*, 49–56.
70. Folstar, P. (1985) in *Coffee (lipids), Vol. 1, Chemistry* (Clarke, R.J., and Macrae, R., eds.), Essex, Elsevier Applied Science, pp. 207–209.
71. Kaufmann, H.P. (1965) *Die Anwendung der GC auf dem Fettgebiet mit besonderer Berucksichtigung der Pharmazeutischen Analyse*, Forschungsbericht des Landes Nordrhein-Westfalen Nr. 1568, 69, Köln, Opladen, Westdeutscher Verlag.
72. Hardon, H., and Zurcher, K. (1967) Beitrag zur gaschromatographischen Analyse von Fetten und Ölen, *Mitt. Gebiete Lebens. Hyg. 58*, 209,236,351.
73. Sykes, P. (1988) *Reaktionsmechanismen der organischen Chemie*, 9th edn. VCH-Verlagsgesellschaft, Weinheim.

74. Speer, K., Sehat, N., and Montag, A. (1993) Fatty Acids in Coffee, *Association Scientifique Internationale sur le Café (ASIC)* Vol. II, Paris, France, pp. 583–591.
75. Sehat, N., Speer, K., and Montag, A. (1993) Lipids in the Coffee Brew, *Association Scientifique Internationale sur le Café (ASIC)* Vol. II, Paris, France, pp. 869–872.
76. Viani, R. (1988) Coffee, *Physiologically Active Substances in Coffee* (Clarke, R.J., and Macrae, R., eds.), Vol. 3, Elsevier Applied Science, pp. 1–31.
77. Van Dusseldrop, M., Katan, M.B., and Van Vliet, T. (1991) Cholesterol Raising Factor Does Not Pass a Paper Filter, *Arteriosclerosis and Thrombosis 11*, 586–593.
78. Arab, L., Kohlmeier, M., and Schlief, G. (1983) Coffee and Cholesterol, *N. Engl. J. Med. 309*, 1250.
79. Bønaa, K., Arnesen, E. Thelle, D.S., and Førde, O.H. (1988) Coffee and Cholesterol: Is It All in the Brewing? The Tromsø Study, *Brit. Med. J. 297*, 1103–1104.
80. Arø, A., Tuomilehto, J., Kostiainen, E., and Pietinen, P. (1988) Boiled Coffee Increases Serum Low Density Lipoprotein Concentration, *Metabolism 36*, 1027–1030.
81. Stensvold, I., Tverdal, A., and Foss, O.P. (1989) The Effect of Coffee on Blood Lipids and Blood Pressure. Results from a Norwegian Cross-Sectional Study, Men and Women 40–42 Years, *J. Clin. Epidemiol. 42*, 877–884.
82. Bak, A.A.A., and Grobbee, D.E. (1989) The Effect on Serum Cholesterol Levels of Coffee Brewed by Filtering or Boiling, *New Engl. J. Med. 321*, 1432–1437.
83. Zock, P.L., Katan, M.B., Merkus, M.P. Van Dusseldrop, M., and Harryvan, J.L. (1990) Effect of a Lipid Rich Fraction from Boiled Coffee on Serum Cholesterol, *The Lancet 335*, 1235–1237.
84. Meyers, M.G., and Basinski, A. (1992) Coffee and Coronary Heart Disease, *Arch. Intern. Med. 152*, 1767–1772.
85. Lancaster, T., Muir, J., and Silagy, C. (1994) The Effects of Coffee on Serum Lipids and Blood Pressure in a UK Population, *J. R. Soc. Med. 87(9)*, 506–507.
86. Wahrburg, U., Martin, H., Schulte, H. Walek, T., and Assmann, G. (1994) Effects of Two Kinds of Decaffeinated Coffee on Serum Lipid Profiles in Healthy Young Adults, *Eur. J. Clin. Nutr. 48(3)*, 172–179.
87. Ranheim, T., Halvorsen, B., Huggett, C.A., Blomhoff, R., and Drevon, A.C.(1995) Effect of a Coffee Lipid (Cafestol) on Regulation of Lipid Metabolism in CaCo-2 Cells, *Journal of Lipid Research 36*, 2079–2089.
88. Burr, M.L., Limb, E.S., Sweetnam, P.M., Fehily, A.M., and Amarah, L., and Hutchings, A. (1995) Instant Coffee and Cholesterol, *Eur. J. Clin. Nutri. 49*, 779–784.
89. Van Rooij, J., van der Stegen, H.D.G., Schoemaker, C.R., Kroon, C., Burggrauf, J., Hollaar, L., Vroon, F.F.P.T., and Smelt, H.M.A., Cohen, F.A.A. (1995) Placebo-Controlled Parallel Study of the Effect of Two Types of Coffee Oil on Serum Lipids and Transaminases, *Am. J. Clin. Nutr. 61*, 1277–1283.
90. Urgert, R., Schulz, G.M.A., and Katan, B.M. (1995) Effects of Cafestol and Kahweol from Coffee Grounds on Serum Lipids and Serum Liver Enzymes in Humans, *Am. J. Clin. Nutr. 61*, 149–154.
91. Mensink, P.R., Lebbink, W.J., Lobezoo, E.I., Weusten-Van der Wouw, P. M.E.M., Zock, L.P., and Katan, B.M. (1995) Diterpene Composition of Oil from *Arabica* and *Robusta* Coffee Beans and Their Effects on Serum Lipids in Man, *Journal of Internal Medicine 237*, 543–550.
92. D'Amicis, A., Scaccini, C., Tomassi, G., Anaclerio, M., Stornelli, R., and Bernini, A. (1996) Italian Style Brewed Coffee, *Int. J. Epidemiol 25(3)*, 513–520.

93. Dawber, T.R., Kannel, W.B., and Gordon ,T. (1974) Coffee and Cardiovascular Disease. Observation from the Framingham Study, *N. Engl. J. Med., 291,* 871–874.
94. Heyden, S., Heiss, G., Manegold, C., Tyroler, H.A., Hames, C.G., Bartel, A.G., and Cooper, G. (1979) The Combined Effect of Smoking and Coffee Drinking on LDL and HDL Cholesterol, *Circulation 60,* 22–25.
95. Hill, C. (1985) Coffee Consumption and Serum Cholesterol Concentrations, *Br. Med. J. 290,* 1590.
96. Burr, M.L., Gallacher, J.E.J., Butland, B.K., Bolton, C.H., and Downs, L.G. (1989) Coffee, Blood Pressure and Plasma Lipids, *Eur. J. Clin. Nutr. 43,* 477–483.
97. Grobbee, D.E., Rimm, E.B., Giovannucci, E., Colditz, G., Stampfer, M., and Willett, W. (1990) Coffee, Caffeine and Cardiovascular Disease in Men, *N. Engl. J. Med. 323,* 1026–1032.
98. Terpstra, A.H.M., Katan, M.B., Weusten-van der Wouw, M.P.M.E., Nicolosi, R.J., and Beynen, A.C. (1995) Coffee Oil Consumption Does Not Affect Serum Cholesterol in Rhesus and Cebus Monkeys, *J. Nutr. 125(9),* 2301–2306.
99. Thelle, D.S., Heyden, S., and Fodor, J.G. (1987) Coffee and Cholesterol in Epidemiological and Experimental Studies, *Atherosclerosis 67,* 97–103.
100. Stavric, B. (1992) An Update on Research with Coffee/Caffeine 1989–1990, *Fd Chem. Toxic. 30, No. 6,* 533–555.

Chapter 18
Improvements in Recovery of Petroleum Hydrocarbons from Marine Fish, Crabs, and Mussels

R.G. Ackman, H. Heras, and S. Zhou

Canadian Institute of Fisheries Technology, Technical University of Nova Scotia, P.O. Box 1000, Halifax, Nova Scotia B3J 2X4

Introduction

Hydrocarbons in foods can come from a variety of contaminating sources. Polycyclic aromatic hydrocarbons (PAHs) are of concern as carcinogens (1,2), but sensory reactions of consumers are more likely from aromatic hydrocarbons of moderate molecular weight (e.g., the familiar benzene, toluene and xylenes). Thus, the folk memory of kerosene leads to that adjective being widely used as a descriptive term for petroleum flavor in foods. The *Braer* oil spill in the Shetland Islands in January 1993 spectacularly demonstrated the visible impact of an oil tanker disaster (3,4), probably before the impact of the *Exxon Valdez* spill in Prince William Sound had worn off in the public mind. Such spills actually occur several times a year, but often in remote parts of the world and without much media attention. For example, 12 major "incidents" between March 1994 and January 1995 required input from the International Tanker Owners Pollution Federation Limited (5).

In the case of the *Braer* wreck the major economic impact was on farmed salmon in nearby waters. The Atlantic salmon, *Salmo salar* is a high-fat ($\leq 16\%$) fish when near market size, and the potential for fat to accumulate and retain hydrocarbons from seawater is obvious. What is less well understood is that this accumulation is not necessarily from physical contact with oil but usually takes the form of assimilating through the gills the hydrocarbons dissolved in seawater. Our concern has been to provide fast, simple, and effective technology for the recovery, identification, and measurement of these particular hydrocarbons from fish and shellfish tissues. Since the disasters tend to occur in remote areas, we felt that the necessary equipment and technology should be found in any nearby modest laboratory.

For the above reasons we eliminated several more traditional approaches to hydrocarbon recovery and recognition, such as exhaustive Soxhlet extraction of fish tissue by organic solvents, isolation by solvent partitioning or by column or high-performance liquid chromatography (HPLC), and in fact all chromatographic fractionation of initial extracts. Some of these more complex techniques are described by Perfetti et al. (6) for the PAHs; at the other extreme, very simple extraction technology was applied to boiled shrimp and crab by Misharina and Golovnya (7). Instead we have revived the simple approach of steam distillation first published over two decades ago (8). We were concerned with analyses for both aliphatic and aromatic hydrocarbons. The latter fluoresce very well, and HPLC can sometimes be used to advantage with ultraviolet absorption or fluorescence detection (1,6). Instead, to also accommodate the aliphatic hydrocarbons seen in the more obvious surface oil slicks or emulsions, we opted for the use of capillary gas-liquid chromatography (GLC), since most laboratories have such equipment. The flame ionization detector (FID) is universally available and suitable for hydrocarbons, since the relative molecular responses are all similar (9). A strictly nonpolar GLC open tubular (capillary) column (polydimethylsiloxane) was also standardized for our research. These are readily available from different suppliers all over the world.

The only luxury we have permitted in our work is the Barrett receiver shown in Fig. 1. This inexpensive and simple-to-use apparatus is combined with standardized volumes of dichloromethane and with an injector split ratio of 1:32 in the

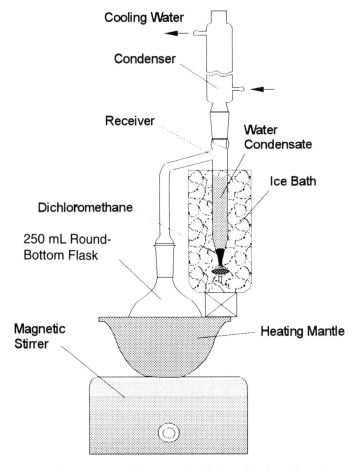

FIG. 1. Equipment used for steam distillation of hydrocarbons from fish or shellfish tissue with recovery in dichloromethane.

GLC injection of 4 µL of the distillate dichloromethane. This easily gave a useful chromatogram profile at a level in fish muscle of 0.1–0.5 ppm or higher total hydrocarbons (10). Often no concentration steps or solvent replacement is required during recovery of hydrocarbons unless needed due to very low levels of hydrocarbons.

Petroleum vs. Natural Hydrocarbons

The petroleum hydrocarbons that could contaminate salmon or other fish and shellfish from seawater solution fall into three simple categories: very volatile ($\leq C_{10}$); mid-range (C_{11}–C_{20}); and nonvolatile ($\geq C_{21}$). Figure 2 shows how a crude petroleum and the water-soluble fraction (WSF) differ. Excluding the alkanes, the very volatile group of hydrocarbons is dominated by the common aromatics, such as benzene, toluene, xylenes, and ethylbenzenes, but can include hydrocarbons with molecular weights as low as that of cyclohexane (MW = 84.2). These are also very soluble in seawater (12–14). The aliphatics so obvious in the GLC of crude petroleums (Fig. 2) or in diesel and fuel oils, are not very water soluble above C_9 or C_{10} (14) and do not significantly enter into long-term low-level exposure of fish.

The obvious group of midrange hydrocarbons embraces the higher alkylated aromatics between 1,3,5-trimethylbenzene and 1- and 2-methylnaphthalenes. The latter are frequently seen as a pair of peaks of equal size among compounds usually present in crude oils and are relatively water-soluble. On the nonpolar polydimethylsiloxane GC column we have used (DB-1, 60 m × 0.25 mm i.d., 0.25 µm film thickness), these elute in approximately the position of the C_{13} n-alkane and well before the C_{21} (n-heneicosane), which we have regularly used as internal standard. Soft tissue of mollusks (such as the blue mussel, *Mytilus edulis,* that has been promoted for international surveillance programs) can accumulate not only polychlorinated biphenyls (15) but a variety of hydrocarbons (16). We always find several natural hydrocarbons, including many of a regular series of n-alkanes of natural origin. Of particular interest is the hydrocarbon pristane derived from algal phytol, that may well be observed adjacent to 17:0 (Fig. 3). The related phytane, and a C_{21} hydrocarbon of algal origin with six ethylenic bonds, may also be observed (17).

Simple steam distillation of minced salmon flesh with a high burden (1–10 ppm) of the WSF of hydrocarbons presented few difficulties, but with crabs and mussels some additional GLC peaks, clearly not hydrocarbons, were often observed (Fig. 4). This pair of peaks—in fact, the whole cluster halfway between pristane and the heneicosane IS—were observed in the condensate from simple steam distillation. Although quantitatively unimportant in the whole picture, they seemed to be oriented to species rather than tissue. We have explored different routes to reduce such interfering materials.

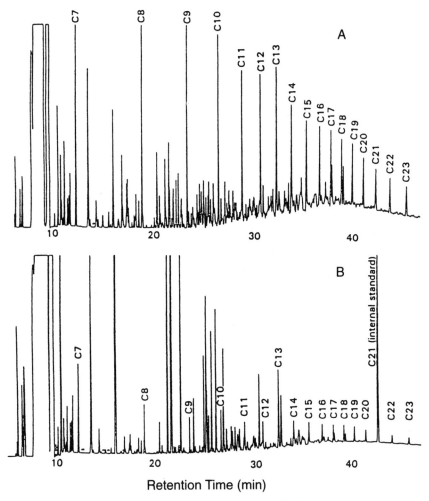

FIG. 2. Comparison of volatiles of (A) whole Flotta North Sea crude petroleum and (B) a hexane extract of the water-soluble hydrocarbon fraction of the same oil recovered from seawater. Adapted from Ref. 11.

Standard Steam Distillation with Sodium Sulfate Addition

Increasing the ionic strength of the aqueous medium seemed to be a reasonable approach to improve hydrocarbon recovery and possibly eliminate unwanted foaming. The steam distillation technology with sodium sulfate as a special addition to the routine technology consisted of fitting a 250-mL round-bottomed flask with a magnetic stirring bar, and a 20-mL Barrett-type distillation receiver with a Teflon stopcock and a glass condenser; all joints were glass standard taper. Double-distilled water (80 mL) and baked sodium sulfate (25 g) were placed in the 250-mL flask and

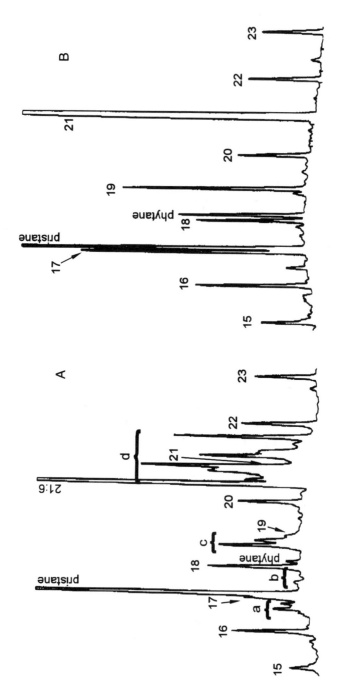

FIG. 3. Comparison of some midrange hydrocarbons recovered from mussels (A) before and (B) after complete hydrogenation. "A" refers to unsaturated C_{17} isomers, "B" to phytadienes etc., "C" to unsaturated C_{19} isomers, and "D" to various isomers of natural all-*cis* 21:6. Reproduced with permission from *Marine Biology*, Ref. 17.

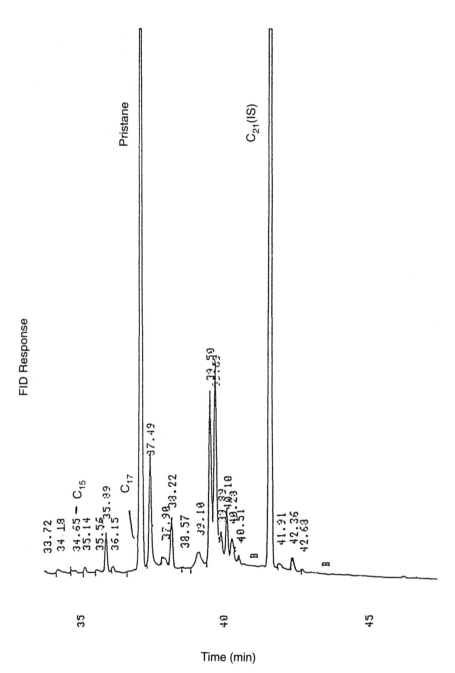

FIG. 4. Partial GLC chromatogram illustrating nonhydrocarbon peaks in steam distillate of fillets from a Pacific fish, *Amphisticus argenteus*. Pristane was a naturally occurring hydrocarbon, C_{21} the internal standard. Adapted from Ref. 23.

heated to boiling using an electric heating mantle. The distillation was conducted at a rate of 1 mL/min and was terminated after 20 mL of condensate was collected. Dichloromethane (1 mL) was added to the receiver through the top of the condenser midway through distillation and also immediately at the end of distillation. The flask was cooled to room temperature, and the water condensate and dichloromethane in the receiver were discarded. The receiver was again rinsed with dichloromethane. Then 20 g of the homogenized fish muscle sample and dichloromethane (2 mL) were added to the flask. The graduated collection receiver was immersed in an ice/water bath before distillation of hydrocarbons from samples. The saturated sodium sulfate solution and homogenized sample in the flask were stirred for 3–5 min to ensure that the sample was well dispersed in the solution before distillation commenced. As soon as 20 mL condensate (which contained the recoverable hydrocarbons) had been collected, the heating mantle was turned off and dichloromethane (1 mL) was added to the receiver through the top of the condenser to rinse the walls of the condenser. After 5–7 min, the condensate was drained to a 50-mL centrifuge tube (borosilicate glass), which had been solvent-rinsed and oven-baked. The receiver was rinsed with dichloromethane (1 mL), which was combined with the condensate in the centrifuge tube. The homogenate and the remaining sodium sulfate solution were discarded.

Dichloromethane (100 µL) containing a known amount of n-heneicosane as internal standard was added to the condensate in the centrifuge tube. The water and dichloromethane in the centrifuge tube were vortexed for 1 min and centrifuged at 1500 rpm for 6 min. The bottom dichloromethane layer was removed by a glass syringe with a long needle to a 5-mL graduated test tube. For heavily contaminated fish this could be used directly as the sample, but for control fish, or fish depurated to low levels of hydrocarbons, the dichloromethane was concentrated to 500 µL under a nitrogen flow of 0.5 L min^{-1}, and 4 µL of this concentrate was injected as the sample for GLC analysis. The concentrate was stored at $-20°C$ before and after GLC analyses in a screw-cap (Teflon-lined) glass container.

The hydrocarbon analyses were conducted with a Perkin Elmer Model 8420 (Norwalk, CT) capillary gas chromatograph equipped with a flame ionization detector (FID) and a split injection system (split ratio 1:32). The analysis was conducted on a fused-silica capillary column with polydimethylsiloxane coating (DB-1 60 m, 0.25 mm i.d., 0.25 µm film thickness, J&W Scientific, Chromatographic Specialties Inc., Brockville, ON). Operating conditions and temperature were: FID, 350°C; injector, 350°C. The column temperature was programmed as follows: initial temperature, 45°C, held for 15 min; increased at a rate of 13°C min^{-1}; final temperature 280°C, held for 25 min. The carrier gas was helium at a pressure of 138 kPa. Hydrogen and air pressures were 90 and 159 kPa, respectively.

Alkali Treatment

For some reason, alkali treatment has been a popular approach to hydrocarbon recovery (16,18–22). Presumably, it is believed that the complete digestion of the matrix should liberate more hydrocarbons. Although we have not investigated the actual recoveries in detail, we do not find any marked advantage in yield from treatment with alkali compared to simply boiling the matrix of fish or shellfish tissue for distillation purposes. Our objective in testing this method was possibly to eliminate foreign materials from among hydrocarbons in the steam distillation. Whether this alkali digestion is better when applied to other animal tissues, or to vegetable matter, is not known.

The alkali treatment of salmon flesh proceeded smoothly, and as part of this work, we examined the recovery of selected hydrocarbons (23) from spiked salmon muscle. Approximately 20 g of salmon muscle tissue from the laboratory-held fish, homogenized in a Sorvall OMNI-Mixer, was weighed into the 500-mL flask; 100 mL of 1 M NaOH solution was then added. The mixture was then spiked with 100 µL of standard solution of mixed hydrocarbons (Table 1) in dichloromethane using a Hamilton microsyringe. The flask was stoppered, and the contents were stirred with a bar magnet at room temperature for 2 h. At the end of this period 100 mL of calcium chloride solution (200 g/L) and 2 mL of dichloromethane were added to the reaction mixture. The mixture was then distilled until 20 mL of condensate was collected. Distilling receivers were kept immersed in ice baths during the entire distillation. The 20 mL of condensate was transferred to 50-mL glass centrifuge tubes, vortexed for 20 s, and centrifuged for 15 min at 2000 rpm. In all of the samples the dichloromethane layer separated cleanly after centrifuging. The bottom dichloromethane layer, containing hydrocarbons, was withdrawn using a glass syringe with a long needle and transferred to a screw-cap glass vial with a Teflon-lined cap for storage at −35°C until GC analysis. The internal standard (IS) n-

TABLE 1
Average Percent Recovery of Standard Hydrocarbons from Spiked Salmon Muscle Homogenate[a]

Hydrocarbon	µg/g Spiked	Recovery (%)
Benzene	15.822	84.7 ± 3.5
Cyclohexane	6.794	40.7 ± 7.9
Isooctane	0.581	21.8 ± 4.5
n-Heptane	0.219	21.6 ± 3.0
Methylcyclohexane	4.153	28.7 ± 2.9
Toluene	62.424	76.3 ± 7.2
Ethylbenzene	12.138	62.8 ± 7.4
m + p-Xylenes	17.260	36.8 ± 1.4
o-Xylene	11.616	60.5 ± 6.1
Propylbenzene	1.724	55.0 ± 4.4
1,3,5-Trimethylbenzene	0.484	68.5 ± 21.6
1,2,4-Trimethylbenzene	1.752	44.8 ± 4.2
1,2,3-Trimethylbenzene	1.430	44.7 ± 3.4
1,2,4,5-Tetramethylbenzene	0.144	85.1 ± 26.4
Naphthalene	0.857	82.6 ± 4.5
2-Methylnaphthalene	0.352	24.3 ± 3.1
1-Methylnaphthalene	0.286	34.7 ± 1.9

[a]Adapted from Ref. 23.

Heneicosane (C_{21}) was added to the sample at this point as a solution in dichloromethane, the amount of IS being adjusted according to the expected concentration of the particular hydrocarbon mixture used. Salmon tissues spiked with the standard hydrocarbon mixture were analyzed in triplicate, and unspiked salmon tissue samples from fish held in the same tank on the same diet were analyzed by the same technology as controls for laboratory blanks.

The results are given in Table 1. It can be seen that recoveries are good, although in a recent paper employing a somewhat similar Dean and Stark distillation approach, the sophisticated technology has given recoveries of up to 95% for some hydrocarbons (3). In our work, foaming (possibly due to free acids, phospholipids, and proteins) had appeared occasionally with whole crab and mussel samples but was partly suppressed by the $CaCl_2$. Such samples include the gut contents and the hepatopancreas, sometimes high in sterols and similar nonhydrocarbon lipid materials.

In an unusual case where long-chain aliphatic hydrocarbons (C_{20}–C_{34}) of presumed algal origin in mussels had to be totally recovered and distinguished from contamination from nearby petroleum operations a more elaborate procedure had to be followed. In contrast to the simplicity of the steam distillation, a procedure of saponification in alkali followed by extraction of all unsaponifiables, with chromatography cleanup was followed (17). This required the use of buffers and backwashing, each sample requiring several days. The mussel unsaponifiables were redissolved twice in hexane to allow removal of the dichloromethane under a nitrogen flow and were finally concentrated to a volume of 0.8 mL in hexane. The replacement of dichloromethane with hexane ensures that only hydrocarbons would be eluted from a final cleanup with Sep-Pak silica cartridges when hexane was used as the eluting solvent. Nonhydrocarbon unsaponifiable components were present, and without the Sep-Pak step and hexane elution, these carried over into the GLC. Chen et al. (1) used a Florisil Sep-Pak cartridge for the same purpose. Granby and Spliid (16) used more conventional column chromatography and reported C_{21}–C_{28} *n*-alkanes from mussels held in Danish waters.

We used Sep-Pak Plus Silica Cartridges (Part No: WAT020520, Waters, Millipore Corporation, 34 Maple Street, Milford, MA; bed volume 1.6 mL). A 2.5-mL glass syringe was connected to the cartridge as a reservoir. The cartridge was cleaned first with 20 mL of dichloromethane and then with 20 mL of *n*-hexane. The concentrated hexane solution of unsaponifiable materials was then quantitatively transferred into the barrel of the syringe, and hexane was passed through the cartridge by gravity. The first 0.5 mL of *n*-hexane to elute was discarded, and the subsequent 6 mL was collected in a graduated glass centrifuge tube. The hexane eluent containing the total hydrocarbons was concentrated to 0.3 mL under a nitrogen flow.

The final result, shown in part in Fig. 3, was a *tour de force* in which we seldom indulge for routine analyses or for urgent environmental samples. Simple steam distillation is effective provided a high-quality GLC analysis is applied to the isolated hydrocarbons. Quantitation through an internal standard (C_{21} or C_{25}) is convenient, but we have found that steam distillation recoveries fell off for PAH materials above naphthalene (Table 1). Fortuitously these PAHs are also not water-soluble, so our recovery of contaminating hydrocarbons from WSF includes most hydrocarbons of immediate interest, especially those necessary to back up sensory panel evaluations of contamination.

In salmon muscle the fat is not evenly spread throughout the tissue. Instead, fat cells (adipocytes) are found grouped along connective tissue (24,25). We expect that the fat cells store the higher-molecular weight hydrocarbons for a long period (Table 2), although the lifespan of the adipocytes is not known. The more water-soluble aromatics are likely present in the fish in the water available throughout the other cells and intercellular spaces, so the fish depurates these rapidly through the circulatory system when placed in clean water (10).

Hydrocarbons are a normal part of the marine environment and are found in edible tissues of most species of marine life. Table 3 shows numerous hydrocarbons in different Pacific salmon species, pristane and squalene being ubiquitous. Other hydrocarbons can be accumulated from normal or useful human activities (22,27), but not all are accumulated equally. The organisms with high fat (e.g., salmon) will presumably store more than lean animals such as mussels. Our work in this respect showed that a high level of exogenous algal hydrocarbons (17) was timely in relation to mussel exposure during a phytoplankton bloom. Some bivalves store lipid temporarily in the gut (28); the gonad ova are also high in fat, and this could affect levels of hydrocarbons (16). The C_{15}–C_{33} n-alkanes are common in many vegetable oils (29,30) but are probably harmless. Packaging materials can provide petroleum contamination of foods—for example, from sisal bags for carrying nuts or coffee beans (31)—and hydraulic fluids and lubricants are obvious sources. It has been said that a ship is apparently designed to contaminate all biological materials with petroleum products.

TABLE 2
Biological Half-Life of Individual Hydrocarbons Accumulated in Adipocytes and White Muscle of Atlantic Salmon Put In Hydrocarbon-Free Seawater After 96 h of Exposure to WSF

Hydrocarbon	White muscle	Adipocytes
Benzene	0.5	0.6
Methylcyclohexane	2.5	12
Toluene	0.6	0.8
Ethylbenzene	0.8	3.9
Xylenes	0.9	6.8
Isopropylbenzene	3.3	17
Propylbenzene	3.8	18
Ethylmethylbenzenes	2.5	15
Trimethylbenzenes	2.9	16
Methylnaphthalenes	3.5	>20

Unfortunately, background levels of hydrocarbons in marine organisms are rarely known (32), but examples from Pacific salmon are given in Table 3. The major response of authorities who should be researching petroleum contamination usually follows disasters (e.g., 4,33). Some petroleum "spills" are long-lived, such as that off Santa Barbara, California (34). The analyst seldom has much control over biological sample collection and should be familiar with hydrocarbon profiles from likely sources of contamination before drawing too firm conclusions. Our research with salmon was, fortunately, carried out under controlled exposure conditions and does not include extensive application of selective ion monitoring, a useful adjunct to the basic GLC in cases of low-level contamination. It is particularly important in the food context to note that GLC alone could measure petroleum contamination in salmon well below the level accurately recognized by sensory panelists (10).

TABLE 3
Hydrocarbon Compositions (w/w%) Naturally Accumulated in Samples of Salmon Flesh from Pacific Species[a]

Hydrocarbon	Sockeye salmon	Chum salmon	Steel head	Coho salmon
13:0	0.09	0.14	0.18	0.06
14:0	0.14	0.19	0.21	0.06
15:0	6.58	2.04	0.93	2.55
16:0	0.26	0.38	0.46	0.15
Pristane	29.60	13.90	7.26	46.36
17:0	1.54	3.22	4.04	1.79
17:1	0.38	0.19	0.81	0.21
Norphytene	1.71	1.35	0.85	3.09
Phytane	0.12	0.17	0.39	0.11
18:0	0.42	0.99	0.97	0.41
19:0	0.33	0.79	0.79	0.28
19:1	2.02	1.99	1.54	1.70
Phytadiene	2.41	0.62	0.50	3.19
20:0	0.24	0.32	0.65	0.22
21:0	0.11	0.13	0.45	0.06
21:1	0.52	2.24	0.32	0.69
22:0	0.14	0.24	0.81	0.15
21:5	0.21	0.21	0.21	0.50
23:0	0.09	0.15	0.49	0.05
21:6	2.13	1.05	0.38	8.18
24:0	0.10	0.15	0.51	0.04
$X\ C_{25}H_{46}$	3.68	3.15	0.34	1.19
25:0	0.14	0.50	0.50	0.08
$Y\ C_{25}H_{44}$	0.59	2.23	0.34	0.66
26:0	0.57	0.28	0.72	0.23
27:0	0.42	1.04	3.82	0.49
28:0	0.09	0.13	0.18	0.07
Squalene	37.42	49.43	52.34	15.97

[a]Adapted from Ref. 35.

References

1. Chen, B.H., Wang, C.Y., and Chiu, C.P. (1996) Evaluation of Analysis of Polycyclic Aromatic Hydrocarbons in Meat Products by Liquid Chromatography, *J. Agric. Food Chem. 44*, 2244–2251.
2. Zabik, M.E., Booren, A., Zabik, M.J., Welch, R., and Humphrey, H. (1996) Pesticide Residues, PCBs and PAHs in Baked, Charbroiled, Salt Boiled and Smoked Great Lakes Lake Trout, *Food Chem. 55*, 231–239.
3. Glegg, G.A., and Rowland, S.J. (1996) The *Braer* Oil Spill—Hydrocarbon Concentrations in Intertidal Organisms, *Mar. Pollut. Bull. 32*, 486–492.
4. Ritchie, W., and O'Sullivan, M. (1994) *The Environmental Impact of the Wreck of the Braer*, The Scottish Office, Edinburgh.
5. *The International Tanker Owners Pollution Federation Limited Review*, Staple Hall, Stonehouse Court, 87–90 Houndsditch, London, 1995.
6. Perfetti, G.A., Nyman, P.J., Fisher, S., Joe, F.L., Jr., and Diachenko, G.W. (1992) Determination of Polynuclear Aromatic Hydrocarbons in Seafood by Liquid Chromatography with Fluorescence Detection, *J. AOAC Internat. 75*, 872–877.
7. Misharina, T.A., and Golovnya, R.V. (1992) Hydrocarbon Contaminants of Boiled Shrimp and Crab Meat, *J. High Resol. Chromatogr. 15*, 332–334.
8. Ackman, R.G., and Noble, D. (1973) Steam Distillation, A Simple Technique for Recovery of Petroleum Hydrocarbons from Tainted Fish, *J. Fish. Res. Board Can. 30*, 711–714
9. Ackman, R.G. (1968) The Flame Ionization Detector, Further Comments on Molecular Breakdown and Fundamental Group Responses, *J. Gas Chromatogr. 6*, 497–501.
10. Heras, H., Zhou, S., and Ackman, R.G. (1993) Uptake and Depuration of Petroleum Hydrocarbons by Atlantic Salmon, Effect of Different Lipid Levels, in *Proceedings of the Sixteenth Arctic and Marine Oil Spill Program (AMOP) Technical Seminar, Calgary, Alberta,* Technology Development Directorate, Environmental Protection Service, Environment Canada, Ottawa, pp. 343–351.
11. Zhou, S., Heras, H., and Ackman, R.G. (1994) Preparation and Characterization of a Water Soluble Fraction of Crude Oil by a Karr Reciprocating-Plate Countercurrent Extraction Column, *Arch. Environ. Contam. Toxicol. 26*, 527–533.
12. Ackman, R.G., and Heras, H. (1992) Tainting by Short-Term Exposure Of Atlantic Salmon to Water Soluble Petroleum Hydrocarbons, in *Proceedings of the Fifteenth Arctic and Marine Oil Spill Program Technical Seminar, Edmonton, Alberta,* Technology Development Directorate, Environmental Protection Service, Environment Canada, Ottawa, pp. 757–762.
13. Heras, H., Ackman, R.G., and Macpherson, E.J. (1992) Tainting of Atlantic Salmon (*Salmo salar*) by Petroleum Hydrocarbons during a Short Term Exposure, *Mar. Pollut. Bull. 24*, 310–315.
14. Zhou, S., and Ackman, R.G. (1994) Deposition Sites for Hydrocarbons in Atlantic Salmon Muscle Tissue, in *Proceedings of the Seventeenth Arctic and Marine Oil Spill Program (AMOP) Technical Seminar,* Vancouver, British Columbia. Technology Development Directorate, Environmental Protection Service, Environment Canada, Ottawa, pp. 479–489.
15. Gilek, M., Björk, M., and Näf, C. (1996) Influence of Body Size on the Uptake, Depuration, and Bioaccumulation of Polychlorinated Biphenyl Congeners by Baltic Sea Blue Mussels, *Mytilus edulis, Mar. Biol. 125*, 499–510.

16. Granby, K., and Spliid, N.H. (1995) Hydrocarbons and Organochlorines in Common Mussels from the Kattegat and the Belts and Their Relation to Condition Indices, *Mar. Pollut. Bull. 30*, 74–82.
17. Zhou, S, Ackman, R.G., and Parsons, J. (1996) Very Long-Chain Aliphatic Hydrocarbons in Lipids of Mussels (*Mytilus edulis*) Suspended in the Water Column Near Petroleum Operations Off Sable Island, Nova Scotia, Canada, *Mar Biol. 126*, 499–507.
18. Ogata, M., Miyake, Y., and Yamasoki, Y. (1979) Identification of Substance Transferred to Fish or Shellfish from Petroleum Suspension, *Water Res. 13*, 613–618.
19. Donkin, P., and Evans, S.V. (1984) Application of Steam Distillation in the Determination of Petroleum Hydrocarbons in Water and Mussels from Dosing Experiments with Crude Oil, *Anal. Chim. Acta 156*, 207–219.
20. Fowler, B., Hamilton, M.C., Chiddell, G., and Sojo, L. (1992) Method Development for the Analysis of Tainting Compounds. Environment Canada, Environmental Protection Directorate, Ottawa, Report EE-136.
21. Shchekaturina, T.L., Khesina, A.L., Mironov, O.G., and Krivosheeva, L.G. (1995) Carcinogenic Polycyclic Aromatic Hydrocarbons in Mussels from the Black Sea, *Mar. Pollut. Bull. 30*, 38–40.
22. Snedaker, S.C., Glynn, P.W., Rumbold, D.G., and Corcoran, E.F. (1995) Distribution of *n*-Alkanes in Marine Samples from Southeast Florida, *Mar. Pollut. Bull. 30*, 83–89.
23. Işiğigür, A., Heras, H., and Ackman, R.G. (1996) An Improved Method for the Recovery of Petroleum Hydrocarbons from Fish Muscle Tissue, *Food Chem. 57*, 457–462.
24. Zhou, S., Ackman, R.G., and Morrison, C. (1995) Storage of Lipids in the Myosepta of Atlantic Salmon (*Salmo salar*), *Fish Physiol. Biochem. 14*, 171–178.
25. Zhou, S, Ackman, R.G., and Morrison, C. (1996) Adipocytes and Lipid Distribution in the Muscle Tissue of Atlantic Salmon (*Salmo salar*), *Can. J. Fish. Aquat. Sci. 53*, 326–332.
26. Zhou, S. (1997) Function of Muscle Adipocytes in Atlantic Salmon (*Salmo salar*) Exposed to Hydrocarbons, Ph.D. Thesis, Technical University of Nova Scotia, Halifax.
27. Olsgard, F., and Gray J.S. (1995) A Comprehensive Analysis of the Effects of Offshore Oil and Gas Exploration and Production on the Benthic Communities of the Norwegian Continental Shelf, *Mar. Ecol. Prog. Ser. 122*, 277–306.
28. Napolitano, G.E., and Ackman R.G. (1993) Fatty Acid Dynamics in Sea Scallops *Placopecten magellanicus* (Gmelin, 1791) from Georges Bank, Nova Scotia, *J. Shellfish Res. 12*, 267–277.
29. McGill, A.S., Moffat, C.F., Mackie, P.R., and Cruickshank, P. (1993) The Composition and Concentration of *n*-Alkanes in Retail Samples of Edible Oils, *J. Sci. Food Agric. 61*, 357–362.
30. Tan, Y.A., and Kuntom A. (1994) Hydrocarbons in Crude Palm Kernel Oil, *J. AOAC Internat. 77*, 67–73.
31. Grob, K., Artho, A., Biedermann, M., and Caramaschi, A. (1992) Batching Oils on Sisal Bags Used for Packaging Foods, Analysis by Coupled LC/GC, *J. AOAC Internat. 75*, 283–287.
32. Johansen, P., Jensen, V.B., and Büchert, A. (1977) in *Hydrocarbons in Marine Organisms and Sediment Off West Greenland* (Ackman, R.G., ed.), *Fish. Mar. Serv. Tech. Rep. 729*, 33 pp.

33. *IFREMER Coût Social de la Pollution par les Hydrocarbures, L'exemple de l'*Amoco Cadiz, *Rapports Economiques et Juridiques de L'IFREMER, No. 1, Université de Bretagne Occidentale—Brest*, (1984) 294 pp. (A translation of *Assessing the Social Costs of Oil Spills, The Amoco Cadiz Case Study*, National Oceanic and Atmospheric Administration (NOAA), U.S. Department of Commerce).
34. Squire, J.L. Jr. (1992) Effects of the Santa Barbara, Calif., Oil Spill on the Apparent Abundance of Pelagic Fishery Resources, *Mar. Fish. Rev. 54(1)*, 7–14.
35. Sasaki, S., Ota, T., and Takagi, T. (1993) Hydrocarbon Composition of Salmon in the Gulf of Alaska, *Nippon Suisan Gakkaishi* 59, 801–806.

Index

A

Acid hydrolysis-capillary column GC method, total fat and saturated fat, 2
Acid index, determination in coffee lipids, 369
Acidity determination, methods for, 3
Active oxygen method (AOM), evaluation of antioxidants, 351
Allylic dihydroxy compounds
 ^{13}C-NMR of, 126
 ^{13}C-NMR of saturated, 126–128
 ^{1}H-NMR of, 125–126
Allylic monohydroxy compounds
 ^{13}C-NMR of, 123–125
 ^{13}C-NMR of saturated, 126–128
 ^{1}H-NMR of, 122–123
Anisidine value, 3
 FTIR measurement of, 311–313
Antioxidants, naturally occurring in foods
 classification of, 341
 determination
  ascorbic acid, 350
  flavonoids, 349
  phospholipids, 349–350
  in spice extracts, 342–348
  tocopherols and tocotrienols, 342–346
 evaluation methods, 350–351
 synergists, 349–350
APCI-MS (Atmosphere-pressure chemical ionization mass spectrometry), triacyglycerol analysis using, 45–76
Argentation chromatography, determination of molecular species of lipids, 242–244
Ascorbic acid, determination of antioxidant activity, 350

Atmosphere-pressure chemical ionization mass spectrometry *see* APCI-MS
Attenuated total reflectance (ATR) sampling technique, *trans* isomers determination using, 13, 14, 239, 240, 289, 290–292, 297–298

B

Bouger-Beer-Lambert (Beer's) law, application to IR spectroscopy, 286–287

C

Cafestol, coffee diterpene, 358–359
Caffeine, Iatroscan TLC-FID analysis of, 334–335
Canola, NIR analysis of, 274–276
Canola oil, triacylglycerol analysis by APCI-MS, 47, 65–72
Capillary supercritical fluid chromatography *see* Supercritical fluid chromatography
Cardiolipin (CL), 81
 LS/ES/MS analysis of, 84–86
 oxidation products in tissues, 95–96
 phospholipase D digestion of oxidized, 93–95, 97
Cardiolipin hydroperoxides
 characterization of, 86–88
 in tissues, 95–96
Cardiolipin ozonides, characterization of, 88–93
Chemometric methods, application to virgin olive oils, 20–21
Chiral-phase HPLC, stereospecific analysis of TAGs by, 100, 102, 105–111

Chlorophyll, NIR analysis in canola seeds, 274
Cholesterol
 capillary GC analysis of, 12
 Iatroscan TLC-FID analysis of, 333–334
Coffee *see also* Coffee Lipids
 economic importance of, 356–357
 16-0-methylcafestol as admixing indicator, 356–374
Coffee beans, lipid analysis of, 372–374
Coffee lipids
 determination in coffee brews, 369–370
 diterpenes, 358–360; *see also* 16-0-Methylcafestol
 extraction and separation, 365–368
 fatty acids and triacylglycerols, 357–358
 identification of free fatty acids, 368–369
 isolation in coffee brews, 370–372
 sterols, 358
 total composition of, 357
Combined FTIR and GC, *trans* fatty acid analysis using, 241
Conjugated linoleic acid (CLA), oxidation products of, 183–211
 analytic chemistry, 189–195

D

Designer oils, positional distribution of long-chain fatty acids in, 100
Diacylglycerols (DGs), isolation of, 104–105
Diene conjugation (DC), formation and physiological effects, 183–186
7,10-Dihydroxy-8(E)-octadecenoic acid, microbial synthesis from oleic acid or olive oil, 121
Diphosphatidylglycerol, LC/ES/MS of oxidized, 81–97

Diterpene
 coffee, 358–360
 in rosemary and sage, 348
Docosahexaenoic acid-rich triacylglycerols
 stereochemical analysis by chiral-phase HPLC, 100–118

E

Edible fats and oils, analysis by FTIR spectroscopy, 283–320
Electrospray mass spectrometry, stereochemical analysis of docosaheanoic acid-rich triacylglycerols, 103–117
Evaporative light-scattering (ELSD) detector, applications, 9

F

Fats and oils *see also* Lipids
 FDA definition for nutritional labeling, 2
 newer methods of analysis in foods, 1–21
Fatty acid isomers *see also Trans* fatty acids
 cis and *trans*, 12
 measurement of *trans*, 234–248, 294–301
Fatty acid methyl esters (FAMES)
 GC analysis of, 11, 245
 GC-FTIR spectroscopy of, 15–17
 GC-MS of, 17
 ozonolysis, 3–4
 preparation, 2
 separation by Ag-HPLC, 256–263
Fatty acid monomers, analysis of major structures of oxidized, 229–230
Fatty acids *see also* Fatty acid methyl esters
 double bond position determination, 3–4

metabolism studies with ^{13}C-labeled, 34–43
Fatty compounds
  NMR characterization of, 121–136
  quantitation of oxidized methyl esters, 223–228
Flame ionization detector (FID), hydrocarbon analysis application, 381
Flavonoids, determination of antioxidant activities, 349
Food labeling, fat analysis methodologies, 2
Fourier transform infrared (FTIR) spectroscopy, 13–14
  applications, 13–14
    bulk characterization of fats and oils, 301–304
    oxidative status and stability measurement, 307–314
    solid fat content, 304–307
    trans content determination, 294–301
  edible-oil analysis by, 283–320
  implementation in the fats and oil industry, 315
  principles of, 13, 283–289
  sample handling techniques, 289–294
Free fatty acids, methods of determination, 3
  in coffee lipids, 368–369
Frying fats and oils
  commercial evaluation by size exclusion chromatography, 7–8
  quantitation of oxidized fatty acids in, 226–228
Furan fatty acids (F-acids),186; *see also* Conjugated linoleic acid

G

Gas chromatography (GC)
  techniques and applications, 11–13
  fatty acid composition, 244–245
  trans isomers, 245–248
GC-combustion isotope ratio mass spectrometry (IRMS), principles and applications, 36–43
GC-FID FA composition method, TAG analysis using, 74–75
GC-Fourier transform infrared (GC-FTIR) spectroscopy, applications, 15–17, 241–242
GC-mass spectrometry (GC-MS), 17
Glycerides, supercritical fluid chromatography of
  containing oxygenated fatty acyl groups, 168–170
  estolide-containing, 171
  polymeric lipids, 171
  polyunsaturated lipids, 171–174

H

HPLC (High-performance liquid chromatography)
  fat analysis, 8–10
  phospholipid separation, 84
HPLC-MS, applications, 10
HPSEC (High-performance size exclusion chromatography)
  applications of, 7–8
  quantitation of oxidized FAMEs, 223–228
  quantitation of oxidized triglycerides, 217–223
HPTLC (High-performance thin-layer chromatography), applications, 6
Hydrocarbons
  contamination, 380, 389–390
  petroleum vs. natural in fish and shellfish, 382–383
Hydroperoxides, preparation of, 82

Hydroxy compounds
  ^{13}C-NMR of symmetrical alkenes, 128, 130, 131–132
  ^{1}H-NMR of symmetrical alkenes, 128, 130

I

Iatroscan TLC-FID
  applications, 6–7, 325–337
    AMPL and polar plant lipids, 330–333
    caffeine, 334–335
    cholesterol, 333–334
    gastric digestate lipids, 326–330
    multiple development for fish muscle phospholipids, 330
    shellfish toxins, 335–336
Infrared (IR) spectroscopy
  principles of, 238–240
  trans fatty acid analysis by, 238–241
Iodine value (IV), FTIR spectroscopy measurement of, 301–304
Isotopes *see* Stable isotopes

K

Kahweol, coffee diterpene, 358–359

L

Lard, triacylglycerol analysis by APCI-MS, 57–65
LC (liquid chromatography)
  on-line electrospray mass spectrometry of oxidized diphospatidylglycerol, 81–97
  TAG analysis, 48
LC/MS (liquid chromatography/mass spectrometry), structural characterization of TAGs, 45
Linoleic acid, *see* Conjugated linoleic acid

Lipase hydrolysis, positional isomer determination, 49
Lipid oxidation products
  chromatographic analysis of, 216–231
  of conjugated linoleic acid and furan fatty acids, 183–211
  FTIR spectroscopy monitoring of, 313–314
Lipids
  analytical methods, 1–21; *see also* specific methods
  natural antioxidants in, 341–351
Liquid chromatography *see* LC
Liquid chromatography/mass spectrometry *see* LC/MS

M

Magnetic resonance imaging (MRI), applications, 20–21
16-0-Methylcafestol (16-OMC)
  analytical methods, 367–368
  extraction and purification methods, 361–364
  indicator of admixing of coffees, 358–373
  synthesis and purification of fatty acid esters of, 364–365
Microcolumns TLC, applications, 6

N

Natural antioxidants *see* Antioxidants
NI (near infrared) spectroscopy
  applications, 14–15
  calibration tools, 270–273
  instrumentation, 267–270
  oilseeds analysis by, 273–280
  theory of, 266–267
Normal phase HPLC (high-performance liquid chromatography), applications, 8–10

NMR (nuclear magnetic resonance) spectroscopy, applications, 17–19
Nutrition Labeling and Education Act of 1990, definition of fat, 2

O

Oil oxidation, monitoring by FTIR, 313–314
Oilseeds, NI analysis of, 273–280
Optothermal window spectroscopy, and *trans* fatty acids analysis, 240–241
Ozonides
  characterization of cardiolipin, 86–93
  preparation of, 92
  triphenylphosphine reduction of, 82–83
Ozonolysis methods, for determination of double bond position, 3–4

P

Partial-least-squares (PLS) calibration model, in FTIR analysis, 287–289
Peroxide value (PV)
  FTIR measurement of, 307–311
  methods for determination, 3
Petroleum hydrocarbons
  recovery from marine fish, crabs and mussels, 380–390
    alkali treatment, 386–390
    steam distillation with sodium sulfate, 383–386
Phospholipids, determination of antioxidant activities, 349–350
Polycyclic aromatic hydrocarbons (PAHs)
  carcinogens and oil spills, 380
  comparison with natural hydrocarbons, 382–383
  extraction, 380–382
Polyunsaturated fatty acids (PUFAs), analysis, 7

R

Randomized fat method, TAG analysis, 74
Reverse phase HPLC, lipid applications, 8–9, 10
Rosemary extract, antioxidant components in, 347–348

S

Salmon, recovery of petroleum hydrocarbons from, 380–390
Saponification number, FTIR measurement of, 303–304, 307
SCIRA *see* Stable carbon isotope ratio analysis
SEC *see* Size-exclusion chromatography
Selenites, NMR characterization of, 130, 133, 135
Selenium dioxide, allylic hydroxylation reagent, 121
Selenium dioxide-based oxidations, NMR characterization of fatty compounds, 121–136
SFC *see* Supercritical fluid chromatography
SFE *see* Supercritical fluid extraction
Shellfish toxins, Iatroscan TLC-FID analysis of, 335–336
Silver ion chromatography, *trans* fatty acids analysis by, 242–244
Silver ion HPLC
  applications, 8, 243–244
  separation of FAMEs and TAGs by, 256–263

Singlet oxygen, oxidation of conjugated dienes, 198–199
Size-exclusion chromatography, applications, 7–8
Solid fat content, FTIR spectroscopy measurement of, 304–307
Solid fat index (SFI), determination by FTIR spectroscopy, 304–307
Solid-phase extraction
　description and applications, 4–5
　of oxidized triglycerides, 217–223
Solvents, lipid extraction in foods, 1–3
Solvent toxicity, and lipid extraction techniques, 2
SPE *see* Solid-phase extraction
Spice extracts
　determination of antioxidants in, 346–348
　as natural antioxidants, 341–342
Stable carbon isotope ratio analysis, techniques and applications, 19–20
Stable isotopes, use in dietary fat metabolism study, 34–43
Supercritical fluid chromatography
　applications of capillary, 148–161
　　analysis of minor lipid constituents, 150–153
　　deformulation of commercial products, 153–156
　　comparison with other techniques, 146–148, 166–168
　　of complex lipid and lipid reaction mixtures, 174–177
　　equipment, 164–165
　　features of, 140–143
　　lipid analysis using, 10–11, 139–161
　　retention trends in capillary, 143–146
　　of unusual triglycerides, 163–177

Supercritical fluid extraction
　applications and advantages, 2–3
　monitoring, 148–150
　properties, 163–164
Supercritical fluid fractionation, monitoring, 148–150

T

TBARS (thiobarbituric acid-reactive substances) method, natural antioxidant evaluation, 351
Thin-layer chromatography (TLC), methods and lipid analysis applications, 5–7
Tocopherols
　determination, 342–346
　　by SFC, 147–148
　structure, 342, 344
Tocotrienols, structure and determination, 342–346
Total fat, methods of determination, 1–3
Trans fatty acids (TFA)
　analytical methods, 234–249
　combined FTIR and GC, 241
　FTIR spectroscopy, 13, 294–301
　GC, 244–248
　GC-MS, 248
　hyphenated GC-FTIR, 241–244
　infrared spectroscopy, 238–241
　and coronary heart disease, 236
　in food products, 234–236, 238
　in hydrogenated oils, 236–238
Triacylglycerols (TAGSs) *see also* Triglycerides
　analysis using APCI-MS spectrometry, 45–76
　determination of fatty composition in coffee lipids, 368–374
　in normal lard composition, 60, 64

partial deacylation of, 101, 104–106
quotient, 46, 53–55, 73–74
in randomized lard composition, 57, 62
separation by Ag-HPLC, 256–263
stereospecific analysis of docosahexaenoic acid-rich, 100–118
Triglycerides (TGs) *see also* Triacylglycerols
quantitation of oxidized, 217–223
supercritical fluid chromatography analysis of unusual, 168–173
Triplet Oxygen, oxidation of conjugated linoleic acid, 199–207

V

Vegetable oils, tocopherols and tocotrienols in, 342–343
Virgin olive oils, chemometric methods applied to, 20–21
Vitamin E *see* Tocopherols and Tocotrienols